Amy Mulnix

Topics in Current Genetics

7

Series Editor: *Stefan Hohmann*

Springer

Berlin
Heidelberg
New York
Hong Kong
London
Milan
Paris
Tokyo

Joris G. Winderickx • Peter M. Taylor (Eds.)

Nutrient-Induced Responses in Eukaryotic Cells

With 48 Figures, 2 in Color; and 1 Table

Springer

Professor Dr. Joris G. Winderickx
Functional Biology
Katholieke Universiteit Leuven
Plantkunde en Microbiologie
Kasteelpark Arenberg 31
3001 Heverlee-Leuven
Belgium

Professor Dr. Peter M. Taylor
Division of Molecular Physiology
School of Life Sciences
University of Dundee
MSI/WTB Complex
Dundee DD1 5EH
UK

The cover illustration depicts pseudohyphal filaments of the ascomycete *Saccharomyces cerevisiae* that enable this organism to forage for nutrients. Pseudohyphal filaments were induced here in a wild-type haploid MATa Σ1278b strain by an unknown readily diffusible factor provided by growth in confrontation with an isogenic petite yeast strain in a sealed petri dish for two weeks and photographed at 100X magnification (provided by Xuewen Pan and Joseph Heitman).

ISSN 1610-2096
ISBN 3-540-20917-4 Springer-Verlag Berlin Heidelberg New York

Cataloging-in-Publication Data applied for

Bibliographic information published by Die Deutsche Bibliothek
Die Deutsche Bibliothek lists this publication in the Deutsche Nationalbibliografie; detailed bibliographic data is available in the Internet at <http:/dnb.ddb.de>

Springer-Verlag is a part of Springer Science + Business Media
springeronline.com

Printed in Germany

Typesetting: Camera ready by editors
Data-conversion: PTP-Berlin, Stefan Sossna e.K.
Cover Design: Design & Production, Heidelberg
39/3150-WI - 5 4 3 2 1 0 - Printed on acid-free paper

Table of contents

List of contributors

Anthony, Tracy G.
Department of Cellular and Molecular Physiology, Mailcode H166, The Pennsylvania State University College of Medicine, Hershey, PA 17033, USA

Best, Leonard
Department of Medicine, University of Manchester, Multipurpose Building, Manchester Royal Infirmary, Oxford Road Manchester, M13 9WL, UK
lbest@man.ac.uk

Cavener, Douglas R.
Department of Biology, The Pennsylvania State University, University Park, PA 16802, USA

Chen, Ching-Yi
Section of Molecular and Cellular Biology, Center for Genetics and Development, Division of Biological Sciences, University of California, Davis, Davis, CA 95616, USA

Cooper, Terrance G.
Department of Molecular Sciences, University of Tennessee, Memphis, Tennessee 38163, USA
tcooper@utmem.edu

Dilova, Ivanka
Section of Molecular and Cellular Biology, Center for Genetics and Development, Division of Biological Sciences, University of California, Davis, CA 95616, USA

Harashima, Toshiaki
Department of Molecular Genetics and Microbiology, Howard Hughes Medical Institute, Duke University Medical Center Durham, NC 27710, USA

Heitman, Joseph
Department of Molecular Genetics and Microbiology, Howard Hughes Medical Institute, Duke University Medical Center Durham, NC 27710, USA
heitm001@duke.edu

Hundal, Harinder S.
Division of Molecular Physiology, School of Life Sciences, University of Dundee, MSI/WTB Complex, Dow Street, Dundee, DD1 5EH, UK

Jefferson, Leonard S.
Department of Cellular and Molecular Physiology, Mailcode H166, The Pennsylvania State University College of Medicine, Hershey, PA 17033, USA

Kilberg, Michael S.
Department of Biochemistry & Molecular Biology, Box 100245, 1600 S.W. Archer Road, University of Florida College of Medicine, Gainesville, Florida 32610-0245, USA
mkilberg@ufl.edu

Kimball, Scot R.
Department of Cellular and Molecular Physiology, Mailcode H166, The Pennsylvania State University College of Medicine, Hershey, PA 17033, USA
skimball@psu.edu

McClellan, Randall
Department of Biochemistry & Molecular Biology, Box 100245, 1600 S.W. Archer Road, University of Florida College of Medicine, Gainesville, Florida 32610-0245, USA

McLaughlin, John
Department of GI Sciences, University of Manchester, Multipurpose Building, Manchester Royal Infirmary, Oxford Road Manchester, M13 9WL, UK

Mosley, Amber L.
Department of Molecular & Cellular Biochemistry, University of Kentucky, College of Medicine, 800 Rose Street, MN 608, Lexington, Kentucky 40536, USA

Narasimhan, Jana
Department of Biochemistry and Molecular Biology, Indiana University School of Medicine, Indianapolis, Indiana 46202, USA

Oesterhelt, Christine
Functional Biology, Katholieke Universiteit Leuven, Kasteelpark Arenberg 31, B-3001, Leuven-Heverlee, Flanders, Belgium

Özcan, Sabire
Department of Molecular & Cellular Biochemistry, University of Kentucky, College of Medicine, 800 Rose Street, MN 608, Lexington, Kentucky 40536, USA
sozcan@uky.edu

Pan, Duojia
Department of Physiology, University of Texas Southwestern Medical Center at Dallas, 5323 Harry Hines Blvd., Dallas, TX 75390-9040, USA
Duojia.Pan@UTSouthwestern.edu

Pan, YuanXiang
Department of Biochemistry & Molecular Biology, Box 100245, 1600 S.W. Archer Road, University of Florida College of Medicine, Gainesville, Florida 32610-0245, USA

Pardons, Katrien
Functional Biology, Katholieke Universiteit Leuven, Kasteelpark Arenberg 31, B-3001, Leuven-Heverlee, Flanders, Belgium

Powers, Ted
Section of Molecular and Cellular Biology, Center for Genetics and Development, Division of Biological Sciences, University of California, Davis, Davis, CA 95616, USA
tpowers@ucdavis.edu

Proud, Christopher G.
Division of Molecular Physiology, School of Life Sciences, University of Dundee, MSI/WTB Complex, Dow Street, Dundee, DD1 5EH, UK

Reinke, Aaron
Section of Molecular and Cellular Biology, Center for Genetics and Development, Division of Biological Sciences, University of California, Davis, Davis, CA 95616, USA

Roosen, Johnny
Functional Biology, Katholieke Universiteit Leuven, Kasteelpark Arenberg 31, B-3001, Leuven-Heverlee, Flanders, Belgium

Sampley, Megan L.
Department of Molecular & Cellular Biochemistry, University of Kentucky, College of Medicine, 800 Rose Street, MN 608, Lexington, Kentucky 40536, USA

Staschke, Kirk A.
Department of Biochemistry and Molecular Biology, Indiana University School of Medicine, Indianapolis, Indiana 46202, USA

Swinnen, Erwin
Functional Biology, Katholieke Universiteit Leuven, Kasteelpark Arenberg 31, B-3001, Leuven-Heverlee, Flanders, Belgium

Taylor, Peter M.
Division of Molecular Physiology, School of Life Sciences, University of Dundee, MSI/WTB Complex, Dow Street, Dundee, DD1 5EH, UK
p.m.taylor@dundee.ac.uk

Wedaman, Karen P.
Section of Molecular and Cellular Biology, Center for Genetics and Development, Division of Biological Sciences, University of California, Davis, CA 95616, USA

Wek, Ronald C.
Department of Biochemistry and Molecular Biology, Indiana University School of Medicine, Indianapolis, Indiana 46202, USA
rwek@iupui.edu

Winderickx, Joris
Functional Biology, Katholieke Universiteit Leuven, Kasteelpark Arenberg 31, B-3001, Leuven-Heverlee, Flanders, Belgium
joris.winderickx@bio.kuleuven.ac.be

Zhong, Can
Department of Biochemistry & Molecular Biology, Box 100245, 1600 S.W. Archer Road, University of Florida College of Medicine, Gainesville, Florida 32610-0245, USA

Introduction

Joris Winderickx and Peter M. Taylor

Cells of all living organisms are able to sense environmental stimuli and respond appropriately. Especially for unicellular organisms, the environment largely controls growth, metabolism, and differentiation. In higher multicellular organisms, most cells experience relative environmental homeostasis. However, growth and metabolism of cells within multicellular organisms require coordination between the cells in a tissue, an organ, and the whole organism. These cells communicate by cell-to-cell contact, gap-junctions, and integrins, or by using molecules such as hormones and growth factors, which allow cell-to-cell signalling.

For unicellular and multicellular organisms alike, nutrients provide the essential building blocks and energy supply to make the necessary cellular components and drive metabolism. Therefore, the availability of nutrients is essential to survive, proliferate, and be productive. Cells have developed mechanisms to sense nutrient availability and produce appropriate responses whereby nutrients are able to influence gene transcription and mRNA processing as well as translation and post-translational modifications. Such mechanisms may, in certain cases, involve direct or near-direct interactions between a nutrient and the regulatory sequences of specific genes involved in its metabolism. There are also reports of metabolite-binding domains in particular mRNA species (so-called "riboswitches"), which serve as metabolite-responsive genetic control elements. Nevertheless, many nutrients appear to affect cell and organismal function largely through intermediate nutrient-responsive signalling pathways. Such nutrient-dependent signalling allows for optimal nutrient consumption in a dynamic integrated manner and particularly in unicellular organisms it enables coordinated induction of a resting phase where the cells cease proliferation upon nutrient limitation, but rapidly resume the process once the conditions are more suitable.

In recent years, our understanding of nutrient sensing and the responses triggered by altered nutrient availability have greatly advanced. The emerging picture is that nutrient signalling mechanisms have evolved early in evolution and that the so-called nutrient-responsive signalling cascades used by microorganisms provide core elements of the more sophisticated regulatory pathways found in multicellular organisms, where hormonal controls have assumed increasingly greater importance. For example, many of the genes regulated by glucose alone in lower eukaryotes are additionally dependent upon the presence of insulin (and to some extent thyroid hormones) in higher eukaryotes. However, perhaps surprisingly, insulin was also found to trigger regulatory effects in microorganisms (Muller et al. 1998). The availability of genomic and proteomic data for an increasing number of eukaryotic species has highlighted the conservation of these basic pathways and, in many cases, conservation has been confirmed by functional complementation of several key proteins. The success of genome and proteome research has led to other exciting and new approaches dealing not only with top-down elucidation

Topics in Current Genetics, Vol. 7
J. Winderickx, P.M. Taylor (Eds) Nutrient-Induced Responses in Eukaryotic Cells

of a single nutrient sensing pathway but with the more global investigation of nutrient signalling networks and the identification of converging effector branches that explain the dynamical but very coordinated nutritional response. Indeed, every step in a particular nutrient pathway represents a potential convergence point for yet another cascade, which may modulate or otherwise alter the final overall response. A nice example of this is the pseudohyphal growth pathway in yeast, which combines modules of the pheromone pathway and the Ras-cAMP cascade (Gancedo 2001).

Several nutritional factors have now been implicated as specific regulators of signal transduction and these include organic nutrients such as glucose, amino acids and fatty acids as well as inorganic compounds like carbon dioxide, ammonia, nitrates, and key micronutrients such as zinc, calcium, and phosphate ions. It has not always been possible to differentiate whether a nutritional stimulus acts really as a signal initiator or whether it merely triggers a local metabolic response. However, in a growing number of cases, specific receptors or transporters of a particular nutrient have been identified which appear to function as sensors. One of the best examples of this has been the discovery of two glucose transporters, i.e. Snf3 and Rgt2, that have been implicated in glucose sensing in yeast (Oscan et al. 1998) and, most recently, a similar function has been recognised for the human sodium/glucose cotransporter SGLT3 (Diez-Sampedro et al. 2003). Other nutrients may bind to GPCR-receptors such as Gpr1, a glucose receptor in yeast (Rolland et al. 2000) and GPR105/P2YX, the UDP-glucose receptor in mammals (Chambers et al. 2000).

Nevertheless, as might be expected there are also some important and fundamental differences in nutrient-induced responses between lower eukaryotes such as yeast and more complex organisms such as mammals. For example, glycogen is stored during glucose abundance in mammals, but in yeast it is only stored at the end of fermentation, before glucose becomes limiting. This is at least in part due to opposing end-point effects of the orthologous signalling pathways in the different organisms. Thus, the nutrient signal itself (in this case glucose availability) does not always induce the same (or even the equivalent) response, depending on the species studied, although the overall response in all cases is regarded as adaptive to the prevailing conditions and the specific biology of the species concerned. Another phylogenetic difference is the apparently unusual importance of free fatty acids as sensed molecules for metabolic regulation in animals. Lipids (fats and oils) are major sources and stores of fuel in higher animals but much less so in most plants and lower eukaryotes. A notable exception occurs during post-germinative growth of oilseeds, which is initially dependent on the breakdown of stored lipid reserves, which can be converted to sugar and other metabolites via the glyoxylate cycle (this is important for seedlings which have large lipid reserves but cannot yet photosynthesize). Recent studies have demonstrated inhibitory effects of sucrose on glyoxylate cycle activity (Borek et al. 2003). This represents an intriguing and important example of a nutrient-induced regulatory response in plants, although the mechanism is not yet known.

A common theme throughout the eukaryotic systems considered here is the link between nutrient availability and cellular energy status (notably as judged by the absolute or relative concentrations of ATP and ADP/AMP). Reduction in cellular ATP levels is also an index of cellular stress and indeed nutritional deprivation *per se* may be considered as a stress. Such stresses are typically associated with activation of one or more "stress-related" signalling or endocrine pathways, which have substantial downstream effects on organismal function, typically involving shut-down of nonessential processes and the induction of the genetic and metabolic program directed towards the protection of the cell.

In this volume on nutritional responses, we brought together experts on nutrient signalling in yeast and animals. The different chapters give an overview of recent advances in the field to guide the reader in the complex but dynamic system of nutrient sensing. We hope that reader will appreciate the dedication of all the scientists whose research has been cited in the reviews presented. Without such expertise, this book would not have been possible.

References

Borek S, Ratajczak W, Ratajczak L (2003) A transfer of carbon atoms from fatty acids to sugars and amino acids in yellow lupine (*Lupinus luteus L.*) seedlings. J Plant Physiol 160:539-545

Chamber JK,Macdonald LE,Sarau HM,Ames RS,Freeman K,Foley JJ,Zhu Y,McLaughlin MM, McMillan L, Trill J, Swift A, Aiyar N, Taylor P, Vawter L, Naheed S, Szekeres P, Hervieu G, Scott C, Watson JM, Murphy AJ, Duzic E, Klein C, Bergsma DJ, Wilson S, Livi GP (2000) A G protein-coupled receptor for UDP-glucose. J Biol Chem 275:10767-10771

Diez-Sampedro A, Hirayama BA, Osswald C, Gorboulev V, Baumgarten K, Volk C, Wright EM, Kopesell H (2003) A glucose sensor hiding in a family of transporters. Proc Natl Acad Sci USA 100:11753-11758

Gancedo JM (2001) Control of pseudohyphae formation in *Saccharomyces cerevisiae*. FEMS Microbiol Rev 25:107-123

Muller G, Rouveyre N, Crecelius A, Bandlow W (1998) Insulin signaling in yeast *Saccharomyces cerevisiae*. 1. Stimulation of glucose metabolism and Snf1 kinase by human insulin. Biochemistry 16:8683-8695

Ozcan S, Dover J, Johnston M (1998) Glucose sensing and signaling by two glucose receptors in the yeast *Saccharomyces cerevisiae*. EMBO J 17:2566-2573

Rolland F, de Winde H, Lemaire K, Boles E, Thevelein J, Winderickx J (2000) Glucose-induced cAMP signalling in yeast requires both a G-protein coupled receptor system for extracellular glucose detection and a separable hexose kinase-dependent sensing process. Mol Microbiol 38:348-358

1 Transcriptional regulatory mechanisms for the response to amino acid deprivation of mammalian cells

Michael S. Kilberg, Can Zhong, Randall McClellan, and YuanXiang Pan

Abstract

Dietary protein is critical to mammalian nutrition and on a cellular level this translates into amino acid availability. Cells monitor amino acids and respond with changes in cellular processes, including gene transcription. Thus, amino acids serve as signal molecules to transmit the nutritional status of the organism to individual cells. Using two target genes, CHOP and asparagine synthetase, this chapter will review the transcriptional control mechanisms triggered by amino acid limitation, a pathway named the amino acid response. The transcription factors, identified thus far, belong to two subfamilies, C/EBP and ATF, of the bZIP superfamily. There is much yet to learn about the signal pathways and the molecular mechanisms responsible for transcriptional regulation by nutrients. Beyond gaining a basic understanding of these biological control mechanisms, characterizing how these processes contribute to the pathology of various disease states represents an exciting aspect of molecular nutrition.

1.1 Introduction

Dietary protein is an important factor in the general nutrition of an entire organism, and on a cellular level this translates into amino acid availability. Although the amino acid content in the bloodstream and protein turnover both act to buffer variation in dietary protein/amino acid intake, fluctuations in the intracellular levels of individual amino acids do occur in response to diet, disease, and metabolic status. Obviously, the cellular metabolic stance must be altered in an attempt to adapt to these changes, and yet, how mammalian cells monitor amino acid levels and respond with changes in fundamental cellular processes is not completely understood. In this context, amino acids are serving as signal transduction messengers to transmit the nutritional status of the organism to individual cells. One of the target mechanisms of amino acid-dependent signaling is altered transcription for specific genes. This chapter will focus on the mechanisms associated with modulation of transcription triggered by amino acid limitation, a signaling pathway that will be referred to as the amino acid response (AAR). Detection of a limiting amount of any single amino acid has been linked to a ribosome-associated

Topics in Current Genetics, Vol. 7
J. Winderickx, P.M. Taylor (Eds) Nutrient-Induced Responses in Eukaryotic Cells

Table 1. Example genes that exhibit increased mRNA content following amino acid limitation

Protein	References
Asparagine synthetase	(Gong et al. 1991; Kilberg and Barbosa-Tessmann 2002; Siu et al. 2002)
Amino Acid Transporters	
CAT1	(Hyatt et al. 1997)
SNAT2	(Bain et al. 2002; Gazzola et al. 2001)
CHOP	(Bruhat et al. 1997; Jousse et al. 1999; Marten et al. 1994)
C/EBPα	(Marten et al. 1994)
C/EBPβ	(Marten et al. 1994)
IGFBP-1	(Jousse et al. 1998; Straus et al. 1993)
Ribosomal proteins	
L17	(Laine et al. 1991)
S25	(Laine et al. 1994)
L35	(Hitomi et al. 1993)
S13	(Hitomi et al. 1993)

kinase, GCN2, that binds and therefore, monitors the level of uncharged tRNAs (Hinnebusch 1997). Starvation-activated GCN2 kinase phosphorylates eIF2α and then the ensuing changes in eIF2α-mediated translation initiation favor increased synthesis of a specific transcription factor. In yeast, this transcription factor is GCN4 (Hinnebusch 1997), which has been reported to alter the transcription rate of up to 1000 genes (Natarajan et al. 2001). This translational detection mechanism has not been as extensively studied in mammalian cells, but a mammalian counterpart to yeast GCN2 has been identified (Berlanga et al. 1999; Sood et al. 2000). It appears from several studies that ATF4 (discussed in more detail below) may represent the mammalian counterpart to GCN4. The translation of pre-existing ATF4 mRNA is rapidly increased following amino acid deprivation (Harding et al. 2000), and ATF4 protein has been shown to mediate the increased transcription of AAR pathway target genes (Siu et al. 2002).

1.2 Examples of mammalian activities altered by amino acid availability

A wide range of enzymatic and transport activities, protein content, mRNA content, and transcription of specific genes have been reported to be regulated by amino acid availability both *in vivo* and *in vitro*. Given current screening technologies using gene arrays, it is anticipated that many more mammalian genes will be identified for which transcription is regulated by amino acid availability. Table 1 presents a partial and ever-increasing representative list of genes for which the corresponding mRNA content is increased following amino acid deprivation of mammalian cells. In some instances, the elevation in mRNA content is known to result from a change in transcription, but for several others, the mechanism remains to be established, and mRNA stabilization may contribute (Abcouwer et al. 1999; Gong et al. 1991). Marten et al. (1994) showed that C/EBPα and C/EBPβ mRNA content was increased by amino acid deprivation of rat hepatoma cells. As discussed in more detail below, C/EBP family members are of particular interest given the participation of C/EBPβ as one of the transcription factors that mediates induction of the human asparagine synthetase gene in response to activation of both the amino acid response (AAR) and the ER stress response (ERSR) nutrient sensing pathways (Siu et al. 2001). Also of note is the amino acid-dependent transcriptional regulation of the C/EBP homologous protein, CHOP (Bruhat et al. 1997; Fafournoux et al. 2000). Future characterization of amino acid-dependent changes in transcription factor expression, especially those (e.g. C/EBPβ, ATF4, and ATF3) that appear to be an integral part of the AAR pathway signaling process, will provide valuable insight into the mechanisms of gene expression following amino acid limitation. Amino acid-dependent regulation of transcription factor synthesis and action is discussed more extensively below.

Although increased transcription of ribosomal protein genes in the face of amino acid limitation appears to be counter-intuitive, the mRNA content for several ribosomal proteins has been shown to be increased (Laine et al. 1994) and for ribosomal proteins L17 and S25 this increase has been demonstrated to be transcriptional in nature (Laine et al. 1994). Interestingly, the newly synthesized mRNA molecules for L17 and S25 are retained within the nucleus for the duration of the amino acid deprivation period, and only released into the cytoplasm for translation following amino acid re-feeding (Laine et al. 1994). Adilakshmi and Laine (2002) have demonstrated that p53 binds to the S25 mRNA in the nucleus and may be associated with this nuclear retention process. Further investigation into how this nuclear retention is controlled should provide mechanistic insight into an interesting and novel cellular process regulated by amino acids.

Amino acid-dependent regulation of insulin-like growth factor binding protein-1 (IGFBP-1) has been reviewed by Bruhat et al. (1999). Among the amino acid-regulated genes identified to date, IGFBP-1 may be of particular importance because nutrient-dependent control of its expression is likely to have significant metabolic effects on a number of tissues and organs. IGFBP-1 may also serve as a prototype for nutrient feedback on metabolism-regulating hormones. A comprehensive examination of how amino acid availability may influence hormone and cytokine expression and/or action has not been undertaken, but such studies will greatly contribute to our understanding of the inter-organ effects that protein nutrition has on cell growth and metabolism. Beyond IGFBP-1, only a limited number of examples are known thus far. It has been demonstrated that histidine deprivation of murine pancreatic cells suppresses the synthesis of glucagon (Paul et al. 1998).

The substrate-dependent regulation of the sodium-dependent zwitterionic amino acid transporter System A activity has been investigated for three decades (Gazzola et al. 1972), and the subject has been reviewed periodically during this period of time (Kilberg et al. 1993; Palacín et al. 1998). The more recent identification of multiple genes encoding this activity has permitted investigators to document that the SNAT2 gene (also known as ATA2, SAT2) is responsible for the modulation of System A transport activity in response to amino acid availability (Gazzola et al. 2001). The laboratory of Hatzoglou and colleagues has described the nutrient control of the cationic amino acid transporter CAT1. An interesting aspect of this work is that CAT1 expression is regulated at both the transcriptional and translational levels (Aulak et al. 1999; Fernandez et al. 2001; Hyatt et al. 1997). The preferential use of an internal ribosome entry site for enhanced translation of the CAT1 mRNA following amino acid limitation is likely to represent a prototypical model for amino acid-dependent translational control of the biosynthesis of selected proteins (Fernandez et al. 2001).

1.3 Nutrient control of C/EBP homology protein

It is well recognized that limitation of an individual metabolite can trigger a cellular response to increase the synthesis of that specific molecule. However, it is also becoming evident that limiting the cellular supply of any single nutrient will also cause a broader "stress response" that activates multiple signal transduction pathways that, subsequently modify the entire metabolic status of the cell through transcriptional and post-transcriptional mechanisms. Included among the many target genes that respond to these more global nutrient stress signals is the transcription factor C/EBP homology protein (CHOP), also known as growth arrest and DNA damage protein 153 (GADD153). Initially, it was believed that CHOP heterodimerized with other basic, leucine zipper (bZip) family members to serve only as a negative regulator (Ron and Habener 1992), but it is now clear that CHOP heterodimers can also function as transcriptional activators (Wang et al. 1998). Originally identified as a gene that was activated following DNA damage, numerous studies have now shown CHOP expression to be increased by a wide array of nutrient stress signals (Sok et al. 1999; Wang et al. 1998; Zinszner et al. 1998).

Glucose starvation of mammalian cells results in accumulation of misfolded glycoproteins in the endoplasmic reticulum (ER) which triggers a signal transduction pathway, the ER Stress Response (ERSR), also called the unfolded protein response (UPR) in yeast (Kaufman, 2002; Patil and Walter 2001). The ERSR pathway leads to increased transcription of a number of target genes, many of which are involved in protein processing within the ER, but several others have been reported for which the connection to ER function is less clear. Nearly all ERSR target genes contain one of two different genomic *cis*-elements. The consensus sequence for the ER stress element (ERSE) is 5'-CCAAT-N_9-CCACG-3' (Yoshida et al. 1998), whereas a second element, the mammalian unfolded protein response element (UPRE), is 5'-TGACGTGG/A-3' (Yoshida et al. 2001, 2003). The human CHOP promoter contains two ERSE sequences (CHOP-ERSE1 and CHOP-ERSE2) (Yoshida et al. 2000). Mutagenesis of these two elements revealed that CHOP-ERSE2 does not link the gene to ER stress, but that CHOP-ERSE1 (nt –93 to –75) mediates activation of the gene by ER stress, for example, following glucose deprivation (Yoshida et al. 2000). Deletion analysis of the human CHOP promoter permitted Jousse et al. (1999) to document that the *cis*-element necessary for activation of transcription by the ERSR pathway was distinct from that responsible for increased transcription following amino acid deprivation. The amino acid response element (AARE) was later shown to reside at nt -302 to -310 (Bruhat et al. 2000; Fafournoux et al. 2000). As detailed below, the presence of two independent genomic elements within the CHOP promoter to respond independently to the AAR or the ERSR pathways contrasts to the single transcriptional control unit within the human asparagine synthetase promoter that mediates activation by both of these two pathways.

Wolfgang et al. (1997) identified an ATF3-responsive element in the CHOP promoter, referred to as a C/EBP-ATF composite site. This *cis*-element mediated inactivation of the gene following enhanced expression of ATF3. Those authors went on to show that following stress activation, for example by arsenite treatment, that the C/EBP-ATF composite site may be occupied first by ATF4 as an activating factor and subsequently, as the transient induction of CHOP transcription declines, ATF3 binding activity was observed to increase (Fawcett et al. 1999). The conclusion was that ATF3 likely functions as a repressor by acting at the C/EBP-ATF site. After establishing that the C/EBP-ATF core sequence (5'-TGATGCAAT-3', nt –302 to –310) also functions as an amino acid response element (AARE), Bruhat et al. (2000) investigated its binding specificity through electrophoresis mobility shift analysis (EMSA). Those authors reported that both C/EBPβ and ATF2 were able to bind to the CHOP AARE sequence *in vitro*, but the absolute amount of these complexes was not increased when extracts from amino acid-deprived cells were tested. Bruhat et al. (2000) also analyzed CHOP mRNA content in mouse embryonic fibroblasts that were deficient for either ATF2 or C/EBPβ and determined that activation of the CHOP gene by the ERSR pathway was functional in both knockout cell types, whereas activation by amino acid limitation occurred only in the C/EBPβ knockout cells, not those deficient in ATF2. In further support of a role for ATF2, transient transfection of the ATF2 deficient cells with an ATF2 cDNA restored amino acid control, and expression of an ATF2 dominant negative isoform suppressed the induction in control fibroblasts.

1.4 Nutrient regulation of the human asparagine synthetase genes

Human asparagine synthetase (Asns) belongs in the class II glutamine amidotransferase superfamily and catalyzes the synthesis of asparagine and glutamate from aspartate and glutamine, with ATP hydrolysis providing energy (Richards and Schuster 1998). The cDNA for Asns has been cloned from a number of species and there is a high degree of conservation (Richards and Schuster 1998). Rat, hamster, and human cells each express a predominant Asns mRNA species of approximately 2.0 kb, whereas hamster cells express a second mRNA of 2.5 kb, and in the rat, two additional Asns mRNA species of 2.5 and 4.0 kb are observed.

Consistent with the presence of a single gene and a single promoter, all three rat mRNAs are coordinately induced by amino acid deprivation (Hutson and Kilberg 1994). Northern analysis using the 3' untranslated region suggested that the species longer than 2.0 kb might result from alternative polyadenylation.

Arfin et al. (1977) showed that after incubation of cells in medium lacking asparagine, the aminoacylation of $tRNA^{Asn}$ decreased and subsequently, the level of Asns activity increased. The data are consistent with work in which Andrulis et al. (1979) analyzed Asns enzymatic activity and documented that it was increased in cells containing defective tRNA synthetase activities for several amino acids. Gong and Basilico (1990) identified asparagine synthetase as the protein that complemented a temperature-sensitive cell line that was blocked at the G1 stage of the cell cycle. As an extension of that research, the same laboratory determined that the Asns mRNA content increased in cells deprived of asparagine, leucine, isoleucine, or glutamine (Gong et al. 1991). Hutson and Kilberg (1994) also demonstrated increased Asns mRNA content following total amino acid deprivation of intact perfused rat liver as well as depletion of a single essential amino acid, such as histidine, threonine, and tryptophan, from the culture medium of rat hepatoma cells. Furthermore, treatment of Fao hepatoma cells with 5 mM of the amino alcohol histidinol, which suppresses the formation of histidinyl-$tRNA^{His}$ by competitively inhibiting the corresponding tRNA synthetase (Hansen et al. 1972), increased Asns mRNA to a level equal to that observed when cells were starved for all amino acids (Hutson and Kilberg 1994). Collectively, these data, from several independent laboratories using different approaches, indicate that the control of Asns mRNA expression is not specific for asparagine limitation, but rather modulated by the availability of many other amino acids as well, indicative of how broadly the AAR pathway in mammalian cells senses amino acid availability. Therefore, the amino acid sensing mechanism in mammalian cells is essentially similar to the general control response in yeast and the mammalian GCN2 kinase, like its yeast counterpart, is likely to bind a range of uncharged tRNAs (Berlanga et al. 1999; Sood et al. 2000).

Guerrini et al. (1993) analyzed the Asns promoter region by a series of sequential deletions followed by scanning mutagenesis and determined that an amino acid response element (AARE) was present near nucleotides –70 to –64 (5′-CATGATG-3′) within the proximal promoter. Barbosa-Tessmann et al. (1999a) later demonstrated that transcription from the human Asns gene is also activated by glucose starvation and that this same promoter region was required. Induction of Asns transcription following glucose starvation is the result of ERSR pathway activation, which was established by showing that other activators of the pathway, such as the protein glycosylation inhibitor tunicamycin and the proline analog, azetidine-2-carboxylate were also effective in causing increase transcription from the Asns gene (Barbosa-Tessmann et al. 1999b). Promoter deletion analysis narrowed the location of the *cis*-element responsible for the ERSR control of the Asns gene to nucleotides –111 to –34 of the Asns promoter, but the previously reported mammalian ERSE consensus sequence (5′-CCAAT-N_9-CCACG-3′), present in other ERSR-inducible genes, was not present in the Asns promoter (Barbosa-Tessmann et al. 1999b). As described below, Barbosa-Tessmann et al.

(2000) went on to demonstrate that activation of Asns gene transcription by either the AAR or the ERSE pathways are mediated through the same pair of unique genomic elements within the Asns proximal promoter.

In vivo footprinting revealed six protein binding sites within the Asns proximal promoter region, from nucleotide –148 to –43 (Barbosa-Tessmann et al. 2000). Of these, five were shown to contribute to nutrient control of the human Asns gene, three GC boxes (GC-I, GC-II, and GC-III) and two novel sequences, originally labeled sites V and VI, and later renamed nutrient sensing response elements (NSRE-1, -2). The collective effect of GC-I (nt –148 t0 –139), GC-II (nt-128 to – 119), and GC-III (nt –107 to –97) is to maintain a high basal transcription rate and to permit maximal activation of the Asns gene by the AAR or ERSR pathway (Leung-Pineda and Kilberg 2002). Deletion or mutagenesis of all three GC boxes simultaneously caused a severe decline in absolute transcription rate, yet induction still occurred following amino acid depletion (Barbosa-Tessmann et al. 2000; Leung-Pineda and Kilberg 2002). While each of the three GC boxes can support some level of transcription, there is not complete redundancy among them as illustrated by a difference in the ability to support both basal and starvation-activated transcription rates, GC-III > GC-II > GC-I. EMSA experiments demonstrated that GC-II and GC-III, but not GC-I, formed protein-DNA complexes with either Sp1 or Sp3 transcription factors. The amount of nuclear Sp1 and Sp3 binding activity, as well as the total cellular content of either Sp1 or Sp3 protein, did not increase following amino acid limitation. *Drosophila* SL2 cells do not express the Sp family of transcription factors (Courey and Tjian 1988) and therefore, are a useful model system for analyzing their effect by transient transfection. Expression of either Sp1 or Sp3 individually in *Drosophila* SL2 cells increased Asns promoter activity, but interestingly, Sp1 maintained basal transcription from the Asns promoter, but did not support increased expression when SL2 cells were amino acid deprived (Leung-Pineda and Kilberg 2002). In contrast, Sp3 expression enhanced both the basal and the starvation-induced Asns-driven transcription. The current view is that these Sp binding sites are permissive in function, but do not directly mediate the nutrient control response.

The NSRE-1 and NSRE-2 binding sites, identified by *in vivo* footprinting, showed increased protein binding following either amino acid or glucose deprivation (Barbosa-Tessmann et al. 2000) and subsequently, single nucleotide mutagenesis across these two sites defined the boundaries of these two elements (Kilberg et al. 2003). The NSRE-1 sequence (5′-TGATGAAAC-3′), nucleotides – 68 to –60 within the Asns promoter, coincides with the site first identified by Guerrini et al. (1993) to have AARE activity. However, from the results illustrating that the NSRE-1 sequence also mediates induction of the Asns gene following activation of the ERSR pathway following glucose limitation (Barbosa-Tessmann et al. 2000), it is clear that this element functions more broadly than simply as an AARE. This broader nutrient detecting capability is the reason that the term nutrient sensing response element was coined. The mutagenesis of the Asns promoter also confirmed the *in vivo* footprinting, which revealed the presence of a second element, NSRE-2 (nucleotides –48 to –43, 5′-GTTACA-3′), positioned eleven nucleotides downstream of NSRE-1. NSRE-2 is also absolutely required for

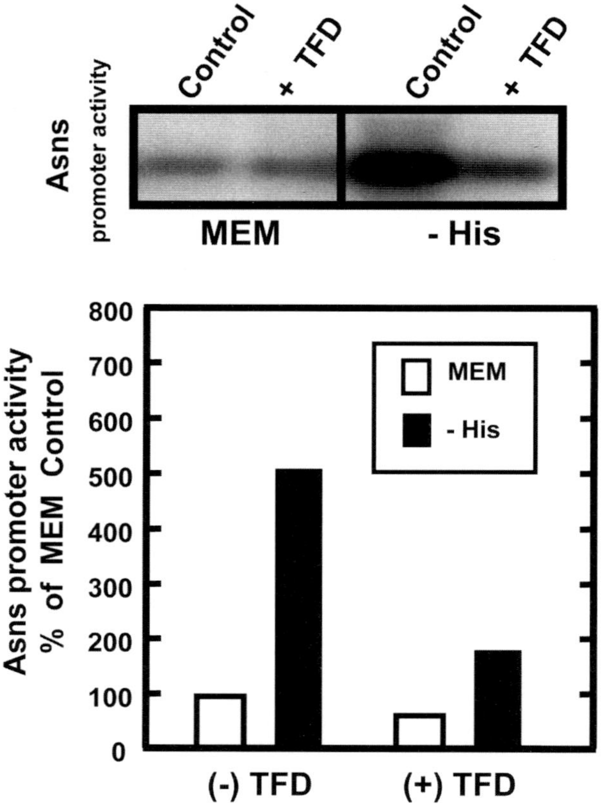

Fig. 1. Activation of Asns promoter activity is blocked by an excess of NSRU binding sites. Human HepG2 hepatoma cells were transiently co-transfected with an Asns promoter/growth hormone reporter construct as described elsewhere (Barbosa-Tessmann et al. 1999b), with (+TFD) or without (Control) a plasmid containing eight copies of the Asns NSRU sequence (nt -79/-35). After a 48 incubation, the cells were transferred to either fresh MEM (MEM) or MEM lacking histidine (- His) for 12 h prior to isolation of RNA and Northern analysis for growth hormone mRNA as described (Barbosa-Tessmann et al. 1999b). The top panel is a representative blot, whereas the bottom panel is the quantified averages of multiple experiments from independent transfections.

induction of the Asns gene by either amino acid or glucose starvation (Barbosa-Tessmann et al. 2000). The core sequence of NSRE-2 does not correspond to any known transcription factor consensus sequence. The term nutrient sensing response unit (NSRU) is used to describe the collective action of NSRE-1 and NSRE-2. To provide further evidence that the NSRU functions *in vivo* to regulate Asns gene expression in response to amino acid availability, HepG2 cells were co-transfected with an Asns expression vector and a plasmid containing multiple copies of the NSRU sequence as a "decoy" for the corresponding transcription factors (Fig. 1). Using this "transcription factor decoy" approach, the induction of Asns promoter activity following amino acid limitation was significantly blocked by an excess of NSRU binding sites.

The NSRE-1 and NSRE-2 sequences, as well as the 11 bp between them, are completely conserved in rat, mouse, hamster, and man (Zhong et al. 2003). Single nucleotide mutagenesis of the 11 bp intervening sequence does not result in loss of nutrient-regulated activity (Kilberg et al. 2003) and *in vitro* binding analysis by EMSA did not reveal formation of protein-DNA complexes when the 11 bp was used as probe (C. Zhong and M.S. Kilberg, unpublished results). However, either reduction (to 6 bp) or extension (to 16 bp) of the 11 bp sequence resulted in complete loss of regulated transcription (Zhong et al. 2003). Interestingly, the addition of 10 bp to extend the spacer region to approximately two turns of DNA also caused a loss of nutrient-controlled transcription (Zhong et al. 2003). These results indicate that NSRE-1 and NSRE-2 must be aligned on the same side of the DNA helix and only one turn away from each other. This spatial relationship is likely the requirement for protein-protein interactions that occur between the transcription factors that bind to these two sites. In support of this hypothesis, *in vitro* protein binding, using the NSRE-2 sequence as probe, does not occur in the absence NSRE-1 (Fig. 2). Consistent with the negative EMSA result, a yeast one-hybrid screen using only the NSRE-2 binding site as bait did not yield identification of the corresponding binding protein (C. Zhong and M.S. Kilberg, unpublished results), whereas a corresponding screen with NSRE-1 was successful (Siu et al. 2001).

As described above, in contrast to Asns, activation of the human CHOP gene by the AAR and ERSR pathways occurs through two completely independent sets of *cis*-acting elements, the AARE at nt -302 to -310 (Bruhat et al. 2000) and the ERSE-1 at nt -93 to -75 (Yoshida et al. 2000). The Asns NSRE-1 sequence differs from the CHOP AARE by only two nucleotides, although the CHOP AARE sequence occurs on the opposite DNA strand (Bruhat et al. 2002). Beyond the sequence difference, the second difference between the two genes is the presence of NSRE-2 in the Asns promoter and its absence in the CHOP promoter (Barbosa-Tessmann et al. 2000). Given the sequence similarity between the Asns NSRE-1 sequence and the CHOP AARE, one might have predicted that mutagenesis or deletion of the Asns NSRE-2 sequence only, with retention of the NSRE-1 site, would block activation of the Asns gene by the ERSR pathway, but permit continued activation by the AAR pathway. However, mutagenesis of any one of the core nucleotides within the NSRE-2 sequence caused a complete loss of Asns responsiveness to both nutrient-regulated pathways (Barbosa-Tessmann et al. 2000). Therefore, the data indicate that the presence of the NSRE-1-like AARE sequence in the context of the CHOP promoter is sufficient to permit transcriptional induction via the AAR pathway, whereas the related Asns NSRE-1 sequence alone is not. Interestingly, insertion of the Asns NSRE-2 sequence 11 nucleotides downstream from the CHOP AARE, such that the physical relation is the same as for NSRE-1 and NSRE-2 in the Asns promoter, conveys responsiveness to the ERSR pathway of the CHOP fragment (Bruhat et al. 2002). Thus, NSRE-2 can transfer ERSR-activated transcription to an AARE-containing promoter, but cannot function alone as a nutrient sensing element.

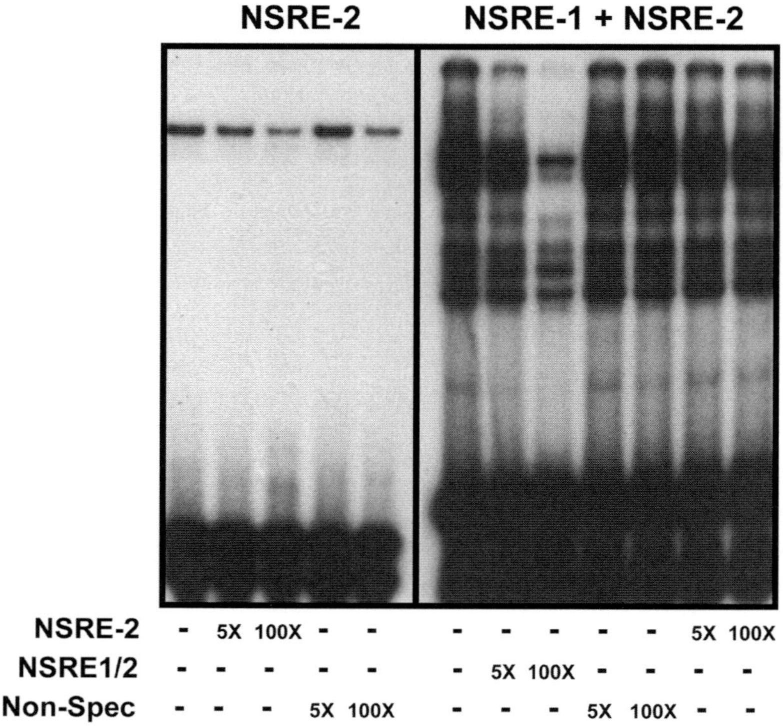

Fig. 2. Electrophoresis mobility shift analysis demonstrates an inability of the Asns promoter NSRE-2 site itself to form specific DNA-protein complexes. Nuclear extracts prepared from HepG2 cells maintained for 16 hours in histidine-free MEM were incubated with 32P-radiolabeled oligonucleotide probes (labeled above the lanes) corresponding to either the NSRE-2 site alone (nt –55/-35 of the Asns proximal promoter) or both NSRE-1 and NSRE-2 sites together (nt -79/-35 of the Asns proximal promoter). The fold-excess of unlabeled competitor oligonucleotides that were included in the incubation is shown below each lane of the gel. An unrelated oligonucleotide (5′-TTGTCGACCTCACAG TGGCTGCTATGTATGC-3′) was used to test for non-specific competition (Non-Spec). The EMSA analysis was performed as described previously (Siu et al. 2002). The results show that the NSRE-2 site by itself does not form specific DNA-protein complexes, whereas when the probe contained both NSRE-1 + NSRE-2 a number of complexes were readily apparent. In addition, an excess of unlabeled NSRE-2 oligonucleotide did not block any of the complexes formed by the NSRE-1 + NSRE-2 probe.

1.5 Transcription factors associated with asparagine synthetase regulation

1.5.1 C/EBP family

The Asns NSRE-1 binding site sequence was used as bait for a yeast one-hybrid screen to identify potential transcription factors specific for this sequence. Those results and subsequent EMSA data indicated that several members of the CCAAT-enhancer binding protein (C/EBP) family could recognize the NSRE-1 sequence (Siu et al. 2001). The C/EBP transcription proteins represent a subclass of the basic leucine zipper (bZIP) family of transcription proteins (Lie-Venema et al. 1998; Takiguchi, 1998), and most have been shown to homodimerize and to heterodimerize with other C/EBP members, or heterodimerize with other bZIP members (Vinson et al. 2002). The C/EBP family includes individual members designated α, β, ϒ δ, ε, and CHOP, which have a wide array of effects on cellular metabolism, differentiation, and growth. The level of expression for each of these family members varies among tissues and cell types. In human HepG2 hepatoma cells, C/EBPβ mRNA is expressed at a high level, but is subject to differential translational start site selection such that both an activating (liver-enriched transcriptional activator protein, LAP) and an inhibitory (liver-enriched transcriptional inhibitory protein, LIP) isoform can be generated from the same mRNA (Descombes and Schibler, 1991). LIP is a shorter protein that corresponds to the C-terminal portion of LAP. LIP lacks the transactivation domain of LAP, but contains the bZIP dimerization region and therefore, acts as a dominant negative repressor of C/EBP function. For example, co-expression of LIP can override the LAP-dependent cell cycle arrest in hepatoma cells (Buck et al. 1994).

Supershift analysis using the NSRE-1 sequence and antibodies specific for individual C/EBP family members demonstrated C/EBPβ binding that was increased when nuclear extracts from cells subjected to amino acid limitation or activation of the ER stress response were tested (Siu et al. 2001). Further evidence that C/EBPβ functions *in vivo* to regulate transcription of the Asns gene was obtained by the observation that basal transcription driven by the Asns promoter was increased after C/EBPβ (LAP) overexpression and conversely, that overexpression of the dominant negative isoform LIP caused a blockade of both basal and starvation-induced transcription. These results are consistent with the report that C/EBPβ mRNA content is increased in rat hepatoma cells following amino acid starvation (Marten et al. 1994), although the Northern analysis in that study does not establish whether LAP or LIP was produced from the increased mRNA.

1.5.2 ATF4

The activating transcription factor (ATF) family of transcription proteins represents a second subclass of the bZIP family and selected ATF members are known to heterodimerize with members of the C/EBP bZIP subgroup. In particular, a

C/EBPβ and ATF4 complex has been detected at cAMP response elements (Vallejo et al. 1993) and the crystal structure of a C/EBPβ–ATF4 complex has been published (Podust et al. 2001). As mentioned above, the proximal promoter region of the human CHOP gene contains an AARE (5′-TGATGCAAT-3′) that differs from the Asns NSRE-1 sequence by only two nucleotides, and has been shown to be a C/EBP-ATF composite site (Fawcett et al. 1999; Wolfgang et al. 1997). Fawcett et al. (Fawcett et al. 1999) reported a transient ATF4 binding to this site in response to arsenite-induced stress, which is then replaced by ATF3 binding which causes a suppression of the gene back towards the basal expression rate. This observation, coupled with increased translation of ATF4 following amino acid deprivation (Harding et al. 2000) and the known characteristics of Asns transcription are consistent with a proposed role for ATF4. *In vitro* binding analysis revealed that ATF4 binding had affinity for the NSRE-1 sequence and binding was increased when nuclear extracts from either histidine-deprived (AAR pathway) or glucose-deprived cells were tested (Siu et al. 2002). Following nutrient limitation, there is a lag of about 4 h prior to a significant increase in Asns mRNA content (Barbosa-Tessmann et al. 1999a; Hutson and Kilberg, 1994) and this increase in Asns mRNA is protein synthesis dependent (Hutson et al. 1996; Hutson and Kilberg, 1994). Those results indicate that synthesis of a regulatory protein is required prior to activation of Asns gene transcription. In support of this proposal, inhibition of protein synthesis blocked the starvation-dependent enhancement in protein-NSRE-1 complex formation as assayed by EMSA and completely prevented the increase in ATF4 binding to NSRE-1 (Siu et al. 2002). Collectively, these results are consistent with the observation that translation of ATF4 from pre-existing mRNA is enhanced following amino acid deprivation (Harding et al. 2000). As further evidence for ATF4 modulation, Asns was identified during a microarray screen as a gene for which stress-induced expression was suppressed (14% of control) in ATF4-deficient cells (Harding et al. 2003). Finally, the basal rate of Asns promoter-driven transcription was induced in ATF4 overexpressing cells (Siu et al. 2002), and expression of a dominant negative ATF4 mutant prevented nutrient control of the Asns promoter.

1.5.3 ATF3

By EMSA, binding to the NSRE-1 sequence by another ATF family member, ATF3, was observed (Siu et al. 2002). The full-length ATF3 protein (ATF3-FL) contains the basic region and leucine zipper motifs characteristic of the bZIP superfamily of transcription factors and it can homodimerize, but can also heterodimerize with c-Jun, JunB, JunD, ATF2, and CHOP to facilitate DNA binding to an ATF/CRE or AP-1 consensus site (Vinson et al. 2002). The transcriptional consequences are different depending on whether ATF3-FL binds as a homodimer, in which case it appears to most often act to repress transcription, or as a heterodimer with other bZIP family members, in which case it can either repress or activate transcription (Hai and Hartman 2001). When ATF3 mRNA and protein content was assayed following activation of either the AAR or ERSR

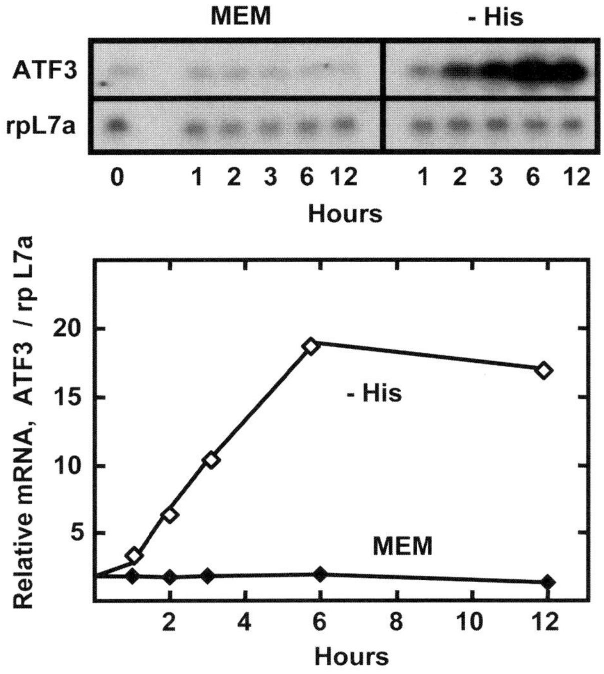

Fig. 3. Time-course of ATF3 mRNA content following histidine deprivation of HepG2 cells. HepG2 cells were incubated for 0–12 hours in MEM (MEM) or MEM lacking histidine (- His). At the times indicated, RNA was isolated and subjected to Northern analysis (20 μg of RNA/lane) for ATF3 or ribosomal protein L7a mRNA content, as described elsewhere (Siu et al. 2002). The results shown in the top panel are a representative blot, whereas the bottom graph shows the quantified data, expressed as a ratio of the value obtained for the loading control, ribosomal protein L7a.

pathways in HepG2 hepatoma cells, it was determined that both nutrient signaling pathways increased ATF3 expression (Fig. 3; Y. Pan and M.S. Kilberg, unpublished results). However, as described below, with regard to ATF3 protein expression this observation is more complicated that it first appears because of the complex synthesis of ATF3 isoforms. The human ATF3 gene covers 55 kilobases on chromosome 1q32.2. The full-length ATF3 mRNA is 1914 bp, is encoded by four exons, and generates a protein of 181 amino acids (Liang et al. 1996). Interestingly, although no experiments have been reported on this subject, the 3′-untranslated region of ATF3-FL mRNA encoded by exon E contains several AUUUA sequences, which is a motif that may help to regulate ATF3 mRNA stability. It has been shown that ATF2 and Jun can activate ATF3 promoter activity (Liang et al. 1996), and recently, Wolfgang et al. (Wolfgang et al. 2000) identified an element (5′-TGATGCAAC-3′) at nt –20 in the ATF3 promoter as an ATF3-binding site responsible for auto-regulating the ATF3 gene by repressing transcription. ATF3, is considered to be a stress response gene (Hai et al. 1999; Hai and Hartman, 2001), is expressed at a low level in quiescent cells, but can be rapidly induced in response to diverse extracellular stress signals (Chen et al. 1996; Hai et al. 1999; Hashimoto et al. 2002). Activation of the AAR pathway by amino acid deprivation can now be included in this list (Fig. 3; Y. Pan and M.S. Kilberg, unpublished results).

Alternative splicing is a mechanism that permits the number of functionally diverse proteins expressed by an organism to exceed the number of genes contained within the genome. Alternative-splicing within the ATF3 gene is known to occur and the Hai laboratory has reported three truncated ATF3 isoforms, ATF3ΔZip (Chen et al. 1994), and ATF3ΔZip2a and ΔZip2b (Hashimoto et al. 2002). ATF3ΔZip expression was enhanced in HeLa cells stimulated by serum (Chen et al. 1994), and represents a truncated protein, of 115 amino acids, lacking much of the leucine zipper domain. In contrast, ATF3ΔZip2a and 2b were detected after treatment of primary human umbilical vein endothelial cells by several stress-associated stimuli (Hashimoto et al. 2002). ATF3ΔZip2a and 2b also encode truncated proteins that lack portions of the leucine zipper dimerization domain. In contrast to ATF3-FL, which is thought to primarily act as a transcriptional repressor (Hai et al. 1999), it has been proposed that the truncated isoforms (ATF3ΔZip, ATF3ΔZip2a, and ATF3ΔZip2b) can function as transcriptional activators indirectly by sequestering co-repressor proteins. This activation is thought to occur because the lack of a functional leucine zipper dimerization domain in the truncated ATF3ΔZip isoforms does not permit DNA binding, but these proteins do retain the binding site for ATF3-associated co-repressors. Therefore, increased expression of these isoforms causes gene activation indirectly by sequestering co-repressor complexes. Evidence for this hypothesis comes from the observation that the ATF3ΔZip isoform can still activate promoters lacking an ATF3 binding site (Chen et al. 1994). Recently, two additional truncated ATF3 isoforms (ATF3ΔZip2c and ATF3ΔZip3) were identified in human HepG2 hepatoma cells (Y. Pan and M.S. Kilberg, unpublished results). The expression of both of these isoforms was increased by activation of either the AAR or ERSR pathways. ATF3ΔZip2c protein, containing 106 amino acids, has a truncated activation domain compared to all other known ATF3 isoforms, but also has a shortened C-terminal that deletes much of the leucine zipper domain. The ATF3ΔZip3 mRNA has an insertion extending from exon C to D′ resulting in a truncated protein of 120 amino acids and a nearly complete loss of the leucine zipper dimerization domain.

Transient expression of ATF3-FL alone caused an inhibition of basal Asns promoter activity in cells maintained in complete MEM medium, consistent with its known repressor activity when present as a homodimer (Hai and Hartman, 2001). However, co-expression of ATF3-FL and ATF4 caused a biphasic response, low ATF3-FL plasmid concentrations (<1 ng/10^5cells) resulted in a further enhancement of ATF4-induced transcription from the Asns promoter, but high ATF3-FL plasmid concentrations (1-100 ng/10^5 cells) caused an antagonism of ATF4 activation. The initial stimulation by ATF3-FL may be due to an "indirect squelching" effect mediated by removing co-repressor molecules from DNA binding sites (Gill and Ptashne, 1988). Chen et al. (Chen et al. 1994) also reported that ATF3-FL can both repress and activate transcription, depending on the expression level. Transient expression of the ATF3ΔZip2c isoform had no detectable effect on Asns promoter-driven transcription, but expression of ATF3ΔZip3 further enhanced ATF4-induced Asns promoter activity in a concentration-dependent manner. The cellular signals and the mechanisms for regulation of the splicing events that determine which ATF3 isoforms are produced are unknown, but these results indicate that mechanisms for nutrient control of gene expression may also include differential gene splicing.

1.6 Summary

Our understanding of how nutrients control gene expression in mammalian cells remains relatively limited, but significant progress has been achieved during the past few years. With regard to amino acid control of transcription, the full complement of target genes has not yet been established, but DNA microarray technology should reveal the remaining candidates. Interestingly, even among the relatively small number of genes that are known to be transcriptionally activated by amino acid limitation, there is already emerging evidence that the mechanisms differ significantly from one gene to another (Bain et al. 2002; Bruhat et al. 2002). Clearly, there is much yet to learn about the signal pathways and the molecular mechanisms responsible for transcriptional regulation by nutrients. Beyond gaining a basic understanding of these important biological control mechanisms, the characterization of how these processes contribute to the pathology of various disease states represents an exciting aspect of future investigations in molecular nutrition.

References

Abcouwer SF, Schwarz C, Meguid RA (1999) Glutamine deprivation induces the expression of *GADD45* and *GADD153* primarily by mRNA stabilization. J Biol Chem 274:28645-28651

Adilakshmi T, Laine R O (2002) Ribosomal protein S25 mRNA partners with MTF-1 and La to provide a p53-mediated mechanism for survival or death. J Biol Chem 277:4147-4151

Andrulis IL, Hatfield GW, Arfin SM (1979) Asparaginyl-tRNA aminoacylation levels and asparagine synthetase expression in cultured Chinese hamster ovary cells. J Biol Chem 254:10629-10633

Arfin SM, Simpson DR, Chiang CS, Andrulis IL, Hatfield GW (1977) A role for asparaginyl-tRNA in the regulation of asparagine synthetase in a mammalian cell line. Proc Natl Acad Sci USA 74:2367-2369

Aulak KS, Mishra R, Zhou L, Hyatt SL, de Jonge W, Lamers W, Snider M, HatzoglouM (1999) Post-transcriptional regulation of the arginine transporter Cat-1 by amino acid availability. J Biol Chem 274:30424-30432

Bain PJ, LeBlanc-Chaffin R, Chen H, Palii SS, Leach KM, Kilberg MS (2002) The mechanism for transcriptional activation of the human ATA2 transporter gene by amino acid deprivation is different than that for asparagine synthetase. J Nutr 132:3023-3029

Barbosa-Tessmann IP, Chen C, Zhong C, Schuster SM, Nick HS, Kilberg MS (1999b) Activation of the unfolded protein response pathway induces human asparagine synthetase gene expression. J Biol Chem 274:31139-31144

Barbosa-Tessmann IP, Chen C, Zhong C, Siu F, Schuster SM, Nick HS, Kilberg MS (2000) Activation of the human asparagine synthetase gene by the amino acid response and the endoplasmic reticulum stress response pathways occurs by common genomic elements. J Biol Chem 275:26976-26985

Barbosa-Tessmann IP, Pineda VL, Nick HS, Schuster SM, Kilberg MS (1999a) Transcriptional regulation of the human asparagine synthetase gene by carbohydrate availability. Biochem J 339:151-158

Berlanga JJ, Santoyo J, De Haro C (1999) Characterization of a mammalian homolog of the GCN2 eukaryotic initiation factor 2alpha kinase. Eur J Biochem 265:754-762

Bruhat A, Averous J, Carraro V, Zhong C, Reimold AM, Kilberg MS, Fafournoux P (2002) Differences in the molecular mechanisms involved in the transcriptional activation of the CHOP and asparagine synthetase genes in response to amino acid deprivation or activation of the unfolded protein response. J Biol Chem 277:48107-48114

Bruhat A, Jousse C, Carraro V, Reimold AM, Ferrara M, Fafournoux P (2000) Amino acids control mammalian gene transcription: activating transcription factor 2 is essential for the amino acid responsiveness of the CHOP promoter. Mol Cell Biol 20:7192-7204

Bruhat A, Jousse C, Fafournoux P (1999) Amino acid limitation regulates gene expression. Proc Nutr Soc 58:625-632

Bruhat A, Jousse C, Wang X-Z, Ron D, Ferrara M, Fafournoux P (1997) Amino acid limitation induces expression of CHOP, a CCAAT/enhancer binding protein-related gene, at both transcriptional and post-transcriptional levels. J Biol Chem 272:17588-17593

Buck M, Turler H, Chojkier M (1994) LAP (NF-IL-6), a tissue-specific transcriptional activator, is an inhibitor of hepatoma cell proliferation. EMBO J 13:851-860

Chen BP, Liang G, Whelan J, Hai T (1994) ATF3 and ATF3 delta Zip. Transcriptional repression versus activation by alternatively spliced isoforms. J Biol Chem 269:15819-15826

Chen BP, Wolfgang CD, Hai T (1996) Analysis of ATF3, a transcription factor induced by physiological stresses and modulated by gadd153/Chop10. Mol Cell Biol 16:1157-1168

Courey AJ, Tjian R (1988) Analysis of Sp1 *in vivo* reveals multiple transcriptional domains, including a novel glutamine-rich activation motif. Cell 55:887-898

Descombes P, Schibler U (1991) A liver-enriched transcriptional activator protein, LAP, and a transcriptional inhibitory protein, LIP, are translated from the same mRNA. Cell 67:569-579

Fafournoux P, Bruhat A, Jousse C (2000) Amino acid regulation of gene expression. Biochem J 351:1-12

Fawcett TW, Martindale JL, Guyton KZ, Hai T, Holbrook NJ (1999) Complexes containing activating transcription factor (ATF)/cAMP-responsive-element-binding protein (CREB) interact with the CCAAT/enhancer-binding protein (C/EBP)-ATF composite

site to regulate *Gadd153* expression during the stress response. Biochem J 339:135-141

Fernandez J, Yaman I, Mishra R, Merrick WC, Snider MD, Lamers WH, Hatzoglou M (2001) Internal ribosome entry site-mediated translation of a mammalian mRNA is regulated by amino acid availability. J Biol Chem 276:12285-12291

Gazzola GC, Franchi R, Saibene V, Ronchi P, Guidotti GG (1972) Regulation of Amino Acid transport in chick embryo heart cells I. Adaptive system of mediation for neutral Amino Acids. Biochim Biophys Acta 266:407-421

Gazzola RF, Sala R, Bussolati O, Visigalli R, Dall'Asta V, Ganapathy V, Gazzola GC (2001) The adaptive regulation of amino acid transport system A is associated to changes in ATA2 expression. FEBS Lett 490:11-14

Gill G, Ptashne M (1988) Negative effect of the transcriptional activator GAL4. Nature 334:721-724

Gong SS, Basilico C (1990) A mammalian temperature-sensitive mutation affecting G1 progression results from a single amino acid substitution in asparagine synthetase. Nucleic Acids Res 18:3509-3513

Gong SS, Guerrini L, Basilico C (1991) Regulation of asparagine synthetase gene expression by amino acid starvation. Mol Cell Biol 11:6059-6066

Guerrini L, Gong SS, Mangasarian K, Basilico C (1993) *Cis*- and *trans*-acting elements involved in amino acid regulation of asparagine synthetase gene expression. Mol Cell Biol 13:3202-3212

Hai T, Hartman MG (2001) The molecular biology and nomenclature of the activating transcription factor/cAMP responsive element binding family of transcription factors: activating transcription factor proteins and homeostasis. Gene 273:1-11

Hai T, Wolfgang CD, Marsee DK, Allen AE, Sivaprasad U (1999) ATF3 and stress responses. Gene Expr 7:321-335

Hansen BS, Vaughan MH, Wang L-J (1972) Reversible inhibition by histidinol of protein synthesis in human cells at the activation of histidine. J Biol Chem 47:3854-3857

Harding HP, Novoa I, Zhang Y, Zeng H, Wek R, Schapira M, Ron D (2000) Regulated translation initiation controls stress-induced gene expression in mammalian cells. Mol Cell 6:1099-1108

Harding HP, Zhang Y, Zeng H, Novoa I, Lu PD, Calfon M, Sadri N, Yun C, Popko B, Paules R, Stojdl DF, Bell JC, Hettmann T, Leiden JM, Ron D (2003) An integrated stress response regulates amino acid metabolism and resistance to oxidative stress. Mol Cell 11:619-633

Hashimoto Y, Zhang C, Kawauchi J, Imoto I, Adachi MT, Inazawa J, Amagasa T, Hai T, Kitajima S (2002) An alternatively spliced isoform of transcriptional repressor ATF3 and its induction by stress stimuli. Nucleic Acids Res 30:2398-2406

Hinnebusch AG (1997) Translational regulation of yeast GCN4. J Biol Chem 72:21661-21664

Hitomi Y, Ito A, Naito Y, Yoshida A (1993) Liver-specific induction of ribosomal protein gene expression by amino acid starvation in rats. Biosci Biotechnol Biochem 57:1216-1217

Hutson RG, Kilberg MS (1994) Cloning of rat asparagine synthetase and specificity of the amino acid-dependent control of its mRNA content. Biochem J 03:745-750

Hutson RG, Warskulat U, Kilberg MS (1996) An example of nutrient control of gene expression: amino acid-dependent regulation of asparagine synthetase. Clin Nutr 5:327-331

Hyatt SL, Aulak KS, Malandro M, Kilberg MS, Hatzoglou M (1997) Adaptive regulation of the cationic amino acid transporter-1 (Cat-1) in Fao cells. J Biol Chem 72:19951-19957

Jousse C, Bruhat A, Ferrara M, Fafournoux P (1998) Physiological concentration of amino acids regulates insulin-like-growth-factor-binding protein 1 expression. Biochem J 34:147-153

Jousse C, Bruhat A, Harding HP, Ferrara M, Ron D, Fafournoux P (1999) Amino acid limitation regulates CHOP expression through a specific pathway independent of the unfolded protein response. FEBS Lett 48:211-216

Kaufman RJ (2002) Orchestrating the unfolded protein response in health and disease. J Clin Invest 110:1389-1398

Kilberg MS, Barbosa-Tessmann IP (2002) Genomic sequences necessary for transcriptional activation by amino acid deprivation of mammalian cells. J Nutr 32:1801-1804

Kilberg MS, Leung-Pineda V, Chen C (2003) Amino acid-dependent control of transcription in mammalian cells, in, "Molecular Nutrition", (H. Daniel and J. Zempleni, eds), CABI Publishing, Wallington, UK, pp 105-119

Kilberg MS, Stevens BR, Novak D (1993) Recent advances in mammalian amino acid transport. Ann Rev Nutr 3:137-165

Laine RO, Laipis PJ, Shay NF, Kilberg MS (1991) Identification of an amino acid-regulated mRNA from rat liver as the mammalian equivalent of bacterial ribosomal protein L22. J Biol Chem 66:16969-16972

Laine RO, Shay NF, Kilberg MS (1994) Nuclear retention of the induced mRNA following amino acid-dependent transcriptional regulation of mammalian ribosomal proteins L17 and S25. J Biol Chem 69:9693-9697

Leung-Pineda V, Kilberg MS (2002) Role of Sp1 and Sp3 in the nutrient-regulated expression of the human asparagine synthetase gene. J Biol Chem 277:16585-16591

Liang G, Wolfgang CD, Chen BP, Chen TH, Hai T (1996) ATF3 gene. Genomic organization, promoter, and regulation. J Biol Chem 271:1695-1701

Lie-Venema H, Hakvoort TB, van Hemert FJ, Moorman AF, Lamers WH (1998) Regulation of the spatiotemporal pattern of expression of the glutamine synthetase gene. Prog Nucleic Acid Res Mol Biol 1:243-308

Marten NW, Burke EJ, Hayden JM, Straus DS (1994) Effect of amino acid limitation on the expession of 19 genes in rat hepatoma cells. FASEB J :538-544

Natarajan K, Meyer MR, Jackson BM, Slade D, Roberts C, Hinnebusch AG, Marton MJ (2001) Transcriptional profiling shows that Gcn4p is a master regulator of gene expression during amino acid starvation in yeast. Mol Cell Biol 1:4347-4368

Palacín M, Estevez R, Bertran J, Zorzano A (1998) Molecular biology of mammalian plasma membrane amino acid transporters. Physiol Rev 8:969-1054

Patil C, Walter P (2001) Intracellular signaling from the endoplasmic reticulum to the nucleus: the unfolded protein response in yeast and mammals. Curr Opin Cell Biol 13:349-355

Paul GL, Waegner A, Gaskins HR, Shay NF (1998) Histidine availability alters glucagon gene expression in murine alphaTC6 cells. J Nutr 128:973-976

Podust LM, Krezel AM, Kim Y (2001) Crystal structure of the CCAAT box/enhancer-binding protein beta activating transcription factor-4 basic leucine zipper heterodimer in the absence of DNA. J Biol Chem 76:505-513

Richards NGJ, Schuster SM (1998) Mechanistic issues in asparagine synthetase catalysis. Adv Enzymol 72:145-198

Ron D, Habener JF (1992) CHOP, a novel developmentally regulated nuclear protein that dimerizes with transcription factors C/EBP and LAP and functions as a dominant-negative inhibitor of gene transcription. Genes Dev 10:439-453

Siu F, Bain PJ, LeBlanc-Chaffin R, Chen H, Kilberg MS (2002) ATF4 is a mediator of the nutrient-sensing response pathway that activates the human asparagine synthetase gene. J Biol Chem 277:24120-24127

Siu FY, Chen C, Zhong C, Kilberg MS (2001) CCAAT/enhancer-binding protein beta (C/EBPb) is a mediator of the nutrient sensing response pathway that activates the human asparagine synthetase gene. J Biol Chem 276:48100-48107

Sok J, Wang XZ, Batchvarova N, Kuroda M, Harding H, Ron D (1999) CHOP-dependent stress-inducible expression of a novel form of carbonic anhydrase VI. Mol Cell Biol 19:495-504

Sood R, Porter AC, Olsen DA, Cavener DR, Wek RC (2000) A mammalian homologue of GCN2 protein kinase important for translational control by phosphorylation of eukaryotic initiation factor-2α. Genetics 154:787-801

Straus DS, Burke EJ, Marten NW (1993) Induction of insulin-like growth factor binding protein-1 gene expression in liver of protein-restricted rats and in rat hepatoma cells limited for a single amino acid. Endocrinology 132:1090-1100

Takiguchi M (1998) The C/EBP family of transcription factors in the liver and other organs. Int J Exp Pathol 79:369-391

Vallejo M, Ron D, Miller CP, Habener JF (1993) C/ATF, a member of the activating transcription factor family of DNA-binding proteins, dimerizes with CAAT/enhancer-binding proteins and directs their binding to cAMP response elements. Proc Natl Acad Sci USA 90:4679-4683

Vinson C, Myakishev M, Acharya A, Mir AA, Moll JR, Bonovich M (2002) Classification of human B-ZIP proteins based on dimerization properties. Mol Cell Biol 22:6321-6335

Wang X-Z, Kuroda M, Sok J, Batchvarova N, Kimmel R, Chung P, Zinszner H, Ron D (1998) Identification of novel stress-induced genes downstream of *chop*. EMBO J 17:3619-3630

Wolfgang CD, Chen BP, Martindale JL, Holbrook NJ, Hai T (1997) gadd153/Chop10, a potential target gene of the transcriptional repressor ATF3. Mol Cell Biol 17:6700-6707

Wolfgang CD, Liang G, Okamoto Y, Allen AE, Hai T (2000) Transcriptional autorepression of the stress-inducible gene ATF3. J Biol Chem 275:16865-16870

Yoshida H, Haze K, Yanagi H, Yura T, Mori K (1998) Identification of the cis-acting endoplasmic reticulum stress response element responsible for transcriptional induction of mammalian glucose-regulated proteins. J Biol Chem 273:33741-33749

Yoshida H, Matsui T, Hosokawa N, Kaufman RJ, Nagata K, Mori K (2003) A time-dependent phase shift in the mammalian unfolded protein response. Dev Cell 4:265-271

Yoshida H, Matsui T, Yamamoto A, Okada T, Mori K (2001) XBP1 mRNA is induced by ATF6 and spliced by IRE1 in response to ER stress to produce a highly active transcription factor. Cell 107:881-891

Yoshida H, Yoshida H, Okada T, Haze K, Yanagi H, Yura T, Negishi M, Mori K (2000) ATF6 activated by proteolysis binds in the presence of NF-Y (CBF) directly to the cis-acting element responsible for the mammalian unfolded protein response. Mol Cell Biol 20:6755-6767

Zhong C, Chen C, Kilberg MS (2003) Characterization of the nutrient sensing response unit in the human asparagine synthetase promoter. Biochem J 372:603-609
Zinszner H, Kuroda M, Wang X, Batchvarova N, Lightfoot RT, Remotti H, Stevens JL, Ron D (1998) CHOP is implicated in programmed cell death in response to impaired function of the endoplasmic reticulum. Genes Dev 12:982-995

Abbreviations

AAR(E): amino acid response (element)
ATF: activating transcription factor
CAT1: cationic amino acid transporter 1
C/EBP: CCAAT enhancer binding protein
CHOP: C/EBP homology binding protein
EMSA: electrophoresis mobility shift analysis
ERSR: endoplasmic reticulum stress response
IGFBP-1: insulin-like growth factor binding protein 1
NSR(E): nutrient sensing response (element)
LAP: liver-enriched activating protein
LIP: liver-enriched inhibitory protein
SNAT2: sodium neutral amino acid transporter 2

2 Nutrient sensing in animal cells and integration of nutrient and endocrine signalling pathways

Christopher G. Proud, Harinder S. Hundal, and Peter M. Taylor

Abstract

Certain nutrients are potent regulators of cell function alongside their essential role in metabolism. Endocrine and nervous systems of higher animals are often regarded as having the dominant role in regulating the responses of tissues to altered nutrient availability, but there is mounting evidence that particular nutrients (perhaps acting through specific receptor or "sensor" mechanisms) have the capability to initiate cell-signalling events and regulate gene expression independently of hormonal influences. Several dietary factors (including glucose, amino acids, and polyunsaturated fatty acids) have now been implicated as specific regulators of gene expression in animal cells. We provide an overview of current understanding of the mechanisms by which these three key groups of macronutrient contribute to regulation of animal cell function and, in addition, how they may also contribute substantially to regulation of endocrine mechanisms.

2.1 Introduction

Certain nutrients are potent regulators of cell function alongside their essential role in metabolism. In lower eukaryotes and prokaryotes, the availability of a specific nutrient may regulate its own acquisition, synthesis, and utilisation as well as modulate general cell metabolism (Fafournoux et al. 2000; Forsberg and Ljungdahl 2001). In contrast, endocrine and neuronal systems of higher eukaryotes are generally considered to have the dominant role in regulating the responses of tissues to altered nutrient availability, at least *in vivo* (Fafournoux et al. 2000; Van Sluijters et al. 2000; Proud 2002). However, there is now mounting evidence (largely from studies in mammalian cells) that certain nutrients, possibly acting through specific nutrient receptor or "sensor" mechanisms, have the capability to initiate cell signalling events and regulate gene expression independently of hormonal influences. Several dietary factors (including vitamins, polyunsaturated fatty acids, cholesterol, minerals, glucose, and amino acids) have now been implicated as specific regulators of gene expression. In addition, certain nutrients may also contribute substantially to regulation of endocrine mechanisms - in particular there is clear evidence of a relationship between nutrient availability and insulin action, al-

Topics in Current Genetics, Vol. 7
J. Winderickx, P.M. Taylor (Eds) Nutrient-Induced Responses in Eukaryotic Cells

though debate continues as to the mechanism(s) by which nutrient availability modulates insulin signalling in health and disease (Proud 2002).

Regulated gene expression in higher eukaryotes requires precise control over the processes of DNA transcription, mRNA translation and associated RNA/protein processing and degradation mechanisms. The availability of substrates for the biosynthetic pathways of transcription and translation (ribonucleotides and amino acids respectively) is not generally believed to be a limiting factor for these processes *in vivo*. Nevertheless, nutrients are able to both directly and indirectly influence gene transcription, mRNA processing, mRNA stability and/or mRNA translation. Such nutrient-gene interactions are now recognised to have an important role in regulation of gene expression. The major nutritional components of the diet in animals are (i) protein, (ii) carbohydrate, and (iii) lipid. The digestion and absorption of these macronutrients generates respectively (i) amino acids, (ii) glucose plus other sugars, and (iii) fatty acids, glycerol and sterols of various types. The aim of this article is to provide a summary of current knowledge on the sensing, signalling and effector mechanisms involved in such regulation. Direct effects include interaction with transcription factors to modulate specific gene expression and the activation of specific intracellular signalling pathways (notably the mTOR pathway in mammals). Indirect effects include stimulation of endocrine mechanisms such as insulin secretion (Patti et al. 1998; Itoh et al. 2003) as well as effects on the general metabolic status of the cell; sugars and fatty acids in particular are major cellular fuels, which when oxidised alter both the energy charge and redox status of the cell. All of the above factors have separate and integrated actions on cellular and whole body function (see Fig. 1). Several types of covalent modification of proteins alter their activity or cellular distribution and these modifications involve chemical groups (e.g. acetyl, methyl, nitro, and phospho moieties) the abundance of which may, at least indirectly, reflect nutrient availability.

Eukaryotic nuclear DNA is tightly associated with nucleoproteins, predominantly the basic histones, to form chromatin. Histone acetylation has a potent effect on transcription because it “slackens” DNA-histone complexes and facilitates access of transcriptional elements to DNA. The dynamic effects of histone acetyltransferases and deacetylases (acting as transcriptional co-activators and repressors respectively) on chromatin structure are proposed as an important mechanism for the reversible activation and repression of transcription, particularly during development (Lehrmann et al. 2002; Schreiber and Bernstein 2002; Fischle et al. 2003). The nutrient-sensitive TOR signalling pathway links nutrient sensing with histone acetylation and cell growth in yeast (Rohde and Cardenas 2003), although there is as yet no direct evidence that this occurs in higher eukaryotes. Transcription may also be modulated directly by interactions of intracellular nutrients with transcription enhancer or repressor elements of specific genes. Transcription and translation are also both regulated by a number of cell-signalling pathways in response to exogenous stimuli such as peptide hormones, mechanical and chemical stresses as well as nutrients (for review see Cohen 2002; de Nadal et al.2002; Ingber 2002). Early "upstream" signalling elements (including receptor activation) frequently involve tyrosine-kinase (or lipid-kinase) activity, whereas later "downstream" elements are generally serine/threonine kinases. Protein phosphorylation

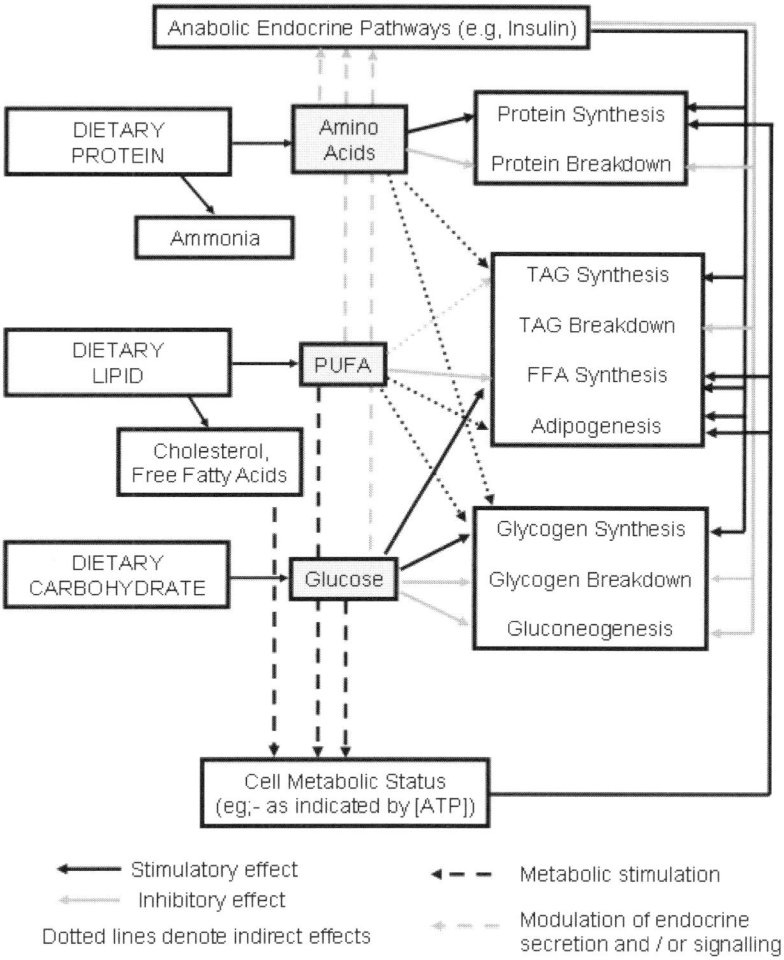

Fig. 1. Interactions between major types of macronutrient, endocrine pathways, and metabolism in mammalian cells.

may have stimulatory or inhibitory effects on downstream targets and protein phosphatases provide an additional level of regulation (Cohen 2002; Klumpp and Krieglstein 2002; Theodosiou and Ashworth 2002; Asante-Appiah and Kennedy 2003).

2.2. Amino acids

2.2.1 Amino acid-induced responses in animal cells

Cells take up or synthesise amino acids as precursors in the synthesis of proteins, whilst simultaneously generating free amino acids from protein breakdown. Numerous studies have demonstrated that elevated amino acid availability generally sustains anabolism and inhibits catabolism in eukaryotic cells (reviewed in Fafournoux et al. 2000; Van Sluijters et al. 2000). It is now well recognised that certain amino acids exert powerful regulatory control over fundamental cellular processes such as the synthesis and degradation of protein, glycogen and lipid both *in vitro* (Krause et al. 2002b) and *in vivo* (Varnier et al. 1995; Patti et al. 1998; Kimball 2002). Thus, global protein synthesis is increased and proteolysis is at least relatively decreased when the precursors for protein synthesis are abundant. Increased amino acid availability can also increase global mRNA abundance, by enhancing both transcriptional activity and mRNA stability (Balavoine et al. 1993; Fafournoux et al. 2000; Proud 2002). When amino acids are scarce the above effects are largely reversed, and it has long been known that starvation or lack of nutrients reduces global rates of mRNA translation (protein synthesis) in mammalian tissues and cells, largely related to the fact that protein synthesis is one of the major energy consuming processes of the cell (Kimball 2002; Proud 2002). Protein kinases that phosphorylate the α-subunit of eukaryotic initiation factor 2 (eIF2α) are also activated in stressed cells and they negatively regulate protein synthesis (Clemens 2001; Kimball 2002). Nevertheless, the synthesis, stability and translation of select gene transcripts (see e.g. Kilberg chapter) may be increased upon amino acid deprivation by specific mechanisms, which oppose the global changes in gene expression (Fafournoux et al. 2000; Fernandez et al. 2002; Kilberg and Barbosa-Tessmann 2002). Absence of a single amino acid may have similar effects to generalised amino acid deprivation, in which regard leucine, alanine, glutamine and histidine appear to be particularly effective in modulating protein turnover *in vitro* (for review see Kimball 2002; Proud 2002). For example, removal of histidine from the culture medium decreases both protein synthesis and albumin mRNA expression in rat hepatocytes (Kimball et al. 1996). An extensive list of genes regulated by amino acid availability is provided in Chapter 5. We are beginning to gain some understanding of the mechanisms underlying these effects, which include modulation of gene transcription, regulation of global and specific mRNA translation, the abundance and activity of ribosomes, as well as the activation state of certain intracellular signalling pathways.

2.2.2 Effects of amino acids on gene transcription

Adaptation to amino acid deficiency is critical for cell survival. It is well established that amino acid deprivation induces expression of specific gene transcripts. Notably those encoding proteins involved in the biosynthesis or transport of amino acids, for example, asparagine synthetase and the System A amino acid transporter (e.g. Kilberg and Barbosa-Tessmann 2002 and Hyde et al. 2003 for review) and their peptide products, despite the generalised global decrease in mRNA translation. Certain genes (e.g. collagenase) (Li et al. 1995) are also upregulated by amino acid supplementation. One of the best-characterised amino acid sensitive processes in animal cells is the transcription-dependent increase in the activity of amino acid transport System A in response to amino acid deprivation (a process referred to as adaptive regulation or derepression). This response to amino acid starvation or replenishment shown by many cell types consists of a derepression of transport activity during amino acid starvation and reversal after replenishment (e.g. Christie et al. 2001 and Hyde et al. 2003 for review). Such processes require *de novo* synthesis of protein and RNA and indeed the adaptive regulation of System A activity involves increases in the expression both of mRNA and protein for the SNAT2 System A transporter (Gazzola et al. 2001; Hyde et al. 2001; Ling et al. 2001).

It is still not known how individual amino acids regulate gene transcription in animal cells. In lower eukaryotes such as yeast, two types of gene regulation in response to changes in amino acid availability have been described (as detailed in Chapters 7, 8, 11): (i) a specific control response in which the abundance of a single amino acid regulates the action of a specific transcriptional activator; and (ii) a general control response in which there is coordinate induction of a specific subset of genes by amino acid starvation. The latter is related to increased phosphorylation of eIF2α by the eIF2α kinase GCN2 leading to increased translation of the mRNA encoding the GCN4 (General Control Nondepressable) transcription factor, which in turn increases the transcription of amino acid biosynthetic genes (Hinnebusch 1997). There is growing evidence that mechanisms, at least analogous to the general control response, may also be operational in cells of higher eukaryotes, notably following the identification of a functional mammalian GCN2 homologue (Sood et al. 2000). For example, although GCN2(-/-) knockout mice exhibit no phenotypic abnormalities under standard growth conditions, prenatal and neonatal mortalities are significantly increased in GCN2(-/-) mice from leucine-, tryptophan-, or glycine-deficient mothers (Zhang et al. 2002). Furthermore, mouse embryonic GCN2(-/-) stem cells do not show the normal induction of eIF2α phosphorylation in cells deprived of leucine (Zhang et al. 2002). Subsequent chapters in this book cover such exciting new developments in our understanding of nutrient-induced responses in animal cells in much greater detail and are summarised below only briefly.

The best-studied example of this phenomenon in higher eukaryotes is the induction of expression of CHOP (a C/EBP-related transcription factor) in several mammalian cell lines after removal of certain amino acids from the culture medium, which involves both transcriptional and post-transcriptional regulatory processes (Fafournoux et al. 2000; Kilberg and Barbosa-Tessmann 2002: and for a review see Chapter 1). CHOP is induced by leucine starvation and activating transcription factors 2 and 4 (ATF-2, 4) are essential for the transcriptional activation of CHOP under these circumstances (Harding et al. 2000; Bruhat et al. 2002). Activation of mammalian GCN2 in amino acid - starved cells selectively increases translation of ATF4, resulting in the induction of CHOP by a signalling pathway that is effectively anagous to the yeast general control response (Harding et al. 2000). Translation of the amino acid transporter CAT1 mRNA also increases during amino acid starvation of mammalian cells by a complex mechanism, which requires GCN2-mediated phosphorylation of eIF2α (Fernandez et al. 2002). This adaptive response to nutritional stress is enabled by the presence of an internal ribosome entry site (IRES) within the CAT1 mRNA leader sequence. Thus, phosphorylation of eIF2α is associated with translation of a small upstream open reading frame (uORF) within this IRES, which unfolds an inhibitory structure in the mRNA leader to reveal the active IRES, enabling translation of the CAT1 ORF (Fernandez et al. 2002; Yaman et al. 2003). It is proposed that eIF2α phosphorylation induces synthesis of an IRES trans-acting factor which stabilizes cat-1 mRNA in this active IRES state (Yaman et al. 2003).

Transcription of the human asparagine synthase gene is also increased in response to deprivation of amino acids (termed the "amino acid response") as well as by deprivation of glucose (through the "endoplasmic reticulum stress/unfolded protein response"; Siu et al. 2001). These two independent response pathways converge on the same set of genomic cis-elements within the asparagine synthase promoter (nutrient sensing response elements (NSRE) 1 and 2), both of which are necessary for gene activation. The transcription factors C/EBPβ and ATF4 bind to the NSRE-1 sequence (Siu et al. 2001, 2002) and this binding is increased in extracts from amino acid-deprived or glucose-deprived cells. A specific amino acid response element (AARE) identified in the CHOP gene is highly similar to the NSRE-1 response element, on which basis the nucleotide sequence 5'-(A/G)-TT(G/t)CATCA-3' has been proposed as an AARE core sequence (Fafournoux et al. 2000). The combination of NSRE-1 and NSRE-2, termed the nutrient-sensing response unit (NSRU), also appears to have enhancer activity (Zhong et al. 2003). Dietary protein restriction or amino acid supplementation is also known to directly affect the quantity or binding-activity of several other liver-enriched transcription factors including hepatocyte nuclear factors (HNF)-1, -3, and -4, CCAAT/enhancer-binding proteins (C/EBP)α and β and liver-enriched transcriptional inhibitory protein (LIP), as well as CHOP and the ubiquitous transcription factor Sp1 (e.g. Marten et al. 1999; Fafournoux et al. 2000).

2.2.3 Effects of amino acids on mRNA translation

Recent advances in understanding of the mechanism of translation and its control have facilitated studies at the molecular level into the regulation of protein synthesis by nutrients, and the interplay between nutrients and hormonal signals (Kimball 2002; Martin and Blenis 2002; Proud 2002). An important finding in the last few years is that a number of components of the translational machinery in mammalian cells are subject to acute regulation by the nutrient status of the cell. The eukaryotic mRNA translation process, as illustrated in Figure 2, consists of three distinct phases (initiation, elongation and termination) and also requires initial "charging" of cytosolic tRNA molecules (e.g. $tRNA^{Met}$) with their respective amino acids (forming e.g. $Met\text{-}tRNA^{Met}$). In higher organisms, proteins

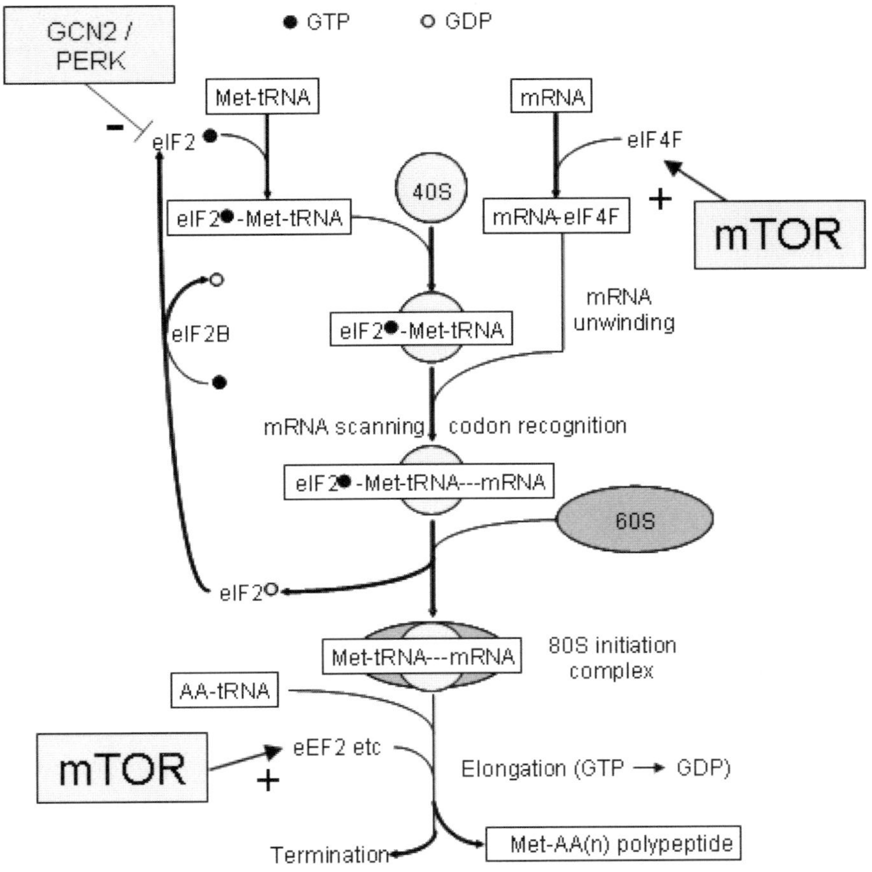

Fig. 2. A schematic illustration of the process of mRNA translation in eukaryotic cells. Steps regulated by the nutrient-sensitive TOR and GCN2/PERK signalling pathways are highlighted. 40S and 60S refer to the ribosomal subunits. Further details are provided in the main text.

constituting eukaryotic initiation factor groups 1 to 5 (eIF1-5) are involved in the initiation phase and specific elongation and releasing/termination factors (eEFs and eRFs respectively) are also required. The charging of tRNA molecules is catalysed by specific aminoacyl-tRNA synthetase enzymes, which have recognition sites for both an individual amino acid and its cognate tRNAs (Francklyn et al. 2002). Initiation involves the binding of both the mRNA and a charged methionyl-tRNA (Met-tRNA^{Met}) molecule to a ribosome (a catalytic complex of 40S and 60S subunits composed of proteins and distinct ribosomal RNAs) to form an 80S initiation complex. This process requires a number of initiation factors, which are involved in binding to the 5' mRNA cap, unwinding of secondary structure in the mRNA and "scanning" it for the initiation codon (AUG), which is recognised by the initiator Met-tRNA^{Met} to correctly align the translational machinery over the mRNA template. Establishment of the initiation complex leads to the elongation phase, which involves complementary "docking" of charged tRNAs to the mRNA template and subsequent peptide chain elongation distal to the initiating Met residue. Components of the 60S ribosomal subunit are involved in the peptidyltransferase reaction of the elongation phase. The ribosome moves along the mRNA until translation is terminated at a stop codon (UAG, UAA, or UGA) in the mRNA.

Eukaryotic messenger RNA translation may be regulated by (a) general activation/inhibition of translation factors or (b) binding of modulators to regulatory regions of individual mRNA species either to inhibit their translation or to (de)stabilize them. The initiation phase includes several key regulated steps. The two major arms of the process initiating messenger RNA translation are the recruitment to the ribosome of the initiator methionyl-tRNA (mediated by eIF2) and the mRNA itself (involving the eIF4 complex). These two arms are under distinct and separate regulatory control by nutrients, although they share some common regulation through endocrine signalling pathways (as detailed in a later section of this chapter).

2.2.3.1 Regulation of eIF2 function by amino acids

The major function of the heteromeric protein eIF2 is in the recruitment of the initiator Met-tRNA^{Met} to the 40S ribosomal subunit to recognise the start codon during translation initiation (Clemens 2001; Erickson et al. 2001). This requires eIF2 to form a complex with GTP; the GTP is hydrolysed following start codon recognition, releasing eIF2 as an inactive complex with GDP. Recycling of eIF2.GDP to active eIF2α.GTP is mediated by a guanine nucleotide exchange factor, the multisubunit protein eIF2B (Erickson et al. 2001; Proud 2001). Binding of eIF2.GTP.Met-tRNA^{Met} complexes to the 40S subunit is required for every initiation event and thus the activity of eIF2B plays a role in regulating global (and also transcript specific) translational control in eukaryotes from yeast to mammals. The activity of eIF2B is regulated by a variety of inputs (Proud 2001) including amino acids and glucose, apparently via several distinct mechanisms (notably, these do not appear to involve signalling via mTOR but may involve TOR proteins in yeast). In L6 myoblasts, regulation of eIF2B (and thus also eIF2 itself) appears to be of more importance than the control of eIF4 components (such as eIF4E) for the activation of protein synthesis by amino acids (Kimball et al. 1998; Kimball 2002). Changes in the activity of eIF2B are also implicated in the overall control of protein synthesis in the liver in response to amino acid imbalance (elevated levels of leucine, glutamine and tyrosine), on the basis of correlations between eIF2B activity (but not levels of eIF4F, for example) and overall protein synthetic rates (Kimball 2002). The effects of nutrients on the control of eIF2B may be mediated via regulatory mechanisms including changes in its phosphorylation state: eIF2B undergoes phosphorylation at multiple sites *in vivo*, one or more of which may be modulated by amino acids/glucose contributing to the regulation of eIF2B activity (Proud 2001). For example, glycogen synthase kinase 3 (GSK3) lies upstream of eIF2B and is reported to be inactivated by amino acids (Peyrollier et al. 2000;

Armstrong et al. 2001). Conversely, activation of the eIF2α kinase GCN2 during amino acid starvation leads to the phosphorylation of the α-subunit of eIF2 to yield eIF2(αP), a potent inhibitor of eIF2B and hence of translation initiation. This contributes to the reduction in global protein synthesis during nutrient starvation, at least in lower eukaryotes such as yeast (and also to the increased translation of the GCN4 mRNA, as discussed above and in Kimball chapter).

2.2.3.2 Regulation of the eIF4F complex by amino acids

The eIF4F complex includes the eIF4E/4G/4A initiation factors and is thought to be of key importance in mediating normal, cap-dependent, translation initiation (Gingras et al. 1999; Proud 2002). The eIF4F complex is involved in attaching the mRNA to the ribosome and in unwinding inhibitory secondary structures in its 5'-untranslated region (5'-UTR). Therefore, it is likely to be of particular importance in modulating translation of mRNAs with structured 5'UTRs. The eIF4G proteins (eIF4G_I and eIF4G_{II}) act as a scaffold for the complex, interacting with proteins such as eIF4E (which binds to the 5'-cap structure of eukaryotic mRNAs), eIF4A (an RNA helicase), PABP (which binds the poly(A)-tail found in almost all mRNAs), eIF3 (which interacts with the 40S ribosomal subunit), and Mnk1 and Mnk2 (eIF4E kinases). It is probable that eIF4E provides the first contact between the translational machinery and the mRNA in *de novo* translation initiation. Its function is regulated by at least two distinct types of protein binding partner: (i) the low molecular mass eIF4E-binding proteins 4E-BP1/2/3 (Proud 2002), and (ii) the nucleocytoplasmic shuttling protein 4E-T, which is believed to help convey eIF4E from the cytoplasm into the nucleus (reviewed by Strudwick and Borden 2002). These partner proteins bind to the same site on eIF4E as eIF4G (or on overlapping ones) (Gingras et al. 1999; Strudwick and Borden 2002). Binding is therefore mutually exclusive, thus eIF4E bound to 4E-BP1 cannot interact with eIF4G to form initiation complexes. 4E-BP1, thus, acts as a repressor of cap-dependent translation (Kimball 2002; Proud 2002).

Of the three 4E-BPs, 4E-BP1 is easily the most intensively studied and best understood. Phosphorylation of several sites in 4E-BP1 is increased by agents that activate protein synthesis, such as amino acids. Phosphorylation of Ser65 and Thr70, and to a lesser extent, Thr37/46, requires activation of the mTOR signalling pathway. The complex nature of the hierarchy of phosphorylation of other sites in 4E-BP1 suggests that they are targets for a range of proline-directed kinases that await identification. It remains to be established which kinases act on 4E-BP1 *in vivo*, although several of these (especially those acting at Ser65 and Thr70) are likely to be regulated by nutrients through the mTOR pathway. In many types of mammalian cell, amino acids exert marked, rapid (15-30 min) effects on the phosphorylation and regulation of 4E-BP1 (reviewed in Gingras et al. 2001a; Kimball 2002). Removal of amino acids quickly causes a marked increase in the amount of 4E-BP1 associated with eIF4E and loss of eIF4F complexes (Proud 2002). These effects are reversed, within minutes, by the re-addition of amino acids. The most effective single amino acid is leucine. Amino acids also exert a positive effect on the phosphorylation of 4E-BP1 *in vivo*, with leucine being effective when given alone (Kimball 2002).

2.2.3.3 Regulation of translation elongation by amino acids

The eukaryotic elongation factor 2 (eEF2), which mediates the translocation step of elongation, is also regulated through the mTOR pathway. Phosphorylation of eEF2 at Thr56 inhibits its activity by preventing it from binding to the ribosome, a process, which is modulated by nutrients (Browne and Proud 2002). Phosphorylation of eEF2 is catalysed by eEF2 kinase, which is itself inactivated by phosphorylation at Ser366 by the downstream mTOR target S6K1 (Wang et al. 2001).

2.2.3.4 The mTOR signalling pathway contributes substantially to translation regulation by amino acids

It is clear that regulation of several of the components of the translational process is linked to the rapamycin-sensitive mTOR (mammalian target of rapamycin) signalling pathway (Gingras et al. 2001b; Proud 2002). The targets of mTOR include translational repressors such as the 4E-binding proteins (4E-BPs) considered above, which are inactivated by mTOR signalling, as well as stimulators of translation such as eEF2 and the S6 kinases (see below), although not eIF2. Elevated amino acid availability signals *via* mTOR and culminates in increased cellular translation rates and *vice versa*. For example, withdrawal of amino acids from Chinese hamster ovary (CHO) cells causes decreased phosphorylation of 4E-BP1 and, in consequence, increased binding of 4E-BP1 to eIF4E and dissociation of eIF4F complexes (Wang et al. 1998). It also results in inactivation of S6K1, decreased phosphorylation of eIF4E and increased phosphorylation of eEF2, all of which are associated with decreased translational activity (Kimball 2002; Proud 2002). These effects are rapidly (<10 min) reversed by resupplying amino acids, except when mTOR is inhibited using rapamycin. Rapamycin itself has similar effects to those of amino acid deprivation.

The TOR kinases link nutrient-sensing to control of ribosomal protein gene expression and cell growth in eukaryotes from yeast to man (Rohde et al. 2001; Rohde and Cardenas 2003). Growth of animal cells to an appropriate size also requires TOR-dependent signals and in fact, rapamycin induces a "small cell size" phenotype (Zhang et al. 2000; Fingar et al. 2002). Current evidence (Fingar et al. 2002) suggests that mTOR signals downstream to both S6K1 and 4EBP1/eIF4E in order to regulate cell size through translational control, as well as regulating essential nutrient uptake (Edinger and Thompson 2002).

2.2.3.5 Amino acid signalling through mTOR

The mTOR protein is large (around 290 kDa) and its primary sequence indicates the presence of a number of potential functional domains (for review see Rohde et al. 2001; Proud 2002), including a series of HEAT domains towards its N-terminus (which are likely to be involved in protein:protein interactions) as well as a domain with similarity to lipid (phosphoinositide) kinases towards its C-terminus. An intact kinase domain is essential for the function of both yeast TOR and mTOR (Mothe-Satney et al. 2000; Rohde et al. 2001), and although mTOR has not been shown to phosphorylate any lipid so far tested, it is able to phosphorylate certain proteins (e.g. 4E-BP1 and S6K1) *in vitro* (Isotani et al. 1999; Mothe-Satney et al. 2000). Nevertheless, it remains to be fully established that mTOR functions as a protein kinase *in vivo*. Proteins non-covalently bound to mTOR, perhaps via its HEAT domains, may provide appropriate protein kinase activities within the intact cellular environment (as described below).

Recent studies (reviewed in detail by Pan in Chapter 3 of this book) are beginning to unravel the complex mechanisms underlying regulation of the mTOR pathway. Dennis et al. (Dennis et al. 2001) were unable to observe any direct effect of amino acid withdrawal on the kinase activity of mTOR. It has very recently been established that a novel 150kDa protein termed raptor (regulatory associated protein of mTOR) interacts with mTOR *in vivo* (Hara et al. 2002; Kim et al. 2002). The binding of raptor to mTOR is necessary for efficient mTOR-catalyzed phosphorylation of 4E-BP1 *in vitro*, and also enhances the mTOR kinase activity toward S6K1. Raptor appears to serve as an mTOR scaffold protein, forming ternary complexes with both 4E-BP1 and S6K1 through their respective TOS (conserved TOR signalling) motifs, a short conserved segment required for mTOR-mediated phosphorylation of its substrates (Nojima et al. 2003; Schalm et al. 2003). Raptor has a positive role in nutrient-stimulated signalling to S6K1, maintenance of cell size, and mTOR protein expression (Hara et al. 2002; Kim et al. 2002). In contrast, conditions that repress the mTOR pathway (e.g. nutrient deprivation and mitochondrial uncoupling) stabilize the mTOR-raptor association and appear to inhibit mTOR kinase activity (Kim et al. 2002). The interaction of raptor with mTOR is stabilized by the protein GβL: binding of GβL to mTOR strongly stimulates mTOR kinase activity towards S6K1 and 4E-BP1 but this effect is reversed by the stable interaction of raptor with mTOR (Kim do et al. 2003). It appears that nutrients (and also rapamycin) regulate the association between mTOR and raptor only in complexes containing GβL (Kim do et al. 2003). It is possible that mTOR makes additional inputs to the regulation of S6K1 through regulation of S6K1 phosphatase activities (Peterson et al. 1999). A potential regulator of the phosphatases acting on S6K1 is α4, which interacts with the catalytic subunit of protein phosphatase (PP) 2A (Chen et al. 1998) and is similar in sequence to the yeast phosphatase partner Tap42p (Jiang and Broach 1999), which in turn is implicated in signalling from yeast TOR to translation. The observations that PP2A interacts with S6K1 and is activated by rapamycin treatment of cells might provide a mechanism by which rapamycin causes dephosphorylation of S6K1 (Peterson et al. 1999; Avruch et al. 2001). The proteins TSC1 and TSC2 (known as hamartin and tuberin respectively) form a complex, which also suppresses signalling via mTOR (Gao et al. 2002; Tee et al. 2002). Loss of TSC1-TSC2 in animal cells prevents the normal inactivation of S6K seen in response to amino acid starvation (Gao et al. 2002). Equally, TSC1 and TSC2 expression blocks the ability of amino acids to activate S6K1 within nutrient-deprived cells (Tee et al. 2002). These studies also indicate that the TSC1-TSC2 complex antagonizes the

TOR-mediated response to amino acid availability upstream of TOR itself (see also section 2.2.5.1) and Pan in Chapter 3 for details of further recent developments in this area).

2.2.3.6 The S6 kinases contribute to mTOR-dependent regulation of translation

The protein kinases that phosphorylate ribosomal protein S6 (the S6 kinases) are an important set of proteins that are regulated via mTOR and are implicated in the control of mRNA translation by nutrients and other stimuli (Avruch et al. 2001; Gingras et al. 2001a; Kimball 2002; Proud 2002). S6Ks phosphorylate a component of the 40S ribosomal subunit (S6) as well as being implicated in the regulation of translation elongation (*via* the elongation factor eEF2) as mentioned above. It has been suggested that the S6 kinases play a particular role in regulating the translation of a set of mRNAs termed the 5'-TOP (terminal oligopyrimidine tract) mRNAs. This group of mRNAs includes those for all of the ribosomal proteins in mammals, plus those for certain other proteins involved in mRNA translation such as eEF2 (Browne and Proud 2002). It is proposed that this amino acid- and hormone-regulated translational control mechanism provides a way in which synthesis of components of the translational machinery can be quickly switched on following treatment of mammalian cells by an anabolic/proliferative stimulus, to increase the cellular capacity for protein synthesis, although the role of S6Ks here has now been challenged (Tang et al. 2001). Activation of the S6Ks involves their phosphorylation at multiple sites, some of which lie in the catalytic domain or its so-called 'extension' or 'linker' while the majority are located in the C-terminal regulatory domain (Avruch et al. 2001). Thr229 in the T-loop of the catalytic domain is phosphorylated by phosphoinositide-dependent kinase 1 (PDK1) *in vitro* and, in addition, mTOR phosphorylates T389 *in vitro* although it may not be the physiological T389 kinase (for review and discussion see Avruch et al. 2001). Nevertheless, the sensitivity of S6K regulation to rapamycin shows that mTOR makes an essential input to the control of the S6Ks. Amino acid deprivation leads to complete dephosphorylation of S6K1 at T389, a major rapamycin-sensitive and thus mTOR-controlled phosphorylation site in S6K1 (for review see Kimball 2002; Proud 2002). An S6K1 mutant resistant to inhibition by rapamycin also proved resistant to the effects of amino acid withdrawal, suggesting that amino acids may signal to S6K1 via mTOR (Pearson et al. 1995). Since S6K2 is also regulated through mTOR, it appears equally likely that the activity of this kinase is also modulated by amino acids.

2.2.3.7 Relative importance of different amino-acid sensitive mechanisms to regulation of mRNA translation in vivo

In skeletal muscle, the activation of protein synthesis elicited by leucine is inhibited only partially by rapamycin (Anthony et al. 2002), consistent with the idea that amino acids stimulate protein synthesis both via mTOR-dependent and -independent pathways (Kimball 2002). As discussed above, regulation of eIF2B (a translation factor not linked to mTOR signalling) may be of more importance than the control of eIF4E (by 4E-BP1) for the activation of protein synthesis by amino acids (Kimball 2002; Kobayashi et al. 2003) in the absence of additional stimuli such as insulin. In L6 myoblasts, histidine, and leucine were each able to activate eIF2B and total protein synthesis, while only leucine was able to modulate 4E-BP1, via the mTOR pathway (Kimball et al. 1998). These data also indicate that the regulation of the activity of eIF2B (conceivably via the GCN2 pathway, at least in part), rather than the control of 4E-BP1 (or the S6Ks), is important for the overall regulation of protein synthesis in these cells. In fact, control of 4E-BP1 and S6Ks via mTOR in response to leucine appears rather to control the translation of specific mRNAs (for review see Kimball 2002).

There is, as yet, no evidence for cross-talk between the nutrient-sensitive TOR and GCN2 pathways in animal cells, but recent work has now established a possible inhibitory effect of TOR on the yeast GCN2 pathway (Cherkasova and Hinnebusch 2003). Studies using rapamycin demonstrate that an active TOR pathway directly inhibits GCN2 activity through phosphorylation of Ser577 in GCN2, which reduces its tRNA binding activity and thus the ability to respond to amino acid deprivation (Cherkasova and Hinnebusch 2003). This process involves the PP2A regulator Tap42p, which already has an established role in signalling from yeast TOR to mechanisms governing mRNA translation (see section 2.2.3.5 above).

Many cell types appear to ensure an adequate cellular amino acid supply by modulating amino acid transport in concert with the rate of protein synthesis. Acute regulation may be achieved at least partly by the use of common signalling elements to modulate translation and amino acid transport (for review see Bode 2001; Proud 2002; Hyde et al. 2003); for example, activation of the mTOR pathway upregulates expression of nutrient transporters such the System L amino acid transporter in mammalian cells (Edinger and Thompson 2002).

2.2.4 Effect of amino acids on protein breakdown

Lysosomal autophagy accounts for the breakdown of the majority of long-lived proteins in animal cells (Blommaart et al. 1997) whereas extra-lysosomal proteolytic pathways (including the ubiquitin-proteasome system) are responsible for the degradation of proteins with high turnover rates (Blommaart et al. 1997; Combaret et al. 2001). An exception is found in skeletal muscle, where degradation of the slow-turnover myofibrillar proteins is normally initiated outside the lysosome because the large size of the myofibrillar apparatus precludes autophagy (Mitch and Price 2003). Macroautophagy is responsible for acceleration of proteolysis in many cell types when amino acid and/or insulin concentrations fall. Amino acids are effective inhibitors of the macroautophagy process, primarily at the initial ATP-dependent sequestration step in which cytosolic material for breakdown becomes membrane-bound prior to fusion with existing lysosomes. Leucine, phenylalanine, and tyrosine in combination with alanine and glutamine, are the most important amino acids involved in control of autophagic proteolysis (for review see Blommaart et al. 1997). Swelling of liver cells has an inhibitory effect on autophagic protein degradation and also increases the sensitivity of the process to inhibition by amino acids: alanine and glutamine may both exert their anti-proteolytic effect (at least in liver cells) *via* cell swelling (see Blommaart et al. 1997 and section 2.2.5.2 below). Leucine and glutamine are also implicated as regulators of proteolysis in skeletal muscle, but the extent to which this regulation is intra and extralysosomal remains unclear. The contribution of extra-lysosomal proteolytic systems to protein degradation in muscle is more important than in liver and accelerated breakdown of myofibrillar protein, for example, during starvation, occurs via the ubiquitin-proteasome system (Mitch and Price 2003). Nutrients contribute to suppression of the rates of ubiquitinylation of protein substrates and of proteasome-dependent proteolysis in skeletal muscle, although factors other than the expression of ubiquitin-proteasome pathway components themselves appear to be responsible here (Combaret et al. 2001). Nevertheless, proteasome-dependent proteolysis in muscle appears much less responsive to nutrients than protein synthesis over the short term (Kee et al. 2003). Amino acids appear to be a key factor required to reduce expression of genes connected with ubiquitin-proteasome dependent proteolysis in the intestine (Adegoke et al. 2003).

2.2.5 Sensing of amino acid availability in animal cells

Our knowledge of the cellular machinery involved in amino acid sensing and signal initiation is much less extensive that that of downstream effector pathways (Van Sluijters et al. 2000; Hyde et al. 2003). For example, although signalling by mTOR is extremely sensitive to amino acid availability, there is no evidence that mTOR kinase activity is regulated directly by amino acids (Dennis et al. 2001). This implies that the initial sensing and signalling events lie upstream of mTOR itself. Precisely how changes in amino acid concentrations are "sensed" by mammalian cells in order to modulate intracellular signalling pathways and/or evoke metabolic responses remains to be elucidated. Mammalian cells have developed discrete chemosensory mechanisms for a variety of nutrients including glucose, fatty acids, and amino acids. These mechanisms may involve either the binding of the nutrient (or a metabolite) to plasma membrane or intracellular receptors, or the detection of physiological signals generated as a result of the cellular uptake of the nutrient (e.g. changes in cell volume or cell membrane potential) (Van Sluijters et al. 2000; Hyde et al. 2003). The exact molecular nature of amino acid sensor mechanisms is not known in the great majority of cases. Nevertheless, it is clear that animal cell types differ in their sensitivities to the omission or addition of specific amino acids; for example, leucine appears to be the only amino acid capable of eliciting an effect on 4E-BP1 and S6K1 in skeletal muscle, while other branched-chain amino acids (isoleucine, valine) are also effective in liver (for review see Kimball 2002). Whether this reflects the operation of different sensing mechanisms in different cell types remains to be discovered.

2.2.5.1 Extracellular amino acid sensors

It is conceivable that the mechanisms for sensing availability of individual amino acids (e.g. leucine) may include "sensors" at the plasma membrane (see Fig. 3). Membrane transporters clearly have the ability to recognise and discriminate between different amino acids and therefore act as sensors (Hyde et al. 2003) and it is also possible to speculate on the existence of extracellular amino acid "sensors" based on the known variety of ionotropic and metabotropic amino acid receptors expressed largely in neural tissue. Amino acid availability may also generate or regulate signal transduction through extracellular receptors such as G-protein coupled taste receptor heterodimer T1R1/T1R3 (Nelson et al. 2002), the CaR calcium receptor and the P2Y purinergic G-protein coupled receptor (for review see Conigrave et al. 2002). More specifically, the leucine oligomer-based membrane impermeant molecule Leu8-MAP (Miotto et al. 1994) and the leucine analogue isovaleryl-L-carnitine (Miotto et al. 1992) both efficiently inhibit hepatic macroautophagy in a similar manner to leucine itself, raising the possibility that Leu8-MAP binds to a plasma-membrane "leucine" receptor with regulatory control over proteolysis. The molecular identity of this putative leucine receptor in mammalian cells has not been established and it does not appear to interact with the mTOR pathway (for further discussion see Van Sluijters et al. 2000).

Eukaryotic solute transporters may be capable of initiating signal transduction pathways and regulating gene expression through mechanisms similar to many receptor-ligand interactions. An excellent example from a lower eukaryote is the yeast (*Saccharomyces cerevisiae*) protein Ssy1p, an amino acid permease homologue (Forsberg et al. 2001), which is considered in more detail in Chapter 9. The mechanisms by which amino acid transporters might contribute to initiation of amino acid-dependent cell signalling fall into two major categories: (**1**) the transporter may initiate cellular signalling in direct response to (or as a direct consequence of) changes in substrate loading or flux or alternatively (**2**) it may regulate the availability of an amino acid to a specific pool of nutrient receptors. It may not even be necessary for amino acids to translocate through a transporter for that protein to initiate signalling: the binding of an amino acid substrate to a transporter may be sufficient to influence cellular signalling proteins that initiate a signal transduction pathway (for review see Hyde et al. 2003). In higher eukaryotes, several amino acid transporters have been implicated directly in the initiation of intracellular signalling in response to altered substrate availability. A representative example is the System A (SNAT2) transporter: during amino acid deprivation, the abundances of SNAT2 mRNA and protein increase by mechanism(s) that are suppressed specifically by Me-AIB (a synthetic, non-metabolisable substrate for System A transporters) as well as natural System A substrates such as asparagine, glutamine and praline (Gazzola et al. 2001; Hyde et al. 2001; Ling et al. 2001; Bain et al. 2002). On this basis, it has been proposed that SNAT2 may act as a substrate receptor for regulation of its own expression (Gazzola et al. 2001; Hyde et al. 2003). SNAT2 exhibits K_M values for its neutral amino acid substrates within the concentration range observed in plasma, consistent with the idea that this System A transporter may act as a nutrient sensor upstream of the (as yet largely

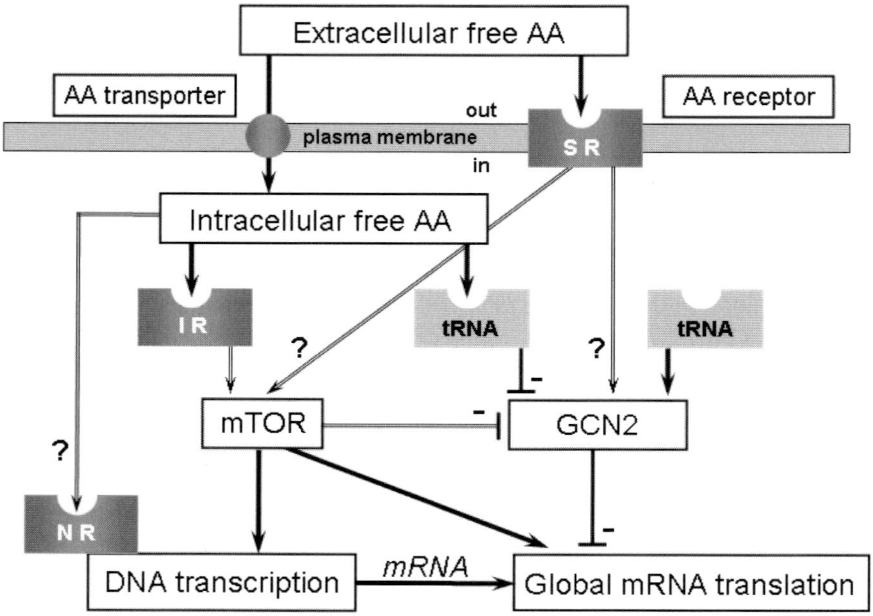

Fig. 3. Major pathways by which amino acids may regulate function of animal cells, illustrating both excitatory and inhibitory effects. Note presence of putative amino acid receptors (SR, extracellular surface receptor; IR, intracellular receptor; NR, nuclear receptor). GCN2 is believed to "sense" amino acid availability from the cellular level of tRNA charging. Solid arrows indicate well-established steps.

unidentified) signalling mechanisms modulating the adaptive regulatory response. Binding of amino acid substrates to transporters of nutritionally-important amino acids may regulate direct interactions of transporters with other proteins or the activity of transporter-associated signalling molecules (for review see Robinson 2002; Hyde et al. 2003).

The activity of amino acid transporters may result in changes in cell volume, transmembrane electrical potential difference and/or pH, which may be detectable by cell sensor mechanisms or may be even of sufficient magnitude to initiate or suppress intracellular signals themselves (Hyde et al. 2003). One significant indirect effect of certain transport mechanisms is cellular swelling and, furthermore, accumulative transport of amino acids is known to promote swelling of both liver and muscle cells during nominally isotonic conditions (Low et al. 1997b; Haussinger et al. 2001; Krause et al. 2002b). This cell swelling in response to increased amino acid loading has general anabolic effects such as the stimulation of protein and glycogen synthesis and suppression of protein breakdown (for review see Haussinger et al. 2001). In particular, the anabolic effects of the amino acid glutamine in skeletal muscle may be at least partly related to cell-swelling, which occurs as an osmotic consequence of increased inward transport of this amino acid (a major osmotic effector in muscle) (Low et al. 1997b). There is evidence in both muscle

(Low et al. 1997a) and liver (Haussinger et al. 2003) that integrin (adhesion)-dependent mechanisms are involved in sensing changes in cell volume and initiating downstream signals and also, in muscle at least, that mTOR is a component of the downstream signalling network (Low et al. 1997b).

2.2.5.2 Intracellular amino acid sensors

The intracellular concentration of a given amino acid is governed by many factors, including the rates of amino acid uptake or release across the plasma membrane, protein synthesis, protein degradation, and amino acyl-tRNA production, plus amino acid catabolism and biosynthesis. Signalling pathways that are regulated by intracellular amino acid concentrations are therefore intrinsically linked to amino acid transporter activity as well as to intracellular amino acid metabolism. Thus, increased extracellular amino acid levels may rapidly lead to increased intracellular amino acid concentrations by virtue of concentrative or equilibrative transmembrane transport. Amino acids regulate a number of intracellular enzymes, which may play a role in nutrient sensing such as glutamate-activated protein phosphatase-2A (Gaussin et al. 1996; Kowluru et al. 2001) and mitochondrial glutamate dehydrogenase, which is allosterically activated by leucine (Xu et al. 2001). Evidence also exists that is highly suggestive of a role for intracellular amino acid sensing upstream of both GCN2 and mTOR signalling pathways, as summarised below (see also Fig. 3).

It is well established that the yeast Gcn2p kinase is activated by amino acid starvation (Hinnebusch 1997; Dong et al. 2000). A key aspect of this activation is the ability of Gcn2p to bind uncharged tRNAs, apparently through a domain homologous to histidinyl-tRNA synthestase juxtaposed to the protein kinase (Dong et al. 2000). It is thought that uncharged tRNA that accumulates during amino acid limitation binds to this sensor domain, activating the kinase (Dong et al. 2000; Sood et al. 2000), which then phosphorylates eIF2α and inhibits translation initiation. Amino acid deprivation also increases eIF2α phosphorylation in mammalian cells and it is likely that the mammalian and yeast GCN2 proteins perform broadly similar roles, possibly including sensing of amino acid availability *via* the level of tRNA charging (Sood et al. 2000; Fernandez et al. 2002; Zhang et al. 2002). This matter is considered in detail in Chapters 5 and 7. Amino acid alcohols can inhibit amino acyl-tRNA synthetases and thus block tRNA charging. Significantly, histidine limitation in the presence of histidinol induced a two-fold increase in the phosphorylation of eIF2α (and a concomitant reduction in eIF2B activity) in perfused livers from wild type mice, but no changes in livers from GCN2(-/-) mice (Zhang et al. 2002), consistent with the uncharged tRNA hypothesis. In contrast to the situation in liver, we (CGP group) have consistently been unable to see any change in the state of phosphorylation of eIF2α in response to amino acid addition or withdrawal in CHO cells (Patel et al. 2001) and only very small changes in eIF2α phosphorylation are seen upon leucine deprivation in L6 myoblasts (Kimball et al, 1998). The relative importance of the mammalian GCN2 pathway in nutrient-induced responses, therefore may depend on the tissue studied. Another mechanism by which uncharged tRNA may influence translation is through an inhibitory effect on the enzyme phosphofructokinase (observed to date only *in vitro*) (Rabinovitz 1995), which would lead to a reduction in intracellular fructose-(1,6)-bisphosphate (an activator of eIF2B; (for review see Rabinovitz 1995; Kimball 2002) in parallel with activation of the GCN2 pathway.

There is very little evidence to suggest that levels of tRNA charging also underlie control of the mTOR pathway. Indeed, neither we (Patel et al. 2001) nor others (Pham et al. 2000) have observed effects of amino acid alcohols on targets of mTOR signalling, although Iiboshi et al. (1999) have reported that treatment of T-lymphoblastoid (Jurkat) cells with amino acid alcohols led to decreased S6K1 activity. In fact, Dennis et al. (2001) reported no significant effect of brief amino acid withdrawal on the cellular level of tRNA charging, whereas there was a marked depletion of intracellular branched chain amino acids (Dennis et al. 2001), consistent with an alternative possibility that vertebrate cells respond directly to changes in intracellular amino acid concentrations. A variety of data now indicates that regulation of S6K1 and other targets for mTOR signalling is substantially influenced by the size of an intracellular pool of amino acids. For example: (i) injection of leucine into the cytoplasm is sufficient to activate TOR signalling in *Xenopus* oocytes (Christie et al. 2002), and (ii) inhibition of protein synthesis in CHO cells by specific inhibitors, which results in increased intracellular amino acid concentrations, both enhances basal mTOR signalling and permits insulin to regulate the mTOR targets 4E-BP1 or p70S6K1 in amino acid-deprived cells (i.e. normalizing the response to that seen in cells supplied with physiological concentrations of amino acids in the medium) (Beugnet et al. 2003). The activity of secondary active amino acid transporters such as System A may therefore fulfil a key (albeit indirect) role in regulating activity of the mTOR pathway, given that their provision of exchange substrates (notably glutamine) for tertiary active transport of branched-chain amino acids through System L should have a marked influence on the intracellular free pool sizes of the latter nutrients (Bevington et al. 2002; Hyde et al. 2003). It appears that full activation of S6K1 requires inputs from both amino acids/glucose <u>and</u> insulin in many certain cell types. In contrast, certain cells (e.g. hepatoma cells - Shigemitsu et al. 1999) appear to contain enough amino acids for effective regulation of S6K1 even when starved for external amino acids. Such cells become dependent upon external amino acids when treated with an inhibitor of autophagy, suggesting that their abundant intracellular supply of amino acids is derived from this form of protein breakdown.

Various products of amino acid metabolism may also be important as intracellular nutrient signals. For example, the intracellular transamination of glutamine accumulated in liver cells to convert intracellular keto acids into amino acids (Holecek 2002) may be a major mechanism for hepatic accumulation of branched-chain amino acids (Leu, Ile, Val) and thus should contribute to activation of the mTOR pathway. In addition, ATP generated following amino acid catabolism may enhance both mTOR signalling and insulin secretion, the former due to increased availability of ATP as a kinase substrate and the latter due to ATP-sensitive membrane depolarisation (see Best chapter). Signalling through mTOR is known to be influenced by changes in intracellular ATP concentration and is inhibited by treatments which either produce radical ATP depletion or mimic high intracellular AMP (Dennis et al. 2001; Bolster et al. 2002; Krause et al. 2002a), thus mTOR may act as an energy sensor showing decreased activity as cellular ATP levels fall. Nevertheless, the K_m value for ATP of mTOR-mediated phosphorylation of 4E-BP1 or S6K1 *in vitro* (around 1 mM - Dennis et al. 2001) is considerably lower than normal cellular ATP concentration, which would therefore have to fall drastically to have a substantial direct effect on the activity of mTOR. A different, less direct mechanism may serve to inhibit the mTOR pathway under conditions of milder energy depletion, involving activation by AMP of the AMP-dependent protein kinase (AMPK). For example, Bolster et al. (2002) have reported that injection of a drug that activates AMPK causes inhibition of mTOR signalling in skeletal muscle (although it should be pointed out that interpretation of such *in vivo* data may be complicated by possible effects of AMPK on the synthesis and/or release of insulin by the pancreas (Leclerc et al. 2002)).

2.3 Carbohydrate

The role of glucose as a nutrient signal in cells of higher animals is closely interwoven with its central position as a regulator of certain endocrine pathways, most notably that of insulin. Glucose is required for full activation of several growth-factor-stimulated signalling pathways and exerts a clear permissive effect with respect to the action of insulin (e.g. Rutter et al. 2000; Vaulont et al. 2000 for perspective). Glucose availability is also directly related to cellular ATP levels, which may also act as a nutrient-related signal as described above and in section 2.2.5.2 Glucose increases the expression of genes involved in lipid synthesis and high dietary intakes of carbohydrate induce the lipogenic capacity of both liver and adipose tissue (see Towle et al. 1997; Foufelle et al. 1998; Rutter et al. 2000; Delzenne et al. 2001). Regulation of many of the genes controlled by glucose is also dependent upon the presence of insulin, and to some extent thyroid hormones and glucocorticoids. Certain genes involved in this type of regulation (e.g. SREBP-1c) are, in effect, purely insulin-responsive genes (Stoeckman and Towle 2002). Another such gene is glucokinase; induction of this gene by insulin stimulates glucose metabolism, thus amplifying the sugar signal (Rutter et al. 2000). In contrast, several genes are responsive to glucose as well as insulin, for example, hepatic phospho-enol pyruvate carboxykinase (PEPCK) and glucose-6-phosphatase. Glucose has also been shown to increase expression of other genes including fatty acid synthase (FAS), acetyl CoA carboxylase (ACC), L-pyruvate kinase (L-PK), and glucose transporter type 2 (GLUT2) (see Towle et al. 1997; Foufelle et al. 1998; Vaulont et al. 2000; Rufo et al. 2001). Genes such as FAS may be activated by glucose at physiological concentrations when insulin is present. The FAS gene includes an E box promoter element, which binds transcription factors such as SREBP (Stoeckman and Towle 2002). Several other transcription factors may bind to distinct carbohydrate response elements on glucose-regulated genes such as L-PK, although ancillary proteins may represent the real receptor for the glucose signal. DNA binding studies and transfection of suspected gene promoter constructs have revealed carbohydrate response elements (ChoRE – see Stoeckman and Towle 2002), which have a core consensus sequence of CACGGG or CACGTG and may bind one of at least two distinct carbohydrate response factors (ChoRF). Although the glucose-dependent DNA-binding proteins involved in these effects are yet to be identified (Rufo et al. 2001; Stoeckman and Towle 2002). ChoRF binds to a distinct site from SREBP-1 on the promoter regions of many lipogenic enzyme genes to activate their expression (Stoeckman and Towle 2002). ChoRF may function synergistically with SREBP to promote lipogenesis in the liver in response to signals generated by glucose and insulin respectively, thus generating an integrated overall response to dietary carbohydrate (Koo et al. 2001).

The insulin gene itself is regulated transcriptionally by glucose in pancreatic β-cells (Melloul et al. 2002). The homeodomain transcription factor PDX1 (pancreatic duodenal homeobox-1) is strongly expressed only in β-cells and binds to a distinct group of *cis* elements on the insulin gene (Macfarlane et al. 2000). Glucose (alongside insulin and T_3) positively regulates the PDX1 gene promoter in pancreatic β-cells (Campbell and Macfarlane 2002). The phosphorylation of PDX1 is believed to be essential for activation of insulin gene expression and indeed inhibition of the stress activated protein kinase SAPK2 appears to block effects of glucose on PDX1 action (Macfarlane et al. 1999). PDX1 is also reported to migrate from the cytosol to the nucleus of glucose stimulated cells (Macfarlane et al. 1999; McKinnon and Docherty 2001). New evidence suggests that sumoylation (Verger et al. 2003) of PDX1 is also associated with its nuclear localization (Kishi et al. 2003). The forkhead transcription factor FOXO1 (also known as FKHR) acts as a repressor of PDX1 expression in the pancreas, an effect that may be relieved by insulin (Kitamura et al. 2002). FOXO1 and PDX1 also exhibit mutually exclusive patterns of nuclear localization in pancreatic β-cells (Kitamura et al. 2002).

Hepatic gluconeogenesis is absolutely required for survival when no exogenous glucose supply is available (i.e. prolonged fasting or starvation). Gluconeogenic enzymes are induced in liver cells during fasting and/or diabetes, a process associated with induction of the transcriptional coactivator PGC-1α (peroxisome proliferative activated receptor-gamma co-activator 1; also known as PPARGC1) (Herzig et al. 2001; Yoon et al. 2001). It is not known whether glucose deprivation *per se* acts as a signal in this situation, although it is clear that both glucocorticoids and glucagon have strong gluconeogenic actions whereas insulin suppresses hepatic gluconeogenesis. Nevertheless, it is noteworthy that PGC-1α binds and coactivates FOXO1 (the PDX1 repressor considered above) to induce gluconeogenic gene expression in hepatocytes (Puigserver et al. 2003). FOXO1 thus appears to have distinct tissue-specific roles related to changes in glucose availability.

It has been suggested (Guillemain et al. 2000) that the facilitated glucose transporter GLUT2 (predominantly expressed in liver, pancreas, intestinal, and renal epithelium) may transduce a glucose signal from the plasma membrane to the nucleus via protein interactions with its large loop. Nevertheless, glucose sensing in most animals cells probably involves metabolism of the sugar: a glucose metabolite may provide the signal for gene regulation and candidates include glucose 6-

phosphate and xylulose 5-phosphate (Foufelle et al. 1998). Increased glucose availability also elevates glycolysis, generating ATP and altering cellular energy status, thus the metabolic stress kinase AMPK (which monitors and responds to changes in AMP/ATP ratio) (Hardie and Hawley 2001) may have a role in some cases. For example, recent results indicate that inhibition of AMPK by glucose (via elevation in ATP levels) is essential for the activation of insulin secretion by the sugar, and may contribute to the transcriptional stimulation of the preproinsulin gene (Da Silva Xavier et al. 2003). Activation of AMP kinase also blocks glucose activation of glucose-responsive genes in liver cells (Leclerc et al. 2002).

2.4 Lipids

Dietary lipids such as polyunsaturated fatty acids (PUFA) and cholesterol are able to regulate gene transcription (Clarke 2001; Jump 2002) and are also increasingly recognised to have additional signalling roles. We will consider the effects of fatty acids and cholesterol separately.

2.4.1 Fatty acid-induced responses in animal cells

Dietary PUFA are important regulators of cell function under physiological conditions; other dietary triacylglycerol components (such as saturated and monounsaturated fatty acids) only have significant effects at very high concentrations. In general, PUFA direct fatty acids away from triglyceride synthesis and towards oxidation and may also stimulate incorporation of glucose into glycogen (Clarke 2001). They do this by upregulation of genes related to fatty acid oxidation whilst downregulating genes associated with lipid synthesis; they also affect RNA processing and stability. For example, increasing the PUFA content of a diet results in a reduction in the rates of fatty acid biosynthesis in rat liver and adipose tissue, due to (n-3) and (n-6) PUFAs inhibiting the expression of genes including FAS, malic enzyme, ACC, L-PK and stearoyl CoA desaturase (Sessler and Ntambi 1998). In contrast, hepatic expression of low-density lipoprotein receptor and lipoprotein lipase are increased by PUFAs (Clarke 2001; Bocher et al. 2002). PUFA and peroxisome proliferators (PPs - e.g. thiazolidinediones and fibrates) have similar effects on FAS gene expression. PUFA are suggested to have a direct effect on the activity or abundance of the PPAR (peroxisome proliferator activated receptor) transcription factors (see Fig. 4). PPARs belong to the superfamily of steroid-thyroid-retinoid nuclear receptors. They act as ligand-dependent transcription factors, with natural ligands including both dietary fat metabolites such as PUFA and products of lipid oxidation (e.g. certain leukotrienes, prostaglandins and their derivatives): PPARs therefore act as fatty acid nuclear receptors (Clarke 2001; Bocher et al. 2002; Jump 2002). In general, PUFA metabolites such as oxidised fatty acids, alongside drugs such as the fibrates and glitazones, appear to be more potent activators of PPARs than PUFA themselves.

There are three types of PPAR, namely PPARα, β (also called δ) and γ, that have different (although overlapping) ligand-sensitivities and are expressed in specific tissues (for review see Bocher et al. 2002; Nosjean and Boutin 2002). PPARα and PPARγ are generally considered as the two main mediators of PUFA effects, whereby PUFA and related compounds bind specifically to PPAR proteins and alter their ability to bind DNA response elements. PPARα is a major regulator of intra and extracellular lipid metabolism and is mainly expressed in tissues having a high metabolic rate such as liver and muscle (also in cells of the atherosclerotic lesion). Ligand-induced activation of PPARα leads to induction of genes involved in lipid oxidation, transport and thermogenesis. Indeed PPARα-/- mice are unable to increase fatty acid oxidation during a fast and display both fatty livers and elevated plasma lipid concentrations (Lee et al. 2002). Upon activation, PPARα alters both synthesis and catabolism of the triglyceride-rich lipoproteins in a way that decreases plasma triglyceride levels (Bocher et al. 2002). PPARα activators also increase the plasma levels of high-density lipoprotein and accelerate the HDL-mediated reverse transport of cholesterol from peripheral tissues to the liver. Activation of PPARβ/δ induces expression of genes required for fatty acid oxidation and utilization in adipocytes and skeletal muscle cells (Wang et al. 2003).

PPARγ is highly enriched in adipose tissue and is also active in epithelial cells and macrophages (Picard and Auwerx 2002). PPARγ agonists (which include the glitazones) are insulin sensitizers that reduce systemic levels of glucose, triglyceride, and free fatty acids in patients with Type 2 diabetes (Celi and Shuldiner 2002). Such antidiabetic actions of PPARγ agonists are the consequence of coordinate effects on gene expression in insulin-sensitive tissues, involving effects on both glucose and fatty acid metabolism, which produce an overall net shift of triglyceride content from liver and muscle to adipose tissue. PPARγ activation stimulates the expression of several genes involved in fatty acid metabolism and lipogenesis in both white and brown adipose tissue. PPARγ activation in liver decreases expression of gluconeogenic genes, whereas in muscle PPARγ activation increases glucose utilization both by derepressing oxidative glucose metabolism (Ciaraldi et al. 2002) and by decreasing expression of genes involved in fatty acid transport and oxidation (Picard and Auwerx 2002). Under appropriate conditions (including the presence of insulin, T_3 and certain growth factors), activation of PPARγ stimulates the transcription of genes responsible for growth and differentiation of adipocytes to produce the mature fat cell phenotype.

Not all the effects of PUFA on gene transcription are via PPARs, as there appear to be at least two mechanisms for PUFA control of gene expression: a PUFA-PPAR-dependent mechanism responsible for upregulation and repression of gene expression, and a PPAR-independent or PUFA-specific mechanism for repression of gene expression. The dominant inhibitory effects of PUFA on lipogenesis mainly involve the latter mechanism (see Fig. 4). Here, PUFA suppress nuclear abundance and expression of transcription factors such as SREBP-1 (sterol regulatory binding protein 1), which contribute to induction of expression of genes involved in de-novo lipogenesis and triglyceride synthesis (e.g. FAS) and also their regulation by sugar and insulin. The effect of PUFA on SREBP-1 results at least

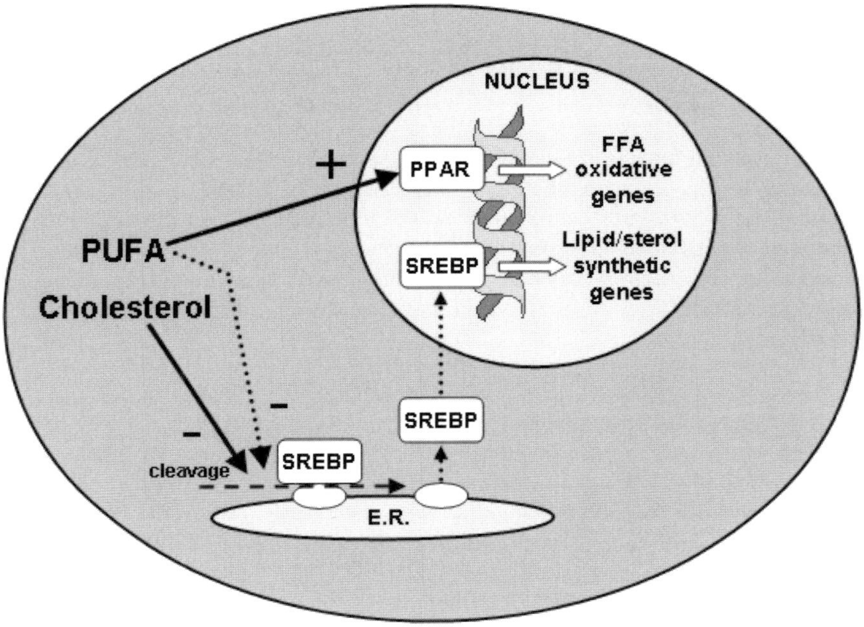

Fig. 4. Simplified illustration of major mechanisms by which polyunsaturated fatty acids (PUFA) and cholesterol regulate expression of genes associated with lipid turnover. Further details of the PPAR and SREBP families of transcription factor are given in the main text.

partly from an inhibition of transcription of the SREBP-1 gene (Duplus and Forest 2002) as well as an indirect inhibition of its proteolytic maturation (Jump 2002; Kersten 2002). The SREBP precursor proteins reside on the endoplasmic reticular membranes and their cleavage liberates mature SREBP forms, which migrate to the nucleus to modulate gene transcription. The intracellular signal involved in the inhibition of this process by PUFA is as yet unknown, although it has been speculated to involve a PUFA-dependent redistribution of either ceramide or cholesterol from the plasma membrane to intracellular compartments (Jump 2002). Both ceramide and cholesterol are associated with the phospholipid sphingomyelin in the plasma membrane: unsaturated (but not saturated) fatty acids such as PUFA stimulate sphingomyelinase with release of both these molecules (Worgall et al. 2002). The fact that such signals derive from a membrane has raised the suggestion that an essential function of SREBP transcription factors is to monitor cell membrane composition and exert feedback control on synthesis of fatty acids and phospholipids (Dobrosotskaya et al. 2002). PUFA also contribute to regulation of the activity/abundance of the LXR and HNF-4a transcription factor families and cross-talk between them (Jump 2002; Kersten 2002; Thomas et al. 2003).

Free fatty acids may also act as signalling molecules in cellular processes including insulin secretion. Both saturated and unsaturated long-chain fatty acids amplify glucose induced insulin secretion by activating the receptor GPR40 (Itoh et al. 2003). This receptor is highly expressed in pancreatic islets, mainly in β-cells. Interaction between free fatty acid and receptor evokes an increase in intracellular calcium ions, which is believed to enhance insulin secretion. Free fatty acids may also acutely stimulate insulin secretion through their metabolism in pancreatic β-cells. Nevertheless, long term exposure to saturated and certain monounsaturated fats reduces insulin secretion and has been implicated in causing insulin resistance, whereas polyunsaturated and omega-3 fatty acids largely do not appear to have adverse effects on insulin action (Lovejoy 2002). One possible reason is that the fatty acid composition of the phospholipids of cell membranes (e.g. in skeletal muscle) is closely related to insulin sensitivity and an increased saturation of membrane fatty acids has been associated with insulin resistance (Vessby 2000).

2.4.2 Cholesterol-induced responses in animal cells

Cholesterol is also known to regulate gene expression, with many of its effects thought to involve the sterol regulatory element binding protein (SREBP) family. A suppression of proteolytic maturation of SREBP-2 (which regulates expression of sterol synthetic genes) by cholesterol occurs in a manner analogous to the situation with SREBP-1 and PUFA described above (see also Fig. 4), a process here involving the sterol-sensing SREBP-activating protein SCAP (Shimano 2001; Jump 2002). SREBP-2 binds to sterol response elements (SREs) in the promoters of genes such as HMG-CoA reductase to activate transcription (Jump 2002). The consensus sequence for the SRE appears to be CACC(C/G)(C/T)AC (Shimano 2001). In cholesterol-depleted cells, SREBP is cleaved from a precursor in membranes of the nuclear envelope and endoplasmic reticulum. The amino-terminal fragment generated is able to enter the nucleus and activate transcription by binding to SREs (Shimano 2001). The binding of cholesterol to the SREBP inhibits the proteolytic cleavage and therefore prevents the activation of transcription. Cholesterol may also directly regulate gene transcription, for example, via the liver-specific transcription factor, hepatocyte nuclear factor-4 (HNF-4) (Shimano 2001; Jump 2002; Kersten 2002).

2.5 Integration between nutrient-sensitive and other intracellular signalling pathways in animal cells

Numerous cell-signalling pathways responsive to a great variety of stimuli share common intermediates and/or end-points with nutrient-responsive pathways. This arrangement provides scope for cross-talk between inputs from various sources through which distinct types of "sensory" input may be integrated into an appropriate cellular or organismal response. Possible roles for nutrients in the development of such responses are considered below and summarised in schematic Fig. 5. The recent discovery of an apparent inhibitory effect of TOR on the yeast GCN2 pathway (as described above in section 2.3.7 - Cherkasova and Hinnebusch 2003) indeed raises the possibility of direct cross-talk between different nutrient-sensitive signalling pathways in animal cells.

2.5.1 Interactions between nutrient and growth factor signalling pathways

We have already noted that glucose and amino acids are both required for full activation of insulin and growth-factor-stimulated signalling pathways. Indeed, many genes regulated by glucose also require insulin (as described above) and glucose itself may exert a permissive effect on the action of insulin. Similarly, plasma concentrations of the growth factor IGF1 (insulin-like growth factor 1) and its binding protein IGFBP are regulated by amino acid availability, providing a mechanism for nutrient feedback to an endocrine pathway involved in metabolic regulation (Bruhat et al. 1999; Takenaka et al. 2000). Equally, amino acids are required for insulin to enhance the formation of complexes between eIF4E and eIF4G and the rate of protein synthesis in skeletal muscle (Kimball 2002). Insulin may serve a permissive function under such circumstances, given that leucine provided orally to rats activates protein synthesis and translation initiation as effectively as a complete meal, but with no rise in plasma insulin concentration (for review see Kimball 2002). Thus, insulin is apparently required for the feeding-induced activation of translation, but an *increase* in circulating insulin level may not be.

Interactions such as those described above are likely to involve effects on processes including gene transcription and mRNA translation. For example, the transcription factor FOXO1 acts as a repressor of the glucose-sensitive PDX1 gene but insulin-induced phosphorylation of FOXO1 by protein kinase B (PKB/Akt) (Puigserver et al. 2003) may overcome such an effect, thus potentially linking insulin and glucose inputs to PDX1-dependent genes such as insulin itself (Campbell and Macfarlane 2002). Phosphatidylinositol 3-kinase (PI3K) dependent activation of PKB by insulin (and other growth factors) also impacts on mRNA translation, through downstream effects on proteins including well-established nutrient-sensitive elements such as TOR and eIF2 (see Rommel et al. 2001 and Fig. 5). GSK3 may also act as a focus for integration of nutrient and endocrine stimuli (see Fig. 5). We have already described how insulin elicits an increase in phosphorylation of the translation inhibitor 4E-BP1 in amino acid-replete cells. Insulin is also able to do this to some extent in cells deprived of amino acids, provided that a metabolisable sugar is also present (Patel et al. 2001). This may reflect an input from metabolic energy to the control of 4E-BP1, perhaps via modulation of the activity of mTOR (see above) and indicates a requirement both for amino acids (especially leucine) and an energy source for activation of this key step in translation initiation. This of course makes excellent physiological sense: amino acids are the precursors for protein synthesis, leucine is an indispensable amino acid, and protein synthesis consumes a large proportion (perhaps 20-25% -

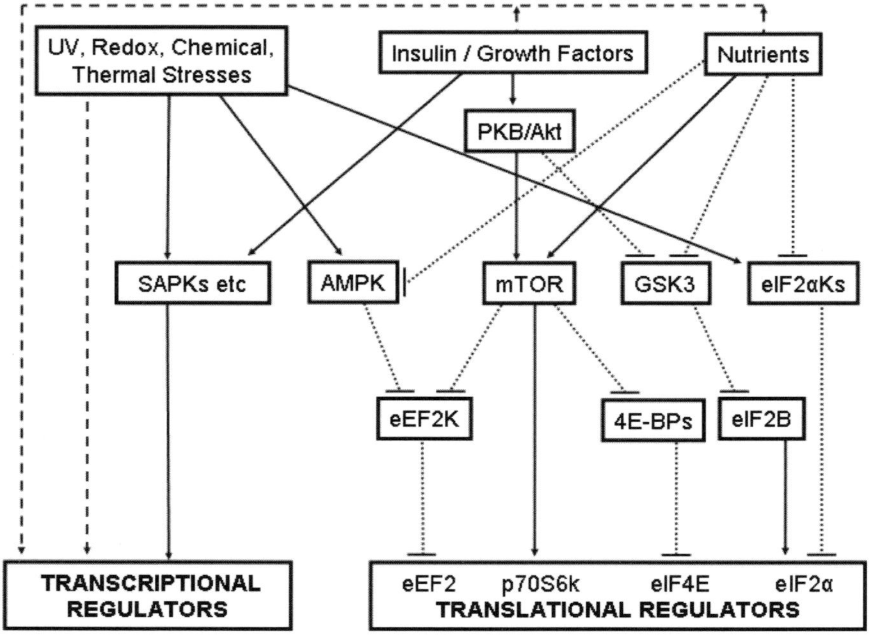

Fig. 5. A schematic diagram showing possible routes for cross-talk between signalling pathways activated by nutrients, growth factors and cellular stresses, which regulate gene transcription and mRNA translation. Only steps recognised as possible foci for cross-talk are shown. Stimulatory signals are shown as solid lines and inhibitory signals as dotted lines. Less direct pathways are indicated by broken arrows. Note a common motif of sequential inhibitory effects leading to a downstream activation. Many such effects result from changes in phosphorylation status of an intermediate resulting from activity of upstream protein kinases or phosphatases. Potential effects may be proscribed or enhanced by cellular or subcellular patterns of distribution of signalling intermediates in particular cell-types under specific conditions. Further details are given in the main text.

Schmidt 1999) of total cellular energy. Other inputs related to cellular energy status are described below in section 2.5.2.

TOR is the central component of a signalling pathway that regulates mammalian cell growth in response to growth factors as well as nutrients (Patti et al. 1999; Proud 2002; Saucedo et al. 2003; Stocker et al. 2003). The TSC2 protein upstream of mTOR (see section 2.2.3.5) is phosphorylated by protein kinase B, thus providing a potential link between phosphatidylinositide 3-kinase signalling and regulation of mTOR. The small GTPase Rheb (Ras homologue enriched in brain) operates downstream of TSC2 in the pathway linking insulin signalling and TOR activation (Saucedo et al. 2003; Stocker et al. 2003; see also Pan chapter). The abundance of Rheb mRNA is rapidly induced in response to protein starvation and overexpressed Rheb promotes cell growth even during starvation (Saucedo et al. 2003), suggesting that Rheb may also function in nutritional, as well as endocrine, control of cell growth (and possibly act as a focus for interplay between the two types of stimuli). It is now clear that the TOR kinases link nutrient-sensing to cell growth in eukaryotes from yeast to man (Rohde et al. 2001) but recent studies suggest that yeast and animal cells may actually control cell growth and division in different ways (Conlon et al. 2001; Conlon and Raff 2003). Most notably, for animal cells it appears that strict cell size checkpoints (as seen in yeast) may not be necessary and proliferation is regulated by the coordinated control of growth and division by diverse stimuli including nutrients and growth factors (Conlon and Raff 2003). The individual cells in multicellular organisms may actually undergo programmed cell death (apoptosis) if deprived of extracellular signals such as growth factors.

Another recognised pathway for nutrient sensing is the hexosamine biosynthetic pathway, which produces the acetylated amino-sugar nucleotide uridine 5'-diphospho-N-acetylglucosamine (UDP-GlcNAc) as its end product. This small molecule includes moieties providing, at least potentially, an index of availability of a remarkable range of nutrients: that is, lipid (acetyl group), carbohydrate (glucose group), protein (amine group), and nucleic acid (UDP group) (Wells et al. 2003). UDP-GlcNAc acts as the donor substrate for modification of nucleocytoplasmic proteins at serine and threonine residues with N-acetylglucosamine (O-GlcNAc). O-GlcNAcylation has regulatory functions in gene transcription and cell-signalling (Vosseller et al. 2002). On several proteins, O-GlcNAc and O-phosphate alternatively occupy the same or adjacent sites, leading to the hypothesis that one function of this sugar is to transiently block phosphorylation, raising the possibility that this posttranslational modification serves as both a nutrient sensor and regulator of insulin signalling (Wells et al. 2001, 2003).

2.5.2 Interactions between nutrient and "stress" signalling pathways

A variety of cellular "stresses", including physical stress (e.g. thermal, mechanical, radiation), chemical stress (e.g. redox, osmotic), and viral infection, also affect cellular functions in ways which potentially influence nutrient – induced responses (for recent reviews see de Nadal et al. 2002; Hoefen and Berk 2002; Johnson and Lapadat 2002; Evans et al. 2003). The mechanisms involved are at least partly related to the action of "stress-activated" signalling pathways. The overall response to these stresses may include systemic components modulated by endocrine and neural system as well as localised components related to specific cytotoxic effects. Major signalling elements activated by cellular stresses (as summarised in Fig. 5) include the SAPK members of the MAP kinase family (for review see de Nadal et al. 2002) and several eIF2α kinases, including the pancreatic endoplasmic reticulum kinase PERK (also called PKR-like ER-resident kinase) and the previously discussed GCN2 (Harding et al. 2000, 2003). Typical responses to stress include a generalised decrease in translation (usually associated with increased phosphorylation of eIF2α) but increased transcription/translation of cyto-protective proteins such as the heat-shock proteins (Wu et al. 2002). Eukaryotic cells respond to unfolded proteins in their endoplasmic reticulum (ER stress), amino acid starvation, or oxidants by phosphorylation of eIF2α (Harding et al. 2003). This adaptation inhibits general protein synthesis while promoting translation and expression of the transcription factor ATF4 under control of the upstream kinase PERK. This signalling pathway initiated by eIF2α phosphorylation protects cells against metabolic consequences of ER oxidation and/or excess cellular amino acid loss through protein secretion by promoting processes geared to improve both resistance to oxidative stress and uptake/assimilation of amino acids (Harding et al. 2003). Thus, mammalian cells have additional mechanisms to regulate eIF2 in response to nutrients in addition to the GCN2 pathway. Such mechanisms may be important in enabling mammalian cells to react to amino acid deficiency before significant accumulation of uncharged tRNAs occurs, which might otherwise compromise translational fidelity (Proud 2002). A cellular glycoprotein known as p67 protects eIF2α from phosphorylation by inhibitory kinases (Chatterjee et al. 1998). The p67 gene includes a heat-shock element (HSE), which promotes transcription after thermal stress, therefore increased p67 expression may facilitate the preferential translation of heat-shock messages during later phases of the response to heat-shock. The pathogenic effects of high nutrient (notably glucose) concentrations in both type 1 and type 2 diabetes may be mediated to a significant extent by oxidative stress related to increased production of reactive oxygen species (ROS) which damage DNA, proteins, and lipids. In such circumstances, ROS also activate cellular stress-sensitive pathways (including those shown in Fig. 5) that contribute to both insulin resistance and impaired insulin secretion characteristic of the diabetic state (Evans et al. 2003).

The AMP-activated protein kinase (AMPK) cascade is activated by cellular stresses that deplete ATP (Hardie and Hawley 2001) and appears to be involved in the glucose signal pathway regulating hepatic gene expression. Stresses such as anoxia activate AMP-activated protein kinase (see Fig. 5), resulting in the inhibition of biosynthetic pathways as a means to conserve cellular ATP. In anoxic rat hepatocytes, AMPK was activated and protein synthesis was inhibited (Horman et al. 2002). The mechanism of inhibition here apparently involves the phosphorylation and inactivation of the translation elongation factor eEF2, possibly due to activation of eEF2 kinase by AMPK-mediated phosphorylation. Much of the energy expended in protein synthesis is consumed during peptide-chain elongation, so its inhibition during conditions of ATP depletion (which may include nutrient deprivation) is highly appropriate (Horman et al. 2002). L-PK gene expression in liver is influenced by cross-talk between oxygen (hypoxia) and glucose stimuli at the level of gene transcription, at least partly because the glucose (carbohydrate) response element (ChoRE) also functions as a hypoxia response element (Kietzmann et al. 2002).

Physical forces producing mechanical stress at the cell surface (e.g. those resulting from swelling or stretch) are increasingly recognised as important factors modulating cellular metabolism, growth and development (Taylor et al. 1999; Ingber 2002; Schliess and Haussinger 2003). For example, physiologically-relevant changes in muscle cell-volume of the sort seen, for example, after exercise or insulin stimulation have coherent effects on muscle metabolism, with swelling stimulating anabolic processes (e.g. protein synthesis) and shrinking having opposite effects (Low et al. 1997b; Taylor et al. 1999). Both muscle and liver cells "sense" volume change by a mechanism requiring integrin-dependent cell adhesion to the extracellular matrix (Low et al. 1997a; Haussinger et al. 2003; Vom Dahl et al. 2003). The ability of plasma-membrane integrins to interact simultaneously with specific extracellular matrix proteins and the cytoskeleton enables them to act both as a "sensor" of cell volume and/or membrane stretch and also as a transducer of these mechanical signals into chemical responses (Parsons 2003). Integrin activation stimulates tyrosine-phosphorylation of focal adhesion kinase (FAK), which promotes interactions between FAK and PI3K (through the p85 regulatory domain - for review see Parsons 2003; Schliess and Haussinger 2003), leading to downstream regulation of effector pathways (see Fig. 5).

Ceramide is a sphingolipid involved in the regulation of diverse cellular processes including apoptosis, cell senescence, the cell cycle, and cellular differentiation (Ruvolo 2003). Ceramide activates a number of enzymes involved in stress signalling cascades including both protein kinases (e.g. SAPKs) and protein phosphatases (PP1; PP2A). Through these protein phosphatases, ceramide may indirectly reverse the action of kinases that are key components of pro-growth signalling processes, for example, protein kinase B (PKB). Ceramide generation is a key feature of the apoptotic process: ceramide activates stress-signal cascades leading to cell death, while simultaneously suppressing growth and survival pathways. Both ceramide and cholesterol are associated with the phospholipid sphingomyelin in the plasma membrane: unsaturated (but not saturated) fatty acids such as PUFA stimulate sphingomyelinase with release of both these molecules but do not in themselves induce apoptosis (Worgall et al. 2002).

Nutrient and stress stimuli may also integrate at the level of gene transcription. For example, asparagine synthase gene transcription is increased in response to deprivation of either amino acids (amino acid response) or glucose (ER stress response) (Siu et al. 2001; see also Kilberg chapter). These two independent pathways converge on the same set of genomic cis-elements within the asparagine synthase promoter (nutrient sensing response elements (NSRE) 1 and 2; together termed NSRU as described in Chapter 1). Transcriptional activation of both NSRE elements is required for gene activation and indeed regulation by amino acid or glucose deprivation as well as ER stress involves common or closely-related transcription factors.

2.6 Summary and perspectives

We have provided a brief overview of current understanding on the mechanisms by which the three key groups of macronutrients in animals contribute to regulation of cell function. Our knowledge in these areas is rapidly expanding at present and several particularly novel aspects of recent findings are highlighted in subsequent chapters of this book. Nevertheless, many important issues related to nutrient-induced responses in animal cells remain to be fully addressed. For example, in relation to the role of mTOR signalling in the control of mRNA translation, we still know relatively little about the machinery through which amino acids are sensed in mammalian cells or by which this information is relayed to mTOR and downstream effectors such as the S6Ks and 4E-BPs. It is also essential that we develop a better understanding of the interplay between signals generated by nutrient, endocrine and stress stimuli which underlie physiological responses and cell fate, as well as the seemingly complex hierarchy of embedded regulatory and permissive mechanisms.

Elucidation of the mechanisms by which cells of higher eukaryotes sense changes in nutrient availability (and respond to them) may have important therapeutic applications, for example, there are many pathological circumstances associated with dysregulation of amino acid and fuel metabolism for which nutritional or pharmacological intervention through such mechanisms may be of clinical benefit. Examples include: (i) obesity and other insulin-resistant states such as diabetes (Delzenne et al. 2001; Celi and Shuldiner 2002; Evans et al. 2003), (ii) protein wasting observed in the musculature upon limb immobilisation, disease, stress, and injury (Hyde et al. 2003), and (iii) increased proliferation and invasiveness of tumour cells resulting from changes in amino acid availability and transport (Singh et al. 1996). Further investigation is needed to fully clarify these issues, with a major long-term goal of developing strategies to modulate sensor and/or signal cascade activation therapeutically as a means to enhance nutrient responsiveness in catabolic disease or, alternatively, to selectively reduce it in tumours.

References

Adegoke OA, McBurney MI, Samuels SE, Baracos VE (2003) Modulation of intestinal protein synthesis and protease mRNA by luminal and systemic nutrients. Am J Physiol Gastrointest Liver Physiol 284:G1017-G1026

Anthony JC, Reiter AK, Anthony TG, Crozier SJ, Lang CH, MacLean DA, Kimball SR, Jefferson LS (2002) Orally administered leucine enhances protein synthesis in skeletal muscle of diabetic rats in the absence of increases in 4E-BP1 or S6K1 phosphorylation. Diabetes 51:928-936

Armstrong JL, Bonavaud SM, Toole BJ, Yeaman SJ (2001) Regulation of glycogen synthesis by amino acids in cultured human muscle cells. J Biol Chem 276:952-956

Asante-Appiah E, Kennedy BP (2003) Protein tyrosine phosphatases: the quest for negative regulators of insulin action. Am J Physiol Endocrinol Metab 284:E663-E670

Avruch J, Belham C, Weng Q, Hara K, Yonezawa K (2001) The p70 S6 kinase integrates nutrient and growth signals to control translational capacity. Prog Mol Subcell Biol 26:115-154

Bain PJ, LeBlanc-Chaffin R, Chen H, Palii SS, Leach KM, Kilberg MS (2002) The mechanism for transcriptional activation of the human ATA2 transporter gene by amino acid deprivation is different than that for asparagine synthetase. J Nutr 132:3023-3029

Balavoine S, Feldmann G, Lardeux B (1993) Regulation of RNA degradation in cultured rat hepatocytes: effects of specific amino acids and insulin. J Cell Physiol 156:56-62

Beugnet A, Tee AR, Taylor PM, Proud CG (2003) Regulation of targets of mTOR signalling by intracellular amino acid availability. Biochem J 372:555-566

Bevington A, Brown J, Butler H, Govindji S, K MK, Sheridan K, Walls J (2002) Impaired system A amino acid transport mimics the catabolic effects of acid in L6 cells. Eur J Clin Invest 32:590-602

Blommaart EF, Luiken JJ, Meijer AJ (1997) Autophagic proteolysis: control and specificity. Histochem J 29:365-385

Bocher V, Pineda-Torra I, Fruchart JC, Staels B (2002) PPARs: transcription factors controlling lipid and lipoprotein metabolism. Ann NY Acad Sci 967:7-18

Bode BP (2001) Recent molecular advances in mammalian glutamine transport. J Nutr 131:2475S-2485S; discussion 2486S-2477S

Bolster DR, Crozier SJ, Kimball SR, Jefferson LS (2002) AMP-activated protein kinase suppresses protein synthesis in rat skeletal muscle through down-regulated mammalian target of rapamycin (mTOR) signaling. J Biol Chem 277:23977-23980

Browne GJ, Proud CG (2002) Regulation of peptide-chain elongation in mammalian cells. Eur J Biochem 269:5360-5368

Bruhat A, Averous J, Carraro V, Zhong C, Reimold AM, Kilberg MS, Fafournoux P (2002) Differences in the molecular mechanisms involved in the transcriptional activation of chop and asparagine synthetase in response to amino acid deprivation or activation of the unfolded protein response. J Biol Chem 25:25

Bruhat A, Jousse C, Fafournoux P (1999) Amino acid limitation regulates gene expression. Proc Nutr Soc 58:625-632

Campbell SC, Macfarlane WM (2002) Regulation of the pdx1 gene promoter in pancreatic beta-cells. Biochem Biophys Res Commun 299:277-284

Celi FS, Shuldiner AR (2002) The role of peroxisome proliferator-activated receptor gamma in diabetes and obesity. Curr Diab Rep 2:179-185

Chatterjee M, Chatterjee N, Datta R, Datta B, Gupta NK (1998) Expression and activity of p67 are induced during heat shock. Biochem Biophys Res Commun 249:113-117

Chen J, Peterson RT, Schreiber SL (1998) Alpha 4 associates with protein phosphatases 2A, 4, and 6. Biochem Biophys Res Commun 247:827-832

Cherkasova VA, Hinnebusch AG (2003) Translational control by TOR and TAP42 through dephosphorylation of eIF2alpha kinase GCN2. Genes Dev 17:859-872

Christie GR, Hajduch E, Hundal HS, Proud CG, Taylor PM (2002) Intracellular sensing of amino acids in *Xenopus laevis* oocytes stimulates p70 S6 kinase in a target of rapamycin-dependent manner. J Biol Chem 277:9952-9957

Christie GR, Hyde R, Hundal HS (2001) Regulation of amino acid transporters by amino acid availability. Curr Opin Clin Nutr Metab Care 4:425-431

Ciaraldi TP, Cha BS, Park KS, Carter L, Mudaliar SR, Henry RR (2002) Free fatty acid metabolism in human skeletal muscle is regulated by PPARgamma and RXR agonists. Ann NY Acad Sci 967:66-70

Clarke SD (2001) Polyunsaturated fatty acid regulation of gene transcription: a molecular mechanism to improve the metabolic syndrome. J Nutr 131:1129-1132

Clemens MJ (2001) Initiation factor eIF2 alpha phosphorylation in stress responses and apoptosis. Prog Mol Subcell Biol 27:57-89

Cohen PT (2002) Protein phosphatase 1--targeted in many directions. J Cell Sci 115:241-256

Combaret L, Taillandier D, Attaix D (2001) Nutritional and hormonal control of protein breakdown. Am J Kidney Dis 37:S108-111

Conigrave AD, Franks AH, Brown EM, Quinn SJ (2002) L-amino acid sensing by the calcium-sensing receptor: a general mechanism for coupling protein and calcium metabolism? Eur J Clin Nutr 56:1072-1080

Conlon I, Raff M (2003) Differences in the way a mammalian cell and yeast cells coordinate cell growth and cell-cycle progression. J Biol 2:7

Conlon IJ, Dunn GA, Mudge AW, Raff MC (2001) Extracellular control of cell size. Nat Cell Biol 3:918-921

Da Silva Xavier G, Leclerc I, Varadi A, Tsuboi T, Moule SK, Rutter GA (2003) Role for AMP-activated protein kinase in glucose-stimulated insulin secretion and preproinsulin gene expression. Biochem J 371:761-774

de Nadal E, Alepuz PM, Posas F (2002) Dealing with osmostress through MAP kinase activation. EMBO Rep 3:735-740

Delzenne N, Ferre P, Beylot M, Daubioul C, Declercq B, Diraison F, Dugail I, Foufelle F, Foretz M, Mace K, Reimer R, Palmer G, Rutter G, Tavare J, Van Loo J, Vidal H (2001) Study of the regulation by nutrients of the expression of genes involved in lipogenesis and obesity in humans and animals. Nutr Metab Cardiovasc Dis 11:118-121

Dennis PB, Jaeschke A, Saitoh M, Fowler B, Kozma SC, Thomas G (2001) Mammalian TOR: a homeostatic ATP sensor. Science 294:1102-1105

Dobrosotskaya IY, Seegmiller AC, Brown MS, Goldstein JL, Rawson RB (2002) Regulation of SREBP processing and membrane lipid production by phospholipids in *Drosophila*. Science 296:879-883

Dong J, Qiu H, Garcia-Barrio M, Anderson J, Hinnebusch AG (2000) Uncharged tRNA activates GCN2 by displacing the protein kinase moiety from a bipartite tRNA-binding domain. Mol Cell 6:269-279

Duplus E, Forest C (2002) Is there a single mechanism for fatty acid regulation of gene transcription? Biochem Pharmacol 64:893-901

Edinger AL, Thompson CB (2002) Akt maintains cell size and survival by increasing mTOR-dependent nutrient uptake. Mol Biol Cell 13:2276-2288

Erickson FL, Nika J, Rippel S, Hannig EM (2001) Minimum requirements for the function of eukaryotic translation initiation factor 2. Genetics 158:123-132

Evans JL, Goldfine ID, Maddux BA, Grodsky GM (2003) Are oxidative stress-activated signaling pathways mediators of insulin resistance and beta-cell dysfunction? Diabetes 52:1-8

Fafournoux P, Bruhat A, Jousse C (2000) Amino acid regulation of gene expression. Biochem J 351:1-12

Fernandez J, Yaman I, Merrick WC, Koromilas A, Wek RC, Sood R, Hensold J, Hatzoglou M (2002) Regulation of internal ribosome entry site-mediated translation by eukaryotic initiation factor-2alpha phosphorylation and translation of a small upstream open reading frame. J Biol Chem 277:2050-2058

Fingar DC, Salama S, Tsou C, Harlow E, Blenis J (2002) Mammalian cell size is controlled by mTOR and its downstream targets S6K1 and 4EBP1/eIF4E. Genes Dev 16:1472-1487

Fischle W, Wang Y, Allis CD (2003) Histone and chromatin cross-talk. Curr Opin Cell Biol 15:172-183

Forsberg H, Gilstring CF, Zargari A, Martinez P, Ljungdahl PO (2001) The role of the yeast plasma membrane SPS nutrient sensor in the metabolic response to extracellular amino acids. Mol Microbiol 42:215-228

Forsberg H, Ljungdahl PO (2001) Sensors of extracellular nutrients in *Saccharomyces cerevisiae*. Curr Genet 40:91-109

Foufelle F, Girard J, Ferre P (1998) Glucose regulation of gene expression. Curr Opin Clin Nutr Metab Care 1:323-328

Francklyn C, Perona JJ, Puetz J, Hou YM (2002) Aminoacyl-tRNA synthetases: versatile players in the changing theater of translation. Rna 8:1363-1372

Gao X, Zhang Y, Arrazola P, Hino O, Kobayashi T, Yeung RS, Ru B, Pan D (2002) Tsc tumour suppressor proteins antagonize amino-acid-TOR signalling. Nat Cell Biol 4:699-704

Gaussin V, Hue L, Stalmans W, Bollen M (1996) Activation of hepatic acetyl-CoA carboxylase by glutamate and Mg2+ is mediated by protein phosphatase-2A. Biochem J 316:217-224

Gazzola RF, Sala R, Bussolati O, Visigalli R, Dall'Asta V, Ganapathy V, Gazzola GC (2001) The adaptive regulation of amino acid transport system A is associated to changes in ATA2 expression. FEBS Lett 490:11-14

Gingras AC, Raught B, Sonenberg N (1999) eIF4 initiation factors: effectors of mRNA recruitment to ribosomes and regulators of translation. Annu Rev Biochem 68:913-963

Gingras AC, Raught B, Sonenberg N (2001a) Control of translation by the target of rapamycin proteins. Prog Mol Subcell Biol 27:143-174

Gingras AC, Raught B, Sonenberg N (2001b) Regulation of translation initiation by FRAP/mTOR. Genes Dev 15:807-826

Guillemain G, Loizeau M, Pincon-Raymond M, Girard J, Leturque A (2000) The large intracytoplasmic loop of the glucose transporter GLUT2 is involved in glucose signaling in hepatic cells. J Cell Sci 113:841-847

Hara K, Maruki Y, Long X, Yoshino K, Oshiro N, Hidayat S, Tokunaga C, Avruch J, Yonezawa K (2002) Raptor, a binding partner of target of rapamycin (TOR), mediates TOR action. Cell 110:177-189

Hardie DG, Hawley SA (2001) AMP-activated protein kinase: the energy charge hypothesis revisited. Bioessays 23:1112-1119

Harding HP, Novoa I, Zhang Y, Zeng H, Wek R, Schapira M, Ron D (2000) Regulated translation initiation controls stress-induced gene expression in mammalian cells. Mol Cell 6:1099-1108

Harding HP, Zhang Y, Zeng H, Novoa I, Lu PD, Calfon M, Sadri N, Yun C, Popko B, Paules R, Stojdl DF, Bell JC, Hettmann T, Leiden JM, Ron D (2003) An integrated stress response regulates amino acid metabolism and resistance to oxidative stress. Mol Cell 11:619-633

Haussinger D, Graf D, Weiergraber OH (2001) Glutamine and cell signaling in liver. J Nutr 131:2509S-2514S; discussion 2523S-2504S

Haussinger D, Kurz AK, Wettstein M, Graf D, Vom Dahl S, Schliess F (2003) Involvement of integrins and Src in tauroursodeoxycholate-induced and swelling-induced choleresis. Gastroenterology 124:1476-1487

Herzig S, Long F, Jhala US, Hedrick S, Quinn R, Bauer A, Rudolph D, Schutz G, Yoon C, Puigserver P, Spiegelman B, Montminy M (2001) CREB regulates hepatic gluconeogenesis through the coactivator PGC-1. Nature 413:179-183

Hinnebusch AG (1997) Translational regulation of yeast GCN4. A window on factors that control initiator-trna binding to the ribosome. J Biol Chem 272:21661-21664

Hoefen RJ, Berk BC (2002) The role of MAP kinases in endothelial activation. Vascul Pharmacol 38:271-273

Holecek M (2002) Relation between glutamine, branched-chain amino acids, and protein metabolism. Nutrition 18:130-133

Horman S, Browne G, Krause U, Patel J, Vertommen D, Bertrand L, Lavoinne A, Hue L, Proud C, Rider M (2002) Activation of AMP-activated protein kinase leads to the phosphorylation of elongation factor 2 and an inhibition of protein synthesis. Curr Biol 12:1419-1423

Hyde R, Christie GR, Litherland GJ, Hajduch E, Taylor PM, Hundal HS (2001) Subcellular localization and adaptive up-regulation of the System A (SAT2) amino acid transporter in skeletal-muscle cells and adipocytes. Biochem J 355:563-568

Hyde R, Taylor PM, Hundal HS (2003) Amino acid transporters: roles in amino acid sensing and signalling in animal cells. Biochem J 373:1-18

Iiboshi Y, Papst PJ, Kawasome H, Hosoi H, Abraham RT, Houghton PJ, Terada N (1999) Amino acid-dependent control of p70(s6k). Involvement of tRNA aminoacylation in the regulation. J Biol Chem 274:1092-1099

Ingber DE (2002) Mechanical signaling and the cellular response to extracellular matrix in angiogenesis and cardiovascular physiology. Circ Res 91:877-887

Isotani S, Hara K, Tokunaga C, Inoue H, Avruch J, Yonezawa K (1999) Immunopurified mammalian target of rapamycin phosphorylates and activates p70 S6 kinase alpha *in vitro*. J Biol Chem 274:34493-34498

Itoh Y, Kawamata Y, Harada M, Kobayashi M, Fujii R, Fukusumi S, Ogi K, Hosoya M, Tanaka Y, Uejima H, Tanaka H, Maruyama M, Satoh R, Okubo S, Kizawa H, Komatsu H, Matsumura F, Noguchi Y, Shinohara T, Hinuma S, Fujisawa Y, Fujino M (2003) Free fatty acids regulate insulin secretion from pancreatic beta cells through GPR40. Nature 422:173-176

Jiang Y, Broach JR (1999) Tor proteins and protein phosphatase 2A reciprocally regulate Tap42 in controlling cell growth in yeast. Embo J 18:2782-2792

Johnson GL, Lapadat R (2002) Mitogen-activated protein kinase pathways mediated by ERK, JNK, and p38 protein kinases. Science 298:1911-1912

Jump DB (2002) Dietary polyunsaturated fatty acids and regulation of gene transcription. Curr Opin Lipidol 13:155-164

Kee AJ, Combaret L, Tilignac T, Souweine B, Aurousseau E, Dalle M, Taillandier D, Attaix D (2003) Ubiquitin-proteasome-dependent muscle proteolysis responds slowly to insulin release and refeeding in starved rats. J Physiol 546:765-776

Kersten S (2002) Effects of fatty acids on gene expression: role of peroxisome proliferator-activated receptor alpha, liver X receptor alpha and sterol regulatory element-binding protein-1c. Proc Nutr Soc 61:371-374

Kietzmann T, Krones-Herzig A, Jungermann K (2002) Signaling cross-talk between hypoxia and glucose via hypoxia-inducible factor 1 and glucose response elements. Biochem Pharmacol 64:903-911

Kilberg MS, Barbosa-Tessmann IP (2002) Genomic sequences necessary for transcriptional activation by amino acid deprivation of mammalian cells. J Nutr 132:1801-1804

Kim DH, Sarbassov DD, Ali SM, King JE, Latek RR, Erdjument-Bromage H, Tempst P, Sabatini DM (2002) mTOR interacts with raptor to form a nutrient-sensitive complex that signals to the cell growth machinery. Cell 110:163-175

Kim do H, Sarbassov dos D, Ali SM, Latek RR, Guntur KV, Erdjument-Bromage H, Tempst P, Sabatini DM (2003) GbetaL, a positive regulator of the rapamycin-sensitive pathway required for the nutrient-sensitive interaction between raptor and mTOR. Mol Cell 11:895-904

Kimball SR (2002) Regulation of global and specific mRNA translation by amino acids. J Nutr 132:883-886
Kimball SR, Horetsky RL, Jefferson LS (1998) Implication of eIF2B rather than eIF4E in the regulation of global protein synthesis by amino acids in L6 myoblasts. J Biol Chem 273:30945-30953
Kimball SR, Yancisin M, Horetsky RL, Jefferson LS (1996) Translational and pretranslational regulation of protein synthesis by amino acid availability in primary cultures of rat hepatocytes. Int J Biochem Cell Biol 28:285-294
Kishi A, Nakamura T, Nishio Y, Maegawa H, Kashiwagi A (2003) Sumoylation of Pdx1 is associated with its nuclear localization and insulin gene activation. Am J Physiol Endocrinol Metab 284:E830-E840
Kitamura T, Nakae J, Kitamura Y, Kido Y, Biggs WH, 3rd, Wright CV, White MF, Arden KC, Accili D (2002) The forkhead transcription factor Foxo1 links insulin signaling to Pdx1 regulation of pancreatic beta cell growth. J Clin Invest 110:1839-1847
Klumpp S, Krieglstein J (2002) Serine/threonine protein phosphatases in apoptosis. Curr Opin Pharmacol 2:458-462
Kobayashi H, Borsheim E, Anthony TG, Traber DL, Badalamenti J, Kimball SR, Jefferson LS, Wolfe RR (2003) Reduced amino acid availability inhibits muscle protein synthesis and decreases activity of initiation factor eIF2B. Am J Physiol Endocrinol Metab 284:E488-E498
Koo SH, Dutcher AK, Towle HC (2001) Glucose and insulin function through two distinct transcription factors to stimulate expression of lipogenic enzyme genes in liver. J Biol Chem 276:9437-9445
Kowluru A, Chen HQ, Modrick LM, Stefanelli C (2001) Activation of acetyl-CoA carboxylase by a glutamate- and magnesium-sensitive protein phosphatase in the islet beta-cell. Diabetes 50:1580-1587
Krause U, Bertrand L, Hue L (2002a) Control of p70 ribosomal protein S6 kinase and acetyl-CoA carboxylase by AMP-activated protein kinase and protein phosphatases in isolated hepatocytes. Eur J Biochem 269:3751-3759
Krause U, Bertrand L, Maisin L, Rosa M, Hue L (2002b) Signalling pathways and combinatory effects of insulin and amino acids in isolated rat hepatocytes. Eur J Biochem 269:3742-3750
Leclerc I, da Silva Xavier G, Rutter GA (2002) AMP- and stress-activated protein kinases: key regulators of glucose-dependent gene transcription in mammalian cells? Prog Nucleic Acid Res Mol Biol 71:69-90
Lee Y, Yu X, Gonzales F, Mangelsdorf DJ, Wang MY, Richardson C, Witters LA, Unger RH (2002) PPAR alpha is necessary for the lipopenic action of hyperleptinemia on white adipose and liver tissue. Proc Natl Acad Sci USA 99:11848-11853
Lehrmann H, Pritchard LL, Harel-Bellan A (2002) Histone acetyltransferases and deacetylases in the control of cell proliferation and differentiation. Adv Cancer Res 86:41-65
Li L, Gotta S, Mauviel A, Varga J (1995) L-tryptophan induces expression of collagenase gene in human fibroblasts: demonstration of enhanced AP-1 binding and AP-1 binding site-driven promoter activity. Cell Mol Biol Res 41:361-368
Ling R, Bridges CC, Sugawara M, Fujita T, Leibach FH, Prasad PD, Ganapathy V (2001) Involvement of transporter recruitment as well as gene expression in the substrate-induced adaptive regulation of amino acid transport system A. Biochim Biophys Acta 1512:15-21

Lovejoy JC (2002) The influence of dietary fat on insulin resistance. Curr Diab Rep 2:435-440

Low SY, Rennie MJ, Taylor PM (1997a) Involvement of integrins and the cytoskeleton in modulation of skeletal muscle glycogen synthesis by changes in cell volume. FEBS Lett 417:101-103

Low SY, Rennie MJ, Taylor PM (1997b) Signaling elements involved in amino acid transport responses to altered muscle cell volume. Faseb J 11:1111-1117

Macfarlane WM, McKinnon CM, Felton-Edkins ZA, Cragg H, James RF, Docherty K (1999) Glucose stimulates translocation of the homeodomain transcription factor PDX1 from the cytoplasm to the nucleus in pancreatic beta-cells. J Biol Chem 274:1011-1016

Macfarlane WM, Shepherd RM, Cosgrove KE, James RF, Dunne MJ, Docherty K (2000) Glucose modulation of insulin mRNA levels is dependent on transcription factor PDX-1 and occurs independently of changes in intracellular Ca2+. Diabetes 49:418-423

Marten NW, Hsiang CH, Yu L, Stollenwerk NS, Straus DS (1999) Functional activity of hepatocyte nuclear factor-1 is specifically decreased in amino acid-limited hepatoma cells. Biochim Biophys Acta 1447:160-174

Martin KA, Blenis J (2002) Coordinate regulation of translation by the PI 3-kinase and mTOR pathways. Adv Cancer Res 86:1-39

McKinnon CM, Docherty K (2001) Pancreatic duodenal homeobox-1, PDX-1, a major regulator of beta cell identity and function. Diabetologia 44:1203-1214

Melloul D, Marshak S, Cerasi E (2002) Regulation of insulin gene transcription. Diabetologia 45:309-326

Miotto G, Venerando R, Khurana KK, Siliprandi N, Mortimore GE (1992) Control of hepatic proteolysis by leucine and isovaleryl-L-carnitine through a common locus. Evidence for a possible mechanism of recognition at the plasma membrane. J Biol Chem 267:22066-22072

Miotto G, Venerando R, Marin O, Siliprandi N, Mortimore GE (1994) Inhibition of macroautophagy and proteolysis in the isolated rat hepatocyte by a nontransportable derivative of the multiple antigen peptide Leu8-Lys4-Lys2-Lys-beta Ala. J Biol Chem 269:25348-25353

Mitch WE, Price SR (2003) Mechanisms activating proteolysis to cause muscle atrophy in catabolic conditions. J Ren Nutr 13:149-152

Mothe-Satney I, Brunn GJ, McMahon LP, Capaldo CT, Abraham RT, Lawrence JC Jr (2000) Mammalian target of rapamycin-dependent phosphorylation of PHAS-I in four (S/T)P sites detected by phospho-specific antibodies. J Biol Chem 275:33836-33843

Nelson G, Chandrashekar J, Hoon MA, Feng L, Zhao G, Ryba NJ, Zuker CS (2002) An amino-acid taste receptor. Nature 416:199-202

Nojima H, Tokunaga C, Eguchi S, Oshiro N, Hidayat S, Yoshino KI, Hara K, Tanaka N, Avruch J, Yonezawa K (2003) The mTOR partner, Raptor binds the mTOR substrates, p70 S6 kinase and 4E-BP1 through their TOS (TOR signaling) motif. J Biol Chem 278:15461-15464

Nosjean O, Boutin JA (2002) Natural ligands of PPARgamma: are prostaglandin J(2) derivatives really playing the part? Cell Signal 14:573-583

Parsons JT (2003) Focal adhesion kinase: the first ten years. J Cell Sci 116:1409-1416

Patel J, Wang X, Proud CG (2001) Glucose exerts a permissive effect on the regulation of the initiation factor 4E binding protein 4E-BP1. Biochem J 358:497-503

Patti ME, Brambilla E, Luzi L, Landaker EJ, Kahn CR (1998) Bidirectional modulation of insulin action by amino acids. J Clin Invest 101:1519-1529

Patti ME, Virkamaki A, Landaker EJ, Kahn CR, Yki-Jarvinen H (1999) Activation of the hexosamine pathway by glucosamine *in vivo* induces insulin resistance of early postreceptor insulin signaling events in skeletal muscle. Diabetes 48:1562-1571

Pearson RB, Dennis PB, Han JW, Williamson NA, Kozma SC, Wettenhall RE, Thomas G (1995) The principal target of rapamycin-induced p70s6k inactivation is a novel phosphorylation site within a conserved hydrophobic domain. Embo J 14:5279-5287

Peterson RT, Desai BN, Hardwick JS, Schreiber SL (1999) Protein phosphatase 2A interacts with the 70-kDa S6 kinase and is activated by inhibition of FKBP12-rapamycinassociated protein. Proc Natl Acad Sci USA 96:4438-4442

Peyrollier K, Hajduch E, Blair AS, Hyde R, Hundal HS (2000) L-leucine availability regulates phosphatidylinositol 3-kinase, p70 S6 kinase and glycogen synthase kinase-3 activity in L6 muscle cells: evidence for the involvement of the mammalian target of rapamycin (mTOR) pathway in the L-leucine-induced up-regulation of system A amino acid transport. Biochem J 350:361-368

Pham PT, Heydrick SJ, Fox HL, Kimball SR, Jefferson LS Jr, Lynch CJ (2000) Assessment of cell-signaling pathways in the regulation of mammalian target of rapamycin (mTOR) by amino acids in rat adipocytes. J Cell Biochem 79:427-441

Picard F, Auwerx J (2002) PPAR(gamma) and glucose homeostasis. Annu Rev Nutr 22:167-197

Proud CG (2001) Regulation of eukaryotic initiation factor eIF2B. Prog Mol Subcell Biol 26:95-114

Proud CG (2002) Regulation of mammalian translation factors by nutrients. Eur J Biochem 269:5338-5349

Puigserver P, Rhee J, Donovan J, Walkey CJ, Yoon JC, Oriente F, Kitamura Y, Altomonte J, Dong H, Accili D, Spiegelman BM (2003) Insulin-regulated hepatic gluconeogenesis through FOXO1-PGC-1alpha interaction. Nature 423:550-555

Rabinovitz M (1995) The phosphofructokinase-uncharged tRNA interaction in metabolic and cell cycle control: an interpretive review. Nucleic Acids Symp Ser 33:182-189

Robinson MB (2002) Regulated trafficking of neurotransmitter transporters: common notes but different melodies. J Neurochem 80:1-11

Rohde J, Heitman J, Cardenas ME (2001) The TOR kinases link nutrient sensing to cell growth. J Biol Chem 276:9583-9586

Rohde JR, Cardenas ME (2003) The tor pathway regulates gene expression by linking nutrient sensing to histone acetylation. Mol Cell Biol 23:629-635

Rommel C, Bodine SC, Clarke BA, Rossman R, Nunez L, Stitt TN, Yancopoulos GD, Glass DJ (2001) Mediation of IGF-1-induced skeletal myotube hypertrophy by PI(3)K/Akt/mTOR and PI(3)K/Akt/GSK3 pathways. Nat Cell Biol 3:1009-1013

Rufo C, Teran-Garcia M, Nakamura MT, Koo SH, Towle HC, Clarke SD (2001) Involvement of a unique carbohydrate-responsive factor in the glucose regulation of rat liver fatty-acid synthase gene transcription. J Biol Chem 276:21969-21975

Rutter GA, Tavare JM, Palmer DG (2000) Regulation of mammalian gene expression by glucose. News Physiol Sci 15:149-154

Ruvolo PP (2003) Intracellular signal transduction pathways activated by ceramide and its metabolites. Pharmacol Res 47:383-392

Saucedo LJ, Gao X, Chiarelli DA, Li L, Pan D, Edgar BA (2003) Rheb promotes cell growth as a component of the insulin/TOR signalling network. Nat Cell Biol 27:27

Schalm SS, Fingar DC, Sabatini DM, Blenis J (2003) TOS motif-mediated raptor binding regulates 4E-BP1 multisite phosphorylation and function. Curr Biol 13:797-806

Schliess F, Haussinger D (2003) Cell volume and insulin signaling. Int Rev Cytol 225:187-228

Schmidt EV (1999) The role of c-myc in cellular growth control. Oncogene 18:2988-2996

Schreiber SL, Bernstein BE (2002) Signaling network model of chromatin. Cell 111:771-778

Sessler AM, Ntambi JM (1998) Polyunsaturated fatty acid regulation of gene expression. J Nutr 128:923-926

Shigemitsu K, Tsujishita Y, Hara K, Nanahoshi M, Avruch J, Yonezawa K (1999) Regulation of translational effectors by amino acid and mammalian target of rapamycin signaling pathways. Possible involvement of autophagy in cultured hepatoma cells. J Biol Chem 274:1058-1065

Shimano H (2001) Sterol regulatory element-binding proteins (SREBPs): transcriptional regulators of lipid synthetic genes. Prog Lipid Res 40:439-452

Singh RK, Rinehart CA, Kim JP, Tolleson-Rinehart S, Lawing LF, Kaufman DG, Siegal GP (1996) Tumor cell invasion of basement membrane *in vitro* is regulated by amino acids. Cancer Invest 14:6-18

Siu F, Bain PJ, LeBlanc-Chaffin R, Chen H, Kilberg MS (2002) ATF4 is a mediator of the nutrient-sensing response pathway that activates the human asparagine synthetase Gene. J Biol Chem 277:24120-24127

Siu F, Chen C, Zhong C, Kilberg MS (2001) CCAAT/enhancer-binding protein-beta is a mediator of the nutrient-sensing response pathway that activates the human asparagine synthetase gene. J Biol Chem 276:48100-48107

Sood R, Porter AC, Olsen DA, Cavener DR, Wek RC (2000) A mammalian homologue of GCN2 protein kinase important for translational control by phosphorylation of eukaryotic initiation factor-2alpha. Genetics 154:787-801

Stocker H, Radimerski T, Schindelholz B, Wittwer F, Belawat P, Daram P, Breuer S, Thomas G, Hafen E (2003) Rheb is an essential regulator of S6K in controlling cell growth in *Drosophila*. Nat Cell Biol 27:27

Stoeckman AK, Towle HC (2002) The role of SREBP-1c in nutritional regulation of lipogenic enzyme gene expression. J Biol Chem 277:27029-27035

Strudwick S, Borden KL (2002) The emerging roles of translation factor eIF4E in the nucleus. Differentiation 70:10-22

Takenaka A, Oki N, Takahashi SI, Noguchi T (2000) Dietary restriction of single essential amino acids reduces plasma insulin-like growth factor-I (IGF-I) but does not affect plasma IGF-binding protein-1 in rats. J Nutr 130:2910-2914

Tang H, Hornstein E, Stolovich M, Levy G, Livingstone M, Templeton D, Avruch J, Meyuhas O (2001) Amino acid-induced translation of TOP mRNAs is fully dependent on phosphatidylinositol 3-kinase-mediated signaling, is partially inhibited by rapamycin, and is independent of S6K1 and rpS6 phosphorylation. Mol Cell Biol 21:8671-8683

Taylor PM, Rennie MJ, Low SY (1999) Biomembrane transport and inter-organ nutrient flows: the amino acids. In: Van Winkle L (ed) Biomembrane Transport. Academic Press, New York, pp 295-325

Tee AR, Fingar DC, Manning BD, Kwiatkowski DJ, Cantley LC, Blenis J (2002) Tuberous sclerosis complex-1 and -2 gene products function together to inhibit mammalian target of rapamycin (mTOR)-mediated downstream signaling. Proc Natl Acad Sci USA 99:13571-13576

Theodosiou A, Ashworth A (2002) MAP kinase phosphatases. Genome Biol 3:REVIEWS3009
Thomas J, Bramlett KS, Montrose C, Foxworthy P, Eacho PI, McCann D, Cao G, Kiefer A, McCowan J, Yu KL, Grese T, Chin WW, Burris TP, Michael LF (2003) A chemical switch regulates fibrate specificity for peroxisome proliferator-activated receptor alpha (PPARalpha) versus liver X receptor. J Biol Chem 278:2403-2410
Towle HC, Kaytor EN, Shih HM (1997) Regulation of the expression of lipogenic enzyme genes by carbohydrate. Annu Rev Nutr 17:405-433
Van Sluijters DA, Dubbelhuis PF, Blommaart EF, Meijer AJ (2000) Amino-acid-dependent signal transduction. Biochem J 351 Pt 3:545-550
Varnier M, Leese GP, Thompson J, Rennie MJ (1995) Stimulatory effect of glutamine on glycogen accumulation in human skeletal muscle. Am J Physiol 269:E309-E315
Vaulont S, Vasseur-Cognet M, Kahn A (2000) Glucose regulation of gene transcription. J Biol Chem 275:31555-31558
Verger A, Perdomo J, Crossley M (2003) Modification with SUMO. EMBO Rep 4:137-142
Vessby B (2000) Dietary fat and insulin action in humans. Br J Nutr 83:S91-S96
Vom Dahl S, Schliess F, Reissmann R, Gorg B, Weiergraber O, Kacalkova M, Dombrowski F, Haussinger D (2003) Involvement of integrins in osmosensing and signalling towards autophagic proteolysis in rat liver. J Biol Chem 28:28
Vosseller K, Sakabe K, Wells L, Hart GW (2002) Diverse regulation of protein function by O-GlcNAc: a nuclear and cytoplasmic carbohydrate post-translational modification. Curr Opin Chem Biol 6:851-857
Wang X, Campbell LE, Miller CM, Proud CG (1998) Amino acid availability regulates p70 S6 kinase and multiple translation factors. Biochem J 334:261-267
Wang X, Li W, Williams M, Terada N, Alessi DR, Proud CG (2001) Regulation of elongation factor 2 kinase by p90(RSK1) and p70 S6 kinase. Embo J 20:4370-4379
Wang YX, Lee CH, Tiep S, Yu RT, Ham J, Kang H, Evans RM (2003) Peroxisome-proliferator-activated receptor delta activates fat metabolism to prevent obesity. Cell 113:159-170
Wells L, Vosseller K, Hart GW (2001) Glycosylation of nucleocytoplasmic proteins: signal transduction and O-GlcNAc. Science 291:2376-2378
Wells L, Vosseller K, Hart GW (2003) A role for N-acetylglucosamine as a nutrient sensor and mediator of insulin resistance. Cell Mol Life Sci 60:222-228
Worgall TS, Johnson RA, Seo T, Gierens H, Deckelbaum RJ (2002) Unsaturated fatty acid-mediated decreases in sterol regulatory element-mediated gene transcription are linked to cellular sphingolipid metabolism. J Biol Chem 277:3878-3885
Wu S, Hu Y, Wang JL, Chatterjee M, Shi Y, Kaufman RJ (2002) Ultraviolet light inhibits translation through activation of the unfolded protein response kinase PERK in the lumen of the endoplasmic reticulum. J Biol Chem 277:18077-18083
Xu G, Kwon G, Cruz WS, Marshall CA, McDaniel ML (2001) Metabolic regulation by leucine of translation initiation through the mTOR-signaling pathway by pancreatic beta-cells. Diabetes 50:353-360
Yaman I, Fernandez J, Liu H, Caprara M, Komar AA, Koromilas AE, Zhou L, Snider MD, Scheuner D, Kaufman RJ, Hatzoglou M (2003) The zipper model of translational control. A small upstream ORF is the switch that controls structural remodeling of an mRNA leader. Cell 113:519-531

Yoon JC, Puigserver P, Chen G, Donovan J, Wu Z, Rhee J, Adelmant G, Stafford J, Kahn CR, Granner DK, Newgard CB, Spiegelman BM (2001) Control of hepatic gluconeogenesis through the transcriptional coactivator PGC-1. Nature 413:131-138

Zhang H, Stallock JP, Ng JC, Reinhard C, Neufeld TP (2000) Regulation of cellular growth by the *Drosophila* target of rapamycin dTOR. Genes Dev 14:2712-2724

Zhang P, McGrath BC, Reinert J, Olsen DS, Lei L, Gill S, Wek SA, Vattem KM, Wek RC, Kimball SR, Jefferson LS, Cavener DR (2002) The GCN2 eIF2alpha kinase is required for adaptation to amino acid deprivation in mice. Mol Cell Biol 22:6681-6688

3 Antagonists of the TOR pathway in animal cells

Duojia Pan

Abstract

The *TSC1* and *TSC2* tumor suppressor genes were initially identified as human disease genes mutated in tuberous sclerosis, a syndrome characterized by the widespread development of benign tumors. TSC2 encodes a putative GTPase activating protein (GAP), while TSC1 encodes a novel protein containing two coiled-coil domains. Recent genetic studies in *Drosophila* have implicated Tsc1 and Tsc2 as negative regulators of TOR signaling in cell growth, with a loss of Tsc1/Tsc2 leading to increased cell size. In animal cells, loss of Tsc1/Tsc2 leads to an increase in S6K activity that is sensitive to rapamycin but resistant to amino acid starvation. Genetic and biochemical studies in *Drosophila* further identified Rheb, a member of the Ras superfamily GTPases, as a direct target of the GAP activity of Tsc2 that functions between Tsc1/Tsc2 and TOR. These exciting developments demonstrate that *Drosophila* is a powerful system to decipher the architecture of the TOR signaling network at molecular, cellular and organismal levels.

3.1 Introduction

The growth of single cell organisms such as yeast is largely regulated by nutrient availability. During metazoan evolution, cells have acquired the ability to respond to intercellular hormones and growth factors, but still retain their ability to directly respond to nutrient levels. Thus in higher eukaryotes, cell growth is dependent on the integration of multiple cell-extrinsic cues including hormones, growth factors and nutrients (reviewed in Rohde et al. 2001; Kimball and Jefferson 2000). Deciphering how these cell-extrinsic signals are integrated to control cell growth is pivotal to our understanding of growth-control mechanisms in normal development as well as pathological conditions such as cancer.

There is increasing evidence that nutrients, in particular amino acids, are sensed by a signaling pathway involving TOR (target of rapamycin, also known as mTOR/FRAP/RAFT1 in mammals), a Ser/Thr protein kinase specifically inhibited by the immunosuppressant rapamycin (Thomas and Hall 1997; Gingras et al. 2001; Kuruvilla and Schreiber 1999). This TOR-dependent amino acid signaling pathway has been shown to control downstream translation initiation regulators including ribosomal subunit S6 kinase (S6K) and initiation factor 4E-binding protein (4E-BP; also known as PHAS) (Thomas and Hall 1997; Gingras et al. 2001). Phosphorylation of S6K increases its kinase activity towards the ribosome S6 sub-

Topics in Current Genetics, Vol. 7
J. Winderickx, P.M. Taylor (Eds) Nutrient-Induced Responses in Eukaryotic Cells

unit, leading to increased translation of 5' TOP (terminal oligopyrimidine tract) mRNAs that largely encode components of the translational apparatus. Phosphorylation of 4E-BP releases the eukaryotic initiation factor 4E (eIF4E) from the inactive eIF4E/4E-BP complex, permitting eIF4E to function in translation initiation. Amino acid starvation results in a rapid dephosphorylation of S6K and 4E-BP, while re-addition of amino acids restores S6K and 4E-BP phosphorylation in a TOR-dependent manner (Hara et al. 1998). TOR has been proposed as a nutrient-dependent "gatekeeper" that couples amino acid availability to cell growth (Thomas and Hall 1997; Gingras et al. 2001). At present, little is known about how amino acids are sensed and transduced to downstream effectors.

The nutrient-sensing pathway mediated by TOR is intimately linked to the insulin signaling pathway (reviewed in Proud and Denton 1997), another pathway that plays a pivotal role in cellular growth. Upon binding of insulin or insulin-like growth factors, insulin receptor (InR) or IGF receptors activates phosphoinositide 3-kinase (PI3K), either directly or through insulin receptor substrate (IRS) protein. Phosphorylation of the membrane lipid phosphatidylinositol 4,5-biphosphate (PIP_2) by PI3K produces the second messenger phosphatidylinositol 3,4,5-triphosphate (PIP_3), which activates phosphoinositide-dependent kinase 1 (PDK1) and Akt (also called PKB). A well-known negative regulator of this pathway is the PTEN (phosphatase and tensin homologue deleted from chromosome 10) tumor suppressor, which functions as a phosphatase to convert PIP3 to PIP2 (reviewed in Cantley and Neel 1999). For simplicity, we will refer to this InR/PI3K/Akt pathway as the "canonical" insulin pathway. In mammalian cells, this pathway is thought to exert its effect on cell growth through phosphorylation of S6K and 4E-BP, the same translation regulators downstream of TOR signaling (reviewed in Thomas and Hall 1997; Sonenberg and Gingras 1998; Schmelzle and Hall 2000). It has been suggested that the TOR signaling pathway is parallel to the canonical insulin pathway, but amino acid starvation or rapamycin treatment can override the insulin signal (Hara et al. 1998). The exact mechanisms by which insulin signaling and TOR signaling intersect with each other, however, are not well understood.

The contribution of insulin and TOR signaling to cell growth has been confirmed by recent genetic studies in *Drosophila* (reviewed in Stocker and Hafen 2000). Loss of TOR or components of the insulin pathway, including *InR*, *IRS*, *PI3K*, *PDK1*, *Akt* and *S6K*, all results in decreased cell size, while upregulation of insulin signaling increases cell size (Weinkove et al. 1999; Böhni et al. 1999; Montagne et al. 1999; Chen et al. 1996; Verdu et al. 1999; Zhang et al. 2000; Oldham et al. 2000; Huang et al. 1999; Gao et al. 2000; Goberdhan et al. 1999; Rintelen et al. 2001; Cho et al. 2001). Studies of TOR and insulin signaling in *Drosophila* have revealed some unexpected findings. In particular, it was shown that in *Drosophila*, S6K activity is PI3K/Akt-independent, but requires PDK1 (Radimerski et al. 2002). These findings suggest a complex interplay between TOR and insulin signaling in *Drosophila* distinct from that postulated in mammalian cells.

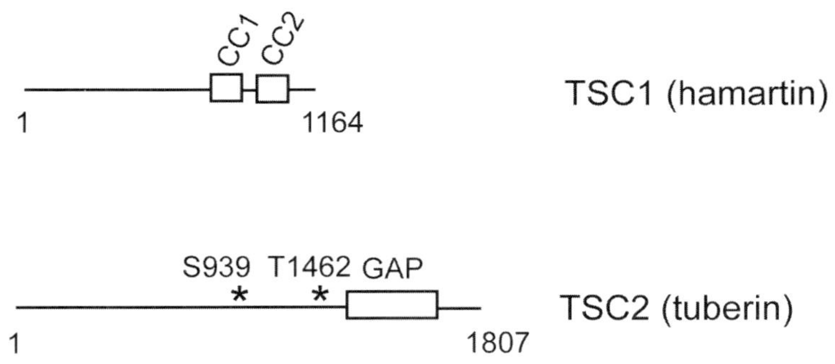

Fig. 1. Schematic structures of the human TSC1 (hamartin) and TSC2 (tuberin) proteins. TSC1 contains two coiled-coil domains (CC1 and CC2), while TSC2 contains a domain with homology to the GAP domain of Rap1GAP. Also marked are the two Akt phosphorylation sites on TSC2 that are conserved between human and *Drosophila*.

At present, little is known about regulators functioning upstream of TOR. In this review, we discuss exciting recent discovery of two tumor suppressors, tuberous sclerosis 1 (TSC1) and tuberous sclerosis 2 (TSC2), as upstream regulators of TOR and its implication for our understanding of the amino acid/TOR signaling.

3.2 TSC1/TSC2 tumor suppressor proteins as antagonists of TOR signaling

3.2.1 Genetic studies of TSC1/TSC2 function in mammalian systems

Tuberous Sclerosis (TSC) is an autosomal dominant disorder that affects 1 in 6000 individuals (reviewed in Young and Povey, 1998). This disease is characterized by the widespread development of benign tumors termed harmatomas, frequently leading to skin rashes, seizures, and mental retardation. TSC is caused by a mutation in either the *TSC1* or *TSC2* tumor suppressor gene. *TSC2* encodes a putative GTPase activating protein (GAP), while *TSC1* encodes a novel protein containing two coiled-coil domains (van Slegtenhorst et al. 1997; The European Chromosome 16 Tuberous Sclerosis Consortium, 1993) (Fig. 1). The TSC1 and TSC2 proteins have also been referred to as hamartin and tuberin, respectively. In mammalian cells, the TSC1 and TSC2 proteins have been shown to form a complex (Nellist et al. 1999; van Slegtenhorst et al. 1998), and have been proposed to control various cellular functions including cell cycle (Soucek et al. 1998; Soucek et al. 1997), endocytosis (Xiao et al. 1997), cell adhesion (Lamb et al. 2000), and transcription (Henry et al. 1998). However, it is not clear how these potential activities relate to the tumor suppressor function of the TSC1/TSC2 proteins.

Studies of mammalian models lacking the *TSC1* or *TSC2* gene have provided certain insight into the function of the *TSC* genes in development. The Eker rat strain contains a germline insertion mutation in the rat *TSC2* gene, which causes a premature truncation of the TSC2 protein (Kobayashi et al. 1997; Rennebeck et al. 1998). Murine models lacking *TSC1* or *TSC2* have also been generated by gene targeting (Onda et al. 1999; Kobayashi et al. 1999; Kobayashi et al. 2001; Kwiatkowski et al. 2002). In these models, homozygous *TSC1* or *TSC2* mutants are embryonic lethal, while heterozygous carriers are prone to tumor formation. While these studies confirmed the importance of *TSC1/TSC2* in tumor suppression, they provided limited insight into their molecular mechanisms.

3.2.2 Genetic studies of TSC1/TSC2 function in *Drosophila* suggest a functional link between TSC1/TSC2 and TOR signaling

A breakthrough in our understanding of TSC came recently from studies of *Drosophila TSC1* and *TSC2* homologues, which will be referred to as *Tsc1* and *Tsc2* to be distinguished from their mammalian counterparts throughout the rest of the review. *Drosophila* mutants of *Tsc1* and *Tsc2* were isolated based on their cell size phenotype: loss of Tsc1 or Tsc2 results in a dramatic increase in cell size (Tapon et al. 2001; Potter et al. 2001; Gao and Pan 2001). This mutant phenotype is strikingly similar to that resulting from activation of the canonical insulin pathway, suggesting an intimate relationship between Tsc1/Tsc2 and insulin signaling. Consistent with this hypothesis, a dosage sensitive genetic interaction is observed between Tsc1/Tsc2 and insulin receptor gene: heterozygosity of *TSC1* or *TSC2* is sufficient to rescue the lethality of certain loss-of-function insulin receptor mutants (Gao and Pan 2001).

Our genetic analyses further revealed that Tsc1/Tsc2 are unlikely to act in a linear pathway within the canonical insulin signaling pathway. Double mutant of *Tsc1* and *PTEN* revealed an additive cell size increase as compared to either mutant alone, suggesting that Tsc1/Tsc2 act in a pathway parallel to the canonical insulin pathway (Gao and Pan 2001). The identity of the Tsc1/Tsc2 pathway was strongly suggested by analyses of the genetic interactions between *Tsc1/Tsc2* and *TOR*: heterozygosity of *TOR* rescued the early lethality and cell size phenotype of a loss-of-function *Tsc1/Tsc2* mutation (Gao et al. 2002). Such dominant dosage sensitive genetic interactions suggest a functional link Tsc1/Tsc2 and TOR. Further genetic analyses placed Tsc1/Tsc2 upstream of TOR in cell size control (Gao et al. 2002). Taken together, these studies suggest that Tsc1/Tsc2 act upstream of and negatively regulate TOR activity in cell growth control.

3.2.3 Biochemical studies of Tsc1/Tsc2 function in TOR signaling using *Drosophila* S2 cells

As powerful as it can be as a genetic model, *Drosophila* as whole animals are usually not amenable to detailed biochemical analyses of cell signaling mechanisms. The *Drosophila* S2 cells provide an ideal system for such biochemical studies. Unlike mammalian cells, *Drosophila* S2 cells can be grown at room temperature without CO2-buffered incubator, yet they are amenable to all the standard techniques typically applied to mammalian cells such as transfection, cell fractionation and most importantly double strand RNA-mediated gene silencing, or RNA interference (RNAi) (Clemens et al. 2000).

To investigate if Tsc1/Tsc2 function upstream of TOR, we used RNAi to specifically knockdown Tsc1 or Tsc2 in S2 cells and followed S6K activity using a phospho-specific antibody against Thr389 of mammalian p70S6K that cross-reacted with the *Drosophila* S6K protein (Gao et al. 2002). Consistent with Tsc1/Tsc2 being negative regulators functioning upstream of TOR, loss of Tsc1/Tsc2 leads to increased S6K activity in a TOR-dependent manner, since RNAi of TOR, or treatment with rapamycin completely abrogate this effect. Most strikingly, knockout of Tsc1/Tsc2 leads to a dramatic effect on the sensitivity of S6K to amino acid levels. While S6K activity in normal S2 cells is dependent on the presence of amino acids such that amino acid starvation leads to rapid inactivation of S6K, cells treated with Tsc1 or Tsc2 RNAi are strongly resistant to amino acid starvation such that S6K activity remains high after long period of amino acid starvation (Gao et al. 2002). A similar resistant to amino acid starvation was also observed in mammalian cells generated from TSC1 or TSC2 mutant animals (Gao et al. 2002). Taken together, these observations provide molecular evidence that Tsc1/Tsc2 are novel regulators of the amino acid/TOR signaling pathway. The exact mechanisms by which Tsc1/Tsc2 regulate amino acid sensitivity remains unknown. Tsc1/Tsc2 could either function as obligatory components between amino acids and TOR in a linear amino acid sensing pathway, or act in a parallel pathway that converges on TOR and negatively regulates TOR activity (Gao et al. 2002).

3.2.4 Regulation of Tsc1/Tsc2 by phosphorylation

Recent studies implicated TSC2 as an Akt substrate (Inoki et al. 2002; Potter et al. 2002; Manning et al. 2002). These studies identified S939 and T1462 of the human TSC2 (or S924 and T1518 of the *Drosophila* TSC2) as Akt phosphorylation sites upon growth factor stimulation or loss of PTEN. Several questions were brought about by these studies. First, the exact regulatory function of this phosphorylation event remains unclear. While some reported that Akt phosphorylation does not affect TSC1/TSC2 complex formation (Manning et al. 2002), others reported that Akt phosphorylation disrupts the TSC1/TSC2 complex (Inoki et al. 2002; Potter et al. 2002). Second, these studies suggest a linear pathway of Akt-TSC2-S6K, which is difficult to reconcile with the findings that Akt does not regulate S6K activity in *Drosophila* (Radimerski et al. 2002) and that Tsc1/Tsc2 functions in parallel to the insulin pathway (Gao and Pan 2001). Further investigations are required to address these discrepancies.

It is worth noting that while the recent Akt/TSC2 studies provided evidence for Akt phosphorylation of TSC2, this observation alone is not sufficient to suggest a physiological role for this phosphorylation event in the context of normal development. A proof for such a role requires engineering TSC2 protein carrying mutations in the identified Akt phosphorylation sites and testing their ability to replace the function of wild type TSC2 in normal development. It is also worth noting that both TSC1 and TSC2 are phosphorylated on Tyr residues as well, although the responsible kinase(s) and the physiological role of such phosphorylation are not clear (Nellist et al. 2001).

3.3 Small GTPase Rheb as a direct target of the tuberous sclerosis tumor suppressor proteins

3.3.1 The GAP function of TSC2 is essential for its biological activity

A further understanding of the mechanisms by which Tsc1/Tsc2 regulates TOR activity has been hampered by a lack of a direct assay for Tsc1/Tsc2 activity. While *TSC1* encodes a novel protein containing two coiled-coil domains, *TSC2* contains a domain with homology to the GTPase activating protein (GAP) domain of Rap1GAP (van Slegtenhorst et al. 1997; The European Chromosome 16 Tuberous Sclerosis Consortium 1993). Small GTPases such as Rap1 cycle between an active GTP-bound form and an inactive GDP-bound form (Boguski and McCormick, 1993). GAP proteins promote the intrinsic GTPase activity of GTPases, facilitating the transition from the GTP-bound active form to the GDP-bound inactive form (Boguski and McCormick 1993). The GAP domain of TSC2 is important for its biological function, since truncation of this domain abolished TSC2's biological function (Kobayashi et al. 1995; Yeung et al. 1994) and missense mutations of this domain were identified in a high proportion of TSC patients (Dabora et al. 2001; Maheshwar et al. 1997). The GAP domain of TSC2 has been shown to exhibit *in vitro* GAP activity against Rab5 and Rap1 (Xiao et al. 1997; Wienecke et al. 1995), though it is unclear how such activity could explain the biology of TSC tumor suppressors. Importantly, inhibition of Rab5 or Rap1 in S2 cells by RNAi does not lead to a downregulation of S6K as predicted for a Tsc2 GAP substrate (Zhang et al. 2003). Thus, the physiological target(s) of the TSC2 GAP activity have been elusive.

3.3.2 Small GTPase Rheb is the direct target of Tsc2 GAP activity

The target GTPase of Tsc2 GAP should meet the following criteria. First, as expected of a substrate of Tsc2 GAP, loss-of-function mutations of the GTPase should lead to a decrease in cell size, while gain-of-function mutations might lead to an increase in cell size. Second, it should behave genetically as being downstream of Tsc1/Tsc2. Third, Tsc2 should show specific GAP activity towards this GTPase, but not other related GTPases. Fourth, inhibition of this GTPase should lead to decrease in S6K activity, a phenotype opposite to that resulting from loss of Tsc2, while activation of this GTPase should lead to increase in S6K activity. Most recent genetic and biochemical studies in *Drosophila* have provided convincing evidence that the small GTPase Rheb (Ras homologue enriched in brain) fulfills these expectations and thus represents the long-sought after target of the tuberous sclerosis tumor suppressor proteins (Saucedo et al. 2003; Stocker et al. 2003; Zhang et al. 2003).

Rheb was initially identified as a member of the Ras superfamily of GTPases that was rapidly induced in hippocampal neurons by synaptic activity (Yamagata et al. 1994). The overall sequence of Rheb does not identify it as a member of any of the known subfamilies of the Ras superfamily of GTPases, although its closest relative is Ras. Unlike Ras and most other Ras superfamily GTPases, which have glycines at the third residue of the G1 box, Rheb has an arginine at this position (residue 15) (Yamagata et al. 1994). Rheb was subsequently found to be widely expressed in human tissues, and is also found in lower eukaryotes such as yeast (Urano et al. 2000; Mach et al. 2000). The molecular function of Rheb has not been defined, and studies in different organisms have yielded conflicting results. In mammalian cells, Rheb has been implicated in positive (Yee and Worley 1997) and negative regulation of Ras/Raf signaling (Clark et al. 1997; Im et al. 2002). In lower eukaryotes, Rheb has been implicated nutrient import in *S. cerevisiae* (Urano et al. 2000) and response to nitrogen levels in *S. pombe* (Mach et al. 2000).

Recent studies in *Drosophila* revealed an essential role for Rheb in growth control (Saucedo et al. 2003; Stocker et al. 2003). Loss of Rheb inhibits cell growth, resulting in a decrease in cell size, while overexpression of Rheb promotes cell growth resulting in a dramatic increase in cell size. Genetic analyses placed Rheb downstream of Tsc1/Tsc2. These observations strongly suggest that Rheb is a target of the Tsc2 GAP activity. Consistent with this hypothesis, loss of Rheb leads to a downregulation of S6K activity, while overexpression of Rheb increases S6K activity.

Additional biochemical analyses provided convincing evidence that Rheb is a direct target substrate of the Tsc2 GAP activity (Zhang et al. 2003). *In vivo*, Tsc2 dramatically affected the GTP-loading status of Rheb. Rheb, like other small GTPases, cycles between the active GTP-bound form and the inactive GDP-bound form. Thus, the steady state GTP/GDP loading status of Rheb can be used as a measurement of its *in vivo* activity. Using an *in vivo* labeling procedure we showed that overexpression of Tsc1/Tsc2 dramatically decrease GTP loading of Rheb, while RNAi inhibition of Tsc2 result in an increase in GTP loading. *In vitro*, Tsc2 displays specific GAP activity towards Rheb, but not the closely related small GTPase Ras, demonstrating the specificity of the Tsc2 GAP activity. Take together, the biochemical analyses (Zhang et al. 2003) and genetic studies (Saucedo et al. 2003; Stocker et al. 2003) provide convincing evidence that Rheb is the direct target of the tuberous sclerosis tumor suppressor proteins in cell growth control.

3.3.3 The molecular relationship between Tsc/Rheb and amino acid sensing

Tsc1/Tsc2 could either function as obligatory components between amino acids and TOR in a linear amino acid sensing pathway, or act in a parallel pathway that converges on TOR and negatively regulates TOR activity (Gao et al. 2002). The former, but not the latter, model predicts that the activity of Tsc1/Tsc2 must be negatively regulated by amino acids. Previously, we could not distinguish between these two models due to the lack of a direct assay for Tsc1/Tsc2 activity (Gao et al. 2002). The identification of Rheb as a direct target of Tsc2 makes it possible to examine the regulation of the TSC/Rheb axis by amino acids using the GTP loading status of Rheb as a readout. Specifically, if TSC/Rheb function as obligatory components between amino acids and TOR in a linear amino acid sensing pathway, we should expect Rheb activity to be dependent on the presence of amino acids and that amino acid starvation should lead to a dramatic decrease of Rheb activity. We observed that the GTP loading status of Rheb is not regulated by amino acid starvation (Zhang et al. 2003). Thus, we favor a model wherein TSC/Rheb acts in a parallel pathway that converges on TOR and negatively regulates TOR activity (Fig. 2). Accordingly to this model, loss of Tsc1/Tsc2 or ectopic activation of Rheb results in constitutive activation of TOR that bypasses the requirement for amino acids, thus rendering S6K activity resistant to amino acid starvation.

3.4 Concluding remarks

Drosophila offers a rich collection of tools for manipulating gene function and its genome shows a high degree of conservation to humans. While orthologs of the *TSC1/TSC2* genes exist in *Drosophila*, such genes do not seem to exist in *C. elegans*, another commonly used invertebrate model system for genetic studies. On the other hand, orthologs of TOR are present in both flies and worms. These observations suggest that the circuitry of the TOR signaling network in *Drosophila* and mammals is different from that in *C. elegans*. Along that line, it is worth noting that perturbation of insulin signaling or TOR function in *C. elegans* does not leads to changes in cell size (Long et al. 2002; Finch and Ruvkun 2001) as seen in *Drosophila* or mammalian cells. Thus some of the effectors of these pathways must be different in worms as compared to flies and mammals. Given the conservation between flies and mammals in insulin signaling and Tsc/TOR signaling, *Drosophila* offers a unique model wherein one can combine genetic and biochemical tools to achieve integrated understanding of these signaling pathways and their cross talk. The recent progress in *Drosophila* linking Tsc1/Tsc2, Rheb and TOR in a signaling pathway testifies the power and elegance of such a combinatorial approach.

The exciting developments in the studies of *Drosophila* Tsc1/Tsc2, TOR and Rheb genes have not only revealed Tsc1/Tsc2 and Rheb as novel regulators of the

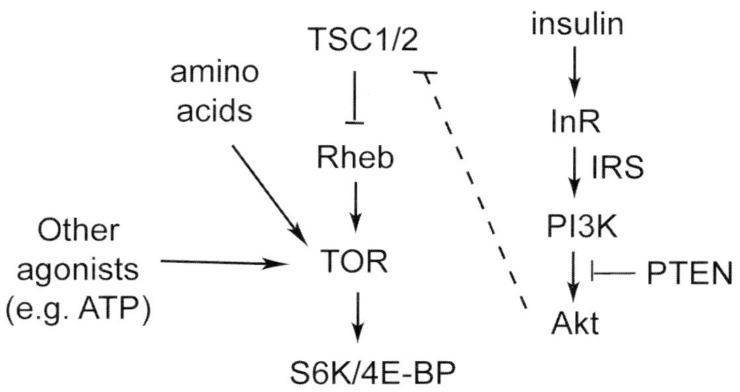

Fig. 2. A working model of the Tsc/Rheb pathway in relationship to amino acid/TOR signaling. Tsc1/Tsc2 antagonize small GTPase Rheb. The Tsc/Rheb pathway converges on and negatively regulate TOR activity. Also shown is the insulin signaling pathway, which could potentially intersect the Rheb pathway through its regulation of TSC2.

TOR pathway, but also provided the first example of tumor suppressors involved in nutrient sensing. While many tumor suppressor genes are involved in aspects of cell signaling such as growth factor receptor signaling and various cell cycle checkpoints (Hanahan and Weinberg 2000), Tsc1/Tsc2 can be viewed as components of an analogous "nutrient checkpoint", and aberrant regulation of this checkpoint can similarly cause neoplasm. The functional link between Tsc1/Tsc2 and TOR further suggests that rapamycin or its derivatives might be used to potential therapeutic advantage against tuberous sclerosis syndrome. The recent identification of the small GTPase Rheb as the direct target of Tsc2 suggests another potential target for the therapeutic intervention of the TSC disease, and provides a new entry point for the regulation of TOR signaling in animal cells. The parallel between Tsc2 tumor suppressor as a Rheb GAP (Zhang et al. 2003) and neurofibromin (NF1) tumor suppressor as a Ras GAP (Boguski and McCormick 1993) also raises the intriguing possibility that Rheb could potentially function as an oncogene. Meanwhile, many questions remain regarding the details of the Tsc/Rheb/TOR axis. Does Tsc/Rheb regulate TOR through changes in its kinase activity or some other aspects of TOR function? How is Tsc1/Tsc2 activity regulated in animal cells? How does the Tsc/Rheb/TOR cross talk to insulin signaling? Are there activators of the small GTPase Rheb such as guanine nucleotide exchange factors? We shall anticipate with great enthusiasm that flies will continue to tell us more about TOR signaling in the coming years.

Acknowledgements

I would like to thank many colleagues in the *Drosophila* research community, in particular Bruce Edgar, Ernst Hafen, and Thomas Neufeld, who have freely exchanged ideas and reagents that make the work described in this chapter possible. D.J.P. is Virginia Murchison Linthicum Endowed Scholar in Medical Science at UT Southwestern Medical Center. This work was supported by grants from National Institutes of Health (GM62323), American Heart Association (0130222N) and American Cancer Society (RSG0303601DDC) to D.J.P.

References

Boguski MS, McCormick F (1993) Proteins regulating Ras and its relatives. Nature 366:643-654

Böhni R, Riesgo-Escovar J, Oldham S, Brogiolo W, Stocker H, Andruss BF, Beckingham K, Hafen E (1999) Autonomous control of cell and organ size by CHICO, a *Drosophila* homolog of vertebrate IRS1-4. Cell 97:865-875

Cantley LC, Neel BG (1999) New insights into tumor suppression:PTEN suppresses tumor formation by restraining the phosphoinositide 3-kinase/AKT pathway. Proc Natl Acad Sci USA 96:4240-4245

Chen C, Jack J, Garofalo RS (1996) The *Drosophila* insulin receptor is required for normal growth. Endocrinology 137:846-856

Cho KS, Lee JH, Kim S, Kim D, Koh H, Lee J, Kim C, Kim J, Chung, J (2001) *Drosophila* phosphoinositide-dependent kinase-1 regulates apoptosis and growth via the phosphoinositide 3-kinase-dependent signaling pathway. Proc Natl Acad Sci USA 98:6144-6149

Clark GJ, Kinch MS, Rogers-Graham K, Sebti SM, Hamilton AD, Der CJ (1997) The Ras-related protein Rheb is farnesylated and antagonizes Ras signaling and transformation. J Biol Chem 272:10608-10615

Clemens JC, Worby CA, Simonson-Leff N, Muda M, Maehama T, Hemmings BA, Dixon JE (2000) Use of double-stranded RNA interference in *Drosophila* cell lines to dissect signal transduction pathways. Proc Natl Acad Sci USA 97:6499-6503

Dabora SL, Jozwiak S, Franz DN, Roberts PS, Nieto A, Chung J, Choy YS, Reeve MP, Thiele E, Egelhoff JC, Kasprzyk-Obara J, Domanska-Pakiela D, Kwiatkowski DJ (2001) Mutational analysis in a cohort of 224 tuberous sclerosis patients indicates increased severity of TSC2, compared with TSC1, disease in multiple organs. Am J Hum Genet 68:64-80

Finch CE, Ruvkun G (2001) The genetics of aging. Annu Rev Genomics Hum Genet 2:435-462

Gao X, Neufeld TP, Pan D (2000) *Drosophila* PTEN regulates cell growth and proliferation through PI3K-dependent and -independent pathways. Dev Biol 221:404-418

Gao X, Pan D (2001) TSC1 and TSC2 tumor suppressors antagonize insulin signaling in cell growth. Genes Dev 15:1383-1392

Gao X, Zhang Y, Arrazola P, Hino O, Kobayashi T, Yeung RS, Ru B, Pan D (2002) Tsc tumour suppressor proteins antagonize amino-acid-TOR signalling. Nat Cell Biol 4:699-704

Gingras AC, Raught B, Sonenberg N (2001) Regulation of translation initiation by FRAP/mTOR. Genes Dev 15:807-826

Goberdhan DCI, Paricio N, Goodman EC, Mlodzik M, Wilson C (1999) *Drosophila* tumor suppressor *PTEN* controls cell size and number by antagonizing the Chico/PI3-kinase signaling pathway. Genes Dev 13:3244-3258

Hanahan D, Weinberg RA (2000) The hallmarks of cancer. Cell 100:57-70

Hara K, Yonezawa K, Weng QP, Kozlowski MT, Belham C, Avruch J (1998) Amino acid sufficiency and mTOR regulate p70 S6 kinase and eIF-4E BP1 through a common effector mechanism. J Biol Chem 273:14484-14494

Henry KW, Yuan X, Koszewski NJ, OndaH, Kwiatkowski DJ, Nordstrom W (1998) Tuberous sclerosis gene 2 product modulates transcription mediated by steroid hormone receptor family members. J Biol Chem 273:20535-20539

Huang H, Potter CJ, Tao W, Li D-M, Brogiolo W, Hafen E, Sun H, Xu T (1999) PTEN affects cell size, cell proliferation and apoptosis during *Drosophila* eye development. Development 126:5365-5372

Im E, von Lintig FC, Chen J, Zhuang S, Qui W, Chowdhury S, WorleyPF, Boss GR, Pilz RB (2002) Rheb is in a high activation state and inhibits B-Raf kinase in mammalian cells. Oncogene 21:6356-6365

Inoki K, Li Y, Zhu T, Wu J, Guan KL (2002) TSC2 is phosphorylated and inhibited by Akt and suppresses mTOR signalling. Nat Cell Biol 4:648-657

Kimball SR, Jefferson LS (2000) Regulation of translation initiation in mammalian cells by amino acids. In Translational control of gene expression, J.W. Sonenberg, J.W.B. Hershey, and M.B. Mathews, eds. (Cold Spring Harbor, NY: Cold Spring Harbor Laboratory Press), pp. 561-579

Kobayashi T, Hirayama Y, Kobayashi E, Kubo Y, Hino O (1995) A germline insertion in the tuberous sclerosis (Tsc2) gene gives rise to the Eker rat model of dominantly inherited cancer. Nat Genet 9:70-74

Kobayashi T, Minowa O, Kuno J, Mitani H, Hino O, Noda T (1999) Renal carcinogenesis, hepatic hemangiomatosis, and embryonic lethality caused by a germ-line Tsc2 mutation in mice. Cancer Res 59:1206-1211

Kobayashi T, Minowa O, Sugitani Y, Takai S, Mitani H, Kobayashi E, Noda T, Hino O (2001) A germ-line Tsc1 mutation causes tumor development and embryonic lethality that are similar, but not identical to, those caused by Tsc2 mutation in mice. Proc Natl Acad Sci USA 98:8762-8767

Kobayashi T, Mitani H, Takahashi R-I, Hirabayashi M, Ueda M, Tamura H, Hino O (1997) Transgenic rescue from embryonic lethality and renal carcinogenesis in the Eker rat model by introduction of a wild-type *Tsc2* gene. Proc Natl Acad Sci USA 94:3990-3993

Kuruvilla FG, Schreiber SL (1999) The PIK-related kinases intercept conventional signaling pathways. Chem Biol 6:R129-R136

Kwiatkowski DJ, Zhang H, Bandura JL, Heiberger KM, Glogauer M, el Hashemite N, Onda H (2002) A mouse model of TSC1 reveals sex-dependent lethality from liver hemangiomas, and up-regulation of p70S6 kinase activity in Tsc1 null cells. Hum Mol Genet 11:525-534

Lamb RS, Roy C, Diefenbach TJ, Vinters HV, Johnson MW, Jay DG, Hall A (2000) The TSC1 tumor suppressor hamartin regulates cell adhesion through ERM proteins and the GTPase Rho. Nature Cell Biol 2:281-287

Long X, Spycher C, Han ZS, Rose AM, Muller F, Avruch J (2002) TOR deficiency in *C. elegans* causes developmental arrest and intestinal atrophy by inhibition of mRNA translation. Curr Biol 12:1448-1461

Mach KE, Furge KA, Albright CF (2000) Loss of Rhb1, a Rheb-related GTPase in fission yeast, causes growth arrest with a terminal phenotype similar to that caused by nitrogen starvation. Genetics 155:611-622

Maheshwar MM, Cheadle JP, Jones AC, Myring J, Fryer AE, Harris PC, Sampson JR (1997) The GAP-related domain of tuberin, the product of the TSC2 gene, is a target for missense mutations in tuberous sclerosis. Hum Mol Genet 6:1991-1996

Manning BD, Tee AR, Logsdon MN, Blenis J, Cantley LC (2002) Identification of the tuberous sclerosis complex-2 tumor suppressor gene product tuberin as a target of the phosphoinositide 3-kinase/akt pathway. Mol Cell 10:151-162

Montagne J, Stewart MJ, Stocker H, Hafen E, Kozma SC, Thomas G (1999) *Drosophila* S6 kinase: a regulator of cell size. Science 285:2126-2129

Nellist M, van Slegtenhorst M, Goedbloed M, van den Ouweland A, Halley D, van der Sluijs P (1999) Characterization of the cytosolic tuberin-hamartin complex. Tuberin is a cytosolic chaperone for hamartin. J Biol Chem 274:35647-35652

Nellist M, Verhaaf B, Goedbloed MA, Reuser AJ, Van Den Ouweland AM, Halley DJ (2001) TSC2 missense mutations inhibit tuberin phosphorylation and prevent formation of the tuberin-hamartin complex. Hum Mol Genet 10:2889-2898

Oldham S, Montagne J, Radimerski T, Thomas G, Hafen E (2000) Genetic and biochemical characterization of dTOR, the *Drosophila* homolog of the target of rapamycin. Genes Dev 14:2689-2694

Onda H, Lueck A, Marks PW, Warren HB, Kwiatkowski DJ (1999) $Tsc2^{+/-}$ mice develop tumors in multiple sites that express gelsolin and are influenced by genetic background. J Clin Invest. 104:687-695

Potter CJ, Huang H, Xu T (2001) *Drosophila* Tsc1 functions with Tsc2 to antagonize insulin signaling in regulating cell growth, cell proliferation, and organ size. Cell 105:357-368

Potter CJ, Pedraza LG, Xu T (2002) Akt regulates growth by directly phosphorylating Tsc2. Nat Cell Biol 4:658-665

Proud CG, Denton RM (1997) Molecular mechanisms for the control of translation by insulin. Biochem J 328:329-341

Radimerski T, Montagne J, Rintelen F, Stocker H, van Der KJ, Downes CP, Hafen E, Thomas G (2002) dS6K-regulated cell growth is dPKB/dPI(3)K-independent, but requires dPDK1. Nat Cell Biol 4:251-255

Rennebeck G, Kleymenova EV, Anderson R, Yeung RS, Artzt K, Walker CL (1998) Loss of function of the tuberous sclerosis 2 tumor suppressor gene results in embryonic lethality characterized by disrupted neuroepithelial growth and development. Proc Natl Acad Sci USA 95:15629-15634

Rintelen F, Stocker H, Thomas G, Hafen E (2001) PDK1 regulates growth through Akt and S6K in *Drosophila*. Proc Natl Acad Sci USA 98:15020-15025

Rohde J, Heitman J, Cardenas ME (2001) The TOR kinases link nutrient sensing to cell growth. J Biol Chem 276:9583-9586

Saucedo LJ, Gao X, Chiarelli DA, Li L, Pan D, Edgar BA (2003) Rehb promotes cell growth as a component of the insulin/TOR signaling network. Nat Cell Biol 5:566-571

Schmelzle T, Hall MN (2000) TOR, a central Controller of cell growth. Cell 103:253-262

Sonenberg N, Gingras A-C (1998) The mRNA 5' cap-binding protein elF4E and control of cell growth. Curr Opin Cell Biol 10:268-275

Soucek T, Pusch O, Wienecke R, DeClue JE, Hengstschläger M (1997) Role of the tuberous sclerosis gene-2 product in cell cycle control. Loss of the tuberous sclerosis gene-2 induces quiescent cells to enter S phase. J Biol Chem 272:29301-29308

Soucek T, Yeung RS, Hengstschläger M (1998) Inactivation of the cyclin-dependent kinase inhibitor p27 upon loss of the tuberous sclerosis complex gene-2. Proc Natl Acad Sci USA 95:15653-15658

Stocker H, Hafen E (2000) Genetic control of cell size. Curr Opin Genet Dev 10:529-535

Stocker H, Radimerski T, Schindelholz B, Wittwer F, Belawat P, Daram P, Breuer S, Thomas G, Hafen E (2003) Rheb is an essential regulator of S6K in controlling cell growth in *Drosophila*. Nat Cell Biol 5:559-566

Tapon N, Ito N, Dickson BJ, Treisman JE, Hariharan IK (2001) The *Drosophila* tuberous sclerosis complex gene homologs restrict cell growth and cell proliferation. Cell 105:345-355

The European Chromosome 16 Tuberous Sclerosis Consortium (1993) Identification and characterization of the tuberous sclerosis gene on chromosome 16. Cell 75:1305-1315

Thomas G, Hall MN (1997) TOR signalling and control of cell growth. Curr Opin Cell Biol 9:782-787

Urano J, Tabancay AP, Yang W, Tamanoi F (2000) The *Saccharomyces cerevisiae* Rheb G-protein is involved in regulating canavanine resistance and arginine uptake. J Biol Chem 275:11198-11206

van Slegtenhorst M, de Hoogt R, Hermans C, Nellist M, Janssen B, Verhoef S, Lindhout D, van den Ouweland A, Halley D, Young J, Burley M, Jeremiah S, Woodward K, Nahmias J, Fox M, Ekong R, Osborne J, Wolfe J, Povey S, Snell RG, Cheadle JP, Jones AC, Tachataki M, Ravine D, Sampson JR, Reeve MP, Richardson P, Wilmer F, Munro C, Hawkins TL, Sepp T, Ali JBM, Ward S, Green AJ, Yates JRW, Kwiatkowska J, Henske EP, Short MP, Haines JH, Jozwiak S, Kwiatkowski DJ (1997) Identification of the tuberous sclerosis gene *TSC1* on chromosome 9q34. Science 277:805-808

van Slegtenhorst M, Nellist M, Nagelkerken B, Cheadle JP, Snell RG, van den Ouweland A, Reuser A, Sampson JR, Halley D, van der Sluijs P (1998) Interaction between hamartin and tuberin, the *TSC1* and *TSC2* gene products. Hum Mol Genet 7:1053-1057

Verdu J, Buratovich MA, Wilder EL, Birnbaum MJ (1999) Cell-autonomous regulation of cell and organ growth in *Drosophila* by Akt/PKB. Nature Cell Biol 1:500-506

Weinkove D, Neufeld TP, Twardzik T, Waterfield MD, Leevers SJ (1999) Regulation of imaginal disc cell size, cell number and organ size by *Drosophila* class I_A phosphoinositide 3-kinase and its adaptor. Curr Biol 9:1019-1029

Wienecke R, Konig A, DeClue JE (1995) Identification of tuberin, the Tuberous Sclerosis-2 product:tuberin possesses specific Rap1GAP activity. J Biol Chem 270:16409-16414

Xiao G-H, Shoarinejad F, Jin F, Golemis EA, Yeung RS (1997) The tuberous sclerosis 2 gene product, tuberin, functions as a Rab5 GTPase activating protein (GAP) in modulating endocytosis. J Biol Chem 272:6097-6100

Yamagata K, Sanders LK, Kaufmann WE, Yee W, Barnes CA, Nathans D, Worley PF (1994) rheb, a growth factor- and synaptic activity-regulated gene, encodes a novel ras-related protein. J Biol Chem 269:16333-16339

Yee WM, Worley PF (1997) Rheb interacts with Raf-1 kinase and may function to integrate growth factor- and protein kinase A-dependent signals. Mol Cell Biol 17:921-933

Yeung RS, Xiao GH, Jin F, Lee WC, Testa JR, Knudson AG (1994) Predisposition to renal carcinoma in the Eker rat is determined by germ- line mutation of the tuberous sclerosis 2 (TSC2) gene. Proc Natl Acad Sci USA 91:11413-11416

Young J, Povey S (1998) The genetic basis of tuberous sclerosis. Mol Med Today 4:313-319

Zhang H, Stallock JP, Ng JC, Reinhard C, Neufeld TP (2000) Regulation of cellular growth by the *Drosophila* target of rapamycin *dTOR*. Genes Dev 14:2712-2724

Zhang Y, Gao X, Saucedo LJ, Ru B, Edgar BA, Pan D (2003) Rheb is a direct target of the tuberous sclerosis tumor suppressor proteins. Nat Cell Biol 5:578-581

Abbreviations:

TSC: tuberous sclerosis complex
TSC1: mammalian TSC1 protein (hamartin)
TSC2: mammalian TSC2 protein (tuberin)
Tsc1: *Drosophila* TSC1 homologue
Tsc2: *Drosophila* TSC2 homologue
TOR: target of rapamycin
PI3K: phosphatidylinositol 3-kinase
S6K: ribosomal subunit S6 kinase
4E-BP: eukaryotic initiation factor 4E-binding protein
PTEN: phosphatase and tensin homologue deleted from chromosome 10
GAP: GTPase activating protein
TOP: terminal oligopyrimidine tract
eIF4E: eukaryotic initiation factor 4E
PDK1: phosphoinositide-dependent kinase 1
PI3K: phosphoinositide 3-kinase
IRS: insulin receptor substrate
RNAi: RNA interference
InR: insulin receptor
Note: Italics and regular fonts are used to designate genes and corresponding proteins, respectively.

4 Nutrients as regulators of endocrine and neuroendocrine secretion

Leonard Best and John McLaughlin

Abstract

Specialised cell types in living organisms are able to monitor continually the levels of key nutrients and produce appropriate metabolic and behavioural responses, typically via altered secretion of certain hormones and neurotransmitters. This review aims to provide a critical evaluation of the molecular mechanisms by which secretory processes in higher animals can be regulated directly by circulating levels of the three major types of macronutrient: sugars, amino acids, and fatty acids. We focus on three organs that play a major role in nutrient sensing; pancreatic islet cells, the gut, and nutrient-sensitive neurones of the hypothalamus. There is evidence for common molecular mechanisms for nutrient sensing in different cells types, involving plasma membrane receptors, electrogenic transport and intracellular metabolism of the nutrient. Nutrient-induced changes in intracellular nucleotide levels have received much attention as a potential coupling mechanism, but the fundamental mechanisms remain elusive. The identification of additional metabolic signals should help to clarify this complex topic.

4.1 Introduction

Two important functions in living organisms are the ability to monitor nutritional status; that is the levels of key nutrients that are available to that organism at any given time, and to produce appropriate metabolic and behavioural responses. In mammals, these processes are regulated to a considerable extent by specialised cells that detect circulating levels of various nutrients. The responses of these cells consist of an increase or decrease in the secretion of certain hormones and neurotransmitters. Such responses are important because they can profoundly influence metabolic processes, and in some cases behaviour, in the organism in accordance with a constantly changing supply of nutrients. Thus, after feeding when nutrient levels are high, metabolism in the organism is directed towards the utilisation and storage of those nutrients, whilst signals are generated to reduce or stop food intake. On the other hand, in times of starvation, metabolism is altered to conserve important nutrients and the feeding response is triggered. Defects in these responses can result in a wide variety of metabolic diseases including diabetes mellitus and obesity.

Topics in Current Genetics, Vol. 7
J. Winderickx, P.M. Taylor (Eds) Nutrient-Induced Responses in Eukaryotic Cells

The purpose of this review is to provide a critical evaluation of the molecular mechanisms by which the secretion of certain hormones and neurotransmitters can be regulated directly by circulating levels of the three types of macronutrient: sugars, amino acids, and fatty acids. We shall primarily consider acute effects of nutrients on exocytosis, though it should be emphasised that secretory activity in virtually any cell type is liable to modification by long-term exposure to altered levels of different nutrients, particularly glucose and fatty acids. Secretory cells that are regulated by acute changes in circulating nutrient levels are widely distributed in the body. We shall focus on three organs that play a major role in nutrient sensing and in which secretory activity is regulated by changes in nutrient levels. Pancreatic islet cells have been most extensively studied, while newer areas of investigation are the endocrine cells of the gut and nutrient-sensitive neurones of the hypothalamus.

4.2 Peptide hormone and neurotransmitter release: exocytosis

The synthesis, trafficking, and secretion of peptide hormones and neurotransmitters have been extensively reviewed elsewhere (Lledo 1997; Lang 1999; Fon and Edwards 2001). For the purpose of this article, the process of exocytosis, which is largely common to peptide hormone-secreting and neuroendocrine cells, will be considered briefly. Peptide hormones and small molecular weight water-soluble neurotransmitters are stored in membrane-bound secretory granules. The movement of these granules through the cytosol is a highly organised process and appears to involve participation of the actin cytoskeleton (Valentijn et al. 1999). The fusion of the secretory granule with the plasma membrane, resulting in ejection of the granule contents into the extracellular medium, is also a highly ordered process, involving a complex system of docking and fusion proteins, notably the SNARE complex (May et al. 2001; Rettig and Neher 2002). Numerous protein components involved in both the actin microfilamentous system and the granule fusion process are subject to regulation, particularly by Ca^{2+} - calmodulin and by cyclic AMP – dependent protein kinase (PKA).

4.3 Theoretical considerations: How can nutrients affect the secretion of hormones and neurotransmitters?

In principle, there are a number of cellular mechanisms whereby cells could detect and respond to changes in ambient levels of nutrients. Here we shall briefly consider some of these mechanisms. Many endocrine and neuroendocrine cells are electrically active; that is stimulation results in depolarisation of the plasma membrane potential leading to electrical activity, consisting of repetitive calcium-dependent action potentials (see Fig. 1). Thus, when the membrane potential

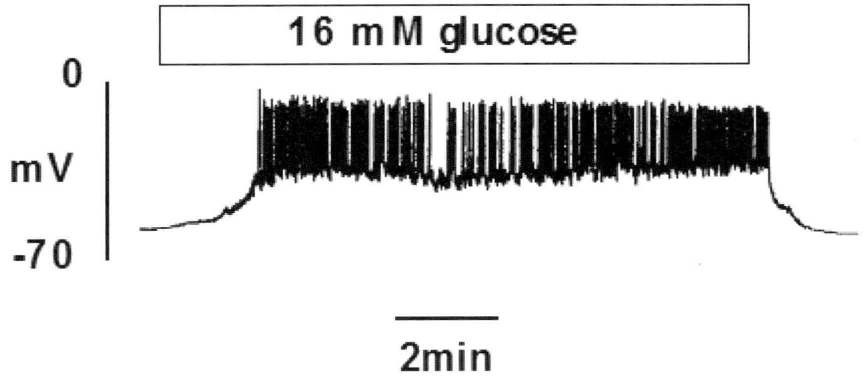

Fig. 1. Glucose-induced electrical activity recorded in a single rat pancreatic β-cell using the perforated patch configuration of the patch-clamp technique. The glucose concentration was raised from 4 to 16 mM for the period shown.

reaches a certain threshold level, voltage-sensitive calcium channels open, resulting in calcium entry into the cell. The resultant increase in cytosolic calcium concentration ($[Ca^{2+}]_i$) is a major factor in triggering exocytosis. In theory, changes in nutrient levels could alter the cell membrane potential, and hence electrical and secretory activity of the cell, in a number of ways. For example, transport across the plasma membrane of a positively charged molecule, or co-transport of a neutral molecule with a positively charged ion, would depolarise the membrane potential. It is also conceivable that a nutrient molecule might bind to a receptor functionally coupled to an ion channel, causing channel activation or inactivation, and thus generating a current carried by the resultant movement of ions. In a number of endocrine and neuroendocrine cell types, membrane potential can be influenced by increased metabolism of nutrients following their transport into the cell. Since exocytosis is subject to regulation by cyclic AMP/PKA and by G-proteins, regulation might also be achieved via a receptor coupled to these systems. Finally, there is the possibility of a more direct 'distal' regulation of the exocytotic machinery by nutrients or their metabolites.

4.4 Regulation of secretion by glucose

Glucose is a major source of fuel for many tissues in the body. As such, maintenance of blood glucose levels is crucial, and numerous regulatory mechanisms exist whereby specialised cell types can detect relatively small changes in ambient glucose concentration and produce an appropriate response.

4.4.1 The pancreatic β-cells

Pancreatic islet cells are of major importance in responding to changes in levels of circulating nutrients and secreting appropriate amounts of hormones, principally insulin and glucagon. Insulin is released from the pancreatic β-cells in response to feeding. The primary stimulus for insulin release is a rise in blood glucose concentration, and the molecular mechanisms underlying this effect have been studied intensively for many years. Until the mid 1970s, two contrasting hypotheses existed concerning the fundamental mechanism of glucose sensing. One theory, the 'glucoreceptor' mechanism, suggested that the glucose molecule was itself responsible for activating the β-cell by binding to a cell surface receptor and thus eliciting an intracellular signal (Matschinsky et al. 1972). An alternative model, the 'fuel' or 'substrate site' hypothesis, stated that the stimulatory effects of glucose on the β-cell were dependent entirely upon metabolism of the sugar within the β-cell (Malaisse et al. 1979). The latter hypothesis was supported by increasingly strong evidence and is now universally accepted. For example, mannoheptulose, an inhibitor of glucose phosphorylation, was shown to be a potent inhibitor of glucose-induced insulin release (Malaisse et al. 1968). Furthermore, several non-hexose substrates, notably certain amino- and keto-acids, could also stimulate secretion in a manner similar to glucose (Malaisse and Malaisse-Lagae 1968).

The first step in glucose-sensing by the pancreatic β-cell is transport of the sugar across the plasma membrane. This process is facilitated by a high capacity, high K_m glucose transporter GLUT-2 (Tal et al. 1992). Activity of this transporter allows changes in extracellular glucose levels to be reflected closely and rapidly by intracellular glucose concentration. Phosphorylation of glucose, the initial step of its metabolism, is also catalysed by a high K_m isoform of hexokinase termed glucokinase (reviewed in Matschinsky et al. 1998). Glucokinase is rate-limiting for glucose metabolism in the β-cell (German 1993) and as such can be considered as the primary glucose-sensor. Indeed, mutations in the glucokinase gene are known to be responsible for one form of maturity-onset diabetes in the young (MODY-2; Velho et al. 1997), emphasizing the importance of this enzyme in glucose-sensing by the β-cell.

4.4.1.1 Electrical activity in the pancreatic β-cell

The mechanisms through which increased glucose metabolism in the pancreatic β-cell is coupled to insulin secretion are not fully understood. However, an important feature of glucose-induced insulin release is electrical activity in the β-cell (see Fig. 1). The ionic events, which underlie this complex phenomenon, have been extensively reviewed elsewhere (Ashcroft and Rorsman 1989; Misler et al. 1992). Briefly, at fasting blood glucose levels (normally 3-5 mM), the β-cell maintains a membrane potential of between –60 and –70 mV. Under these conditions, the cell is electrically 'silent' and the rate of insulin release is low. A rise in glucose concentration to stimulatory levels (6-25 mM) causes a gradual depolarisation of the membrane potential. When the threshold potential (-40 to –50 mV) is reached, voltage-sensitive Ca^{2+} channels open, resulting in Ca^{2+} entry into the cell. This process is manifest as rapid repetitive Ca^{2+}-dependent action potentials or 'spikes', which are important because they result in a rise in $[Ca^{2+}]_i$ that activates the exocytotic machinery. The major question is how increased glucose metabolism depolarises the β-cell.

4.4.1.2 Inhibition of K_{ATP} channel activity by glucose

It has been known for many years that glucose reduces islet cell K^+ permeability, assessed from the rate of efflux of $^{86}Rb^+$ (a radioactive marker for K^+) from preloaded pancreatic islets (Henquin 1978). The ion channel responsible for this effect was subsequently identified (Ashcroft et al. 1984), and later designated the ATP-sensitive K^+ or K_{ATP} channel. The structure and function of this channel have been extensively reviewed (for example, see Ashcroft 2000). Briefly, the functional K_{ATP} channel consists of two subunits; the pore-forming $K_{IR6.2}$ and the sulphonylurea receptor SUR1. The K_{ATP} channel is inhibited by ATP (Cook and Hales 1984) and activated by Mg^{2+}-ADP (Dunne and Petersen 1986), leading to the idea that the intracellular ATP/ADP ratio could be an important determinant of channel regulation by glucose. Glucose has been reported to increase this ratio in insulin-secreting cells, predominantly by lowering the concentration of ADP (Malaisse and Sener 1987; Ronner et al. 2001). In contrast, studies in intact pancreas suggest that β-cell adenine nucleotide levels change little in response to glucose over the physiological concentration range (Ghosh et al. 1991). Despite these inconsistencies, the current 'consensus model' for glucose-stimulated electrical activity suggests that a rise in glucose concentration leads to increased oxidative metabolism of the hexose, resulting in a rise in ATP/ADP ratio, K_{ATP} channel inhibition and hence depolarization of the plasma membrane (see Fig. 2). In accordance with this model, β-cell mitochondria have been suggested to play a major role in the regulation of insulin secretion by glucose (Malaisse 1992; Maechler and Wollheim 2000). The stimulatory effects on the β-cell of several cell-permeable esters of mitochondrial substrates would appear to be consistent with this idea (Malaisse 1995). However, the fact that pyruvate is a very weak stimulus, despite being effectively oxidised islet cells, implies a signal derived from the glycolytic pathway in addition to mitochondrial metabolism may be required (German 1993).

K_{ATP} channel activity is inhibited not only by glucose but also by hypoglycaemic sulphonylureas (reviewed in Panten et al. 1996). These drugs interact with the SUR1 subunit of the channel, and are commonly used in the treatment of type 2 diabetes mellitus. This mechanism is thought to be primarily responsible for the stimulatory effect of sulphonylureas on insulin release. It should be borne in mind that, whatever the mechanism of inhibition of the K_{ATP} channel, the presence of an inward (depolarizing) current is necessary to produce a depolarisation of the plasma membrane. The nature of this current is unknown, and is generally assumed to be a non-specific, non-regulated current.

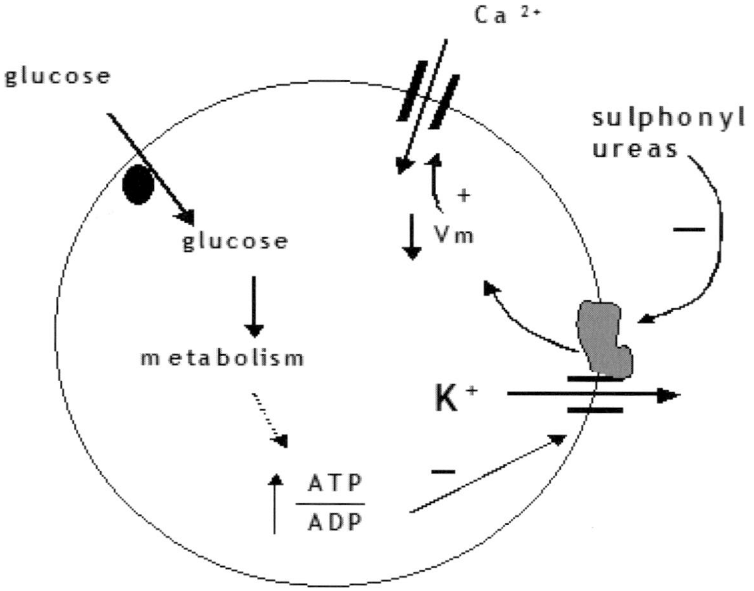

Fig. 2. The consensus model for glucose regulation of β-cell membrane potential and electrical activity. Increased oxidative metabolism of glucose raises the cytosolic ATP/ADP ratio, resulting in K_{ATP} channel inhibition, depolarisation of the plasma membrane (membrane potential V_m becomes more positive) leading to activation of voltage-sensitive Ca^{2+} channels and Ca^{2+} entry. Hypoglycaemic sulphonylureas directly inhibit K_{ATP} channel activity via interaction with the SUR1 subunit of the channel.

An important role for the K_{ATP} channel in the normal functioning of pancreatic β-cells is indicated by studies of transgenic mice. As noted above, expression of both $K_{IR6.2}$ and SUR1 subunits is required for functional K_{ATP} channels. In $K_{IR6.2}$ knockout mice, β-cell $[Ca^{2+}]_i$ was reported to be raised in the presence of fasting glucose concentrations due to a depolarised membrane potential (Miki et al. 1998). Furthermore, neither these parameters nor insulin release were significantly affected by either a rise in glucose concentration or by the sulphonylurea tolbutamide. These findings suggest that the K_{ATP} channel is essential for normal β-cell activation by glucose and sulphonylureas. In SUR1 knockout mice, β-cell membrane potential was again depolarised, although the effect of raised glucose concentration was not reported (Seghers et al. 2000). However, an interesting finding in these mice was a positive insulinotropic effect of glucose, albeit at a reduced level compared to wild type (control) animals, indicating a K_{ATP} channel-independent pathway for glucose-induced insulin release. This study also provided evidence that an important function of the K_{ATP} channel was in switching off insulin release when glucose returned towards fasting levels. The clinical condition of familial hyperinsulinism (also known as persistent hyperinsulinaemic hypoglycaemia in infancy; PHHI) also provides evidence for a role of K_{ATP} channels in normal β-cell function (reviewed in Sharma et al. 2000). The disease often results from mutations in either $K_{IR6.2}$ or SUR1, causing dysregulation of the K_{ATP} channel, continuous β-cell electrical activity and hence inappropriately high rates of insulin release in the presence of hypoglycaemia.

4.4.1.3 A glucose-regulated, K_{ATP} channel – independent ionic mechanism

Despite the considerable evidence outlined above supporting a central role for K_{ATP} channels in modulating the effects of glucose on β-cell electrical activity and insulin release, several observations have been made which argue strongly for the existence of an additional ionic mechanism sensitive to changes in glucose concentration. For example, the inhibitory effect of glucose on $^{86}Rb^{+}$ efflux from islets (often used as an index of K_{ATP} channel activity) is minimal at substimulatory concentrations of glucose (~ 5 mM); above this concentration, the hexose actually stimulates efflux (Carpinelli and Malaisse 1981). Consistent with this finding, several studies have shown that glucose progressively inhibits β-cell K_{ATP} channel activity at the single channel level over the range 0-5 mM (e.g. Ashcroft et al. 1988), whilst channel activity is virtually undetectable at glucose concentrations over the stimulatory range (5-20 mM; Best 2002a). Similarly, β-cell input conductance, which largely reflects whole-cell K_{ATP} channel activity, is also minimal at glucose concentrations of ~ 5mM; above this concentration, input conductance actually shows small increases, presumably due to activation of other ionic conductances (Best 2000). As mentioned above, hypoglycaemic sulphonylureas inhibit K_{ATP} channel activity. However, tolbutamide requires the presence of glucose to mimic the sugar's action on β-cell electrical activity, suggesting that inhibition of K_{ATP} channel activity is not the sole mechanism of action of glucose in depolarising the β-cell (Henquin 1998). This conclusion is strengthened by the finding that glucose evokes electrical activity and insulin release even when K_{ATP} channel activity is completely inhibited by a maximal concentration of tolbutamide (Best et al. 1992b; Best 2002a) or activated by diazoxide (Henquin 1992). These observations appear, at least at first sight, to be at variance with those obtained from $K_{IR6.2}$/SUR1 knockout mice and in PHHI patients. However, it is possible that any genetic abnormality – whether experimental or spontaneous - might lead to compensatory mechanisms during development. Furthermore, SUR1 could be involved in processes other than K_{ATP} channel regulation. Thus, taken together, the above findings strongly suggest the involvement of additional ion channel(s) in the modulation by glucose of β-cell electrical activity. Reports from several laboratories suggest that anionic mechanisms could be at least a component of the K_{ATP} channel-independent glucose-sensing mechanism.

4.4.1.4 A glucose-regulated anion channel

Evidence for the involvement of an anion channel in the regulation by nutrients of islet cell function was first obtained by Sehlin (1978, 1987), who demonstrated that glucose stimulated $^{36}Cl^-$ efflux from pre-loaded mouse islets. More recently, glucose has been shown to increase β-cell Cl^- permeability, based on the fluorimetric measurement of Cl^- fluxes (Eberhardson et al. 2000). The conductance likely to be responsible for these effects has been functionally identified as a volume-sensitive anion channel (VSAC; Kinard and Satin 1995; Best et al. 1996b). This type of channel is ubiquitously expressed in animal cells and plays an important role in cell volume regulation, together with other purported functions (reviewed in Eggermont et al. 2001). Activation of the VSAC in intact β-cells by hypotonic cell swelling causes depolarisation of the cell membrane potential (presumably due to Cl^- efflux) and thus leads to electrical and secretory activity (Best et al. 1996a; Drews et al. 1998). Furthermore, there is evidence that glucose and other nutrient stimuli can also activate the VSAC, both at the whole-cell (Best 1997, 2000) and single channel levels (Best 1999, 2002b). The mechanism by which glucose activates the VSAC is unknown, although glucose has been shown to cause β-cell swelling in a concentration-dependent manner (Miley et al. 1997). Metabolism of glucose is required for both cell swelling and VSAC activation since 3-O-methylglucose, which is transported but not metabolised, is ineffective (Miley et al. 1997; Best 2002b). The simplest mechanism to explain these findings would be an increase in intracellular ionic and/or osmotic strength resulting from glucose metabolism, leading to water entry into the cell, cell swelling and hence VSAC activation. In this regard, it is of interest that glycolysis in islet cells generates considerable amounts of lactate from glucose in a concentration-dependent manner (Sener and Malaisse 1976; Best et al. 1989). Furthermore, activity and expression of the lactate (monocarboxylate) transporter MCT are low or absent in native β-cells (Best et al. 1992a; Zhao et al. 2001) so that raising the glucose concentration might be predicted to result in a progressive intracellular accumulation of lactate. Since glucose metabolism is likely to be accompanied by Na^+/H^+ and Cl^-/HCO_3^- exchange, accumulation of Na^+ and Cl^- could contribute to an increase in intracellular ionic/osmotic strength (see Best et al. 1997 for discussion of this topic). An additional caveat of this model is that the VSAC shows significant permeability to lactate ($P_{lact}/P_{Cl} \sim 0.3$; Best et al. 2001), and could therefore function as an efflux pathway for lactate. In contrast to MCT, which transports $lactate^-/H^+$ and is therefore electroneutral, a channel-mediated pathway would be electrogenic, generating an inward current, which could contribute towards glucose-induced depolarisation of the β-cell.

Further evidence in support of a role for the VSAC in the β-cell response to glucose arises from the use of channel inhibitors. All such drugs examined so far, including DIDS (4,4'-diisothiocyanatostilbene-2,2'-disulphonic acid; Best 1997, 2002a; Best et al. 2000), NPPB (5-nitro-2-(3-phenylpropylamino) benzoic acid; Best 1997), 4-hydroxytamoxifen (Best 2002c) and DCPIB (4-(2-butyl-6,7-dichloro-2-cyclopentyl-indan-1-on-5-yl) oxobutyric acid; Best et al. unpublished findings) have been shown to inhibit glucose-induced electrical activity and insulin release. However, it should be pointed out that the selectivity of the above compounds as VSAC blockers is poor or unproven.

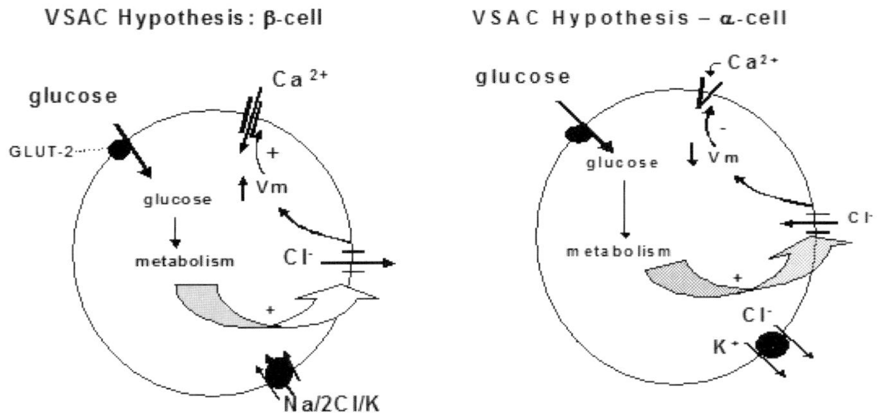

Fig. 3. Volume-sensitive anion channel (VSAC) model for glucose regulation of membrane potential and electrical activity in the pancreatic β-cell (left) and α-cell (right). In both cases, the VSAC is activated by glucose metabolism, possibly due to intracellular accumulation of glucose metabolites, water entry, and cell swelling. In the β-cell, $Na^+/K^+/2Cl^-$ (NKCC1) activity maintains high intracellular $[Cl^-]$ so that VSAC activation results in Cl^- efflux, thus generating an inward (depolarising) current (i.e. the membrane potential V_m becomes more positive), leading to Ca^{2+} channel activation. In contrast, KCC expression in the α-cell maintains low intracellular $[Cl^-]$ so that VSAC activation results in Cl^- influx, generating an outward (hyperpolarising) current (V_m more negative), resulting in Ca^{2+} channel inactivation.

4.4.1.5 β-cell chloride transporters

As mentioned above, activation of the β-cell VSAC by hypotonic cell swelling generates an inward current, presumably due to Cl^- efflux. This implies that intracellular $[Cl^-]$ must be maintained above its electrochemical equilibrium; that is E_{Cl} is positive with respect to the resting membrane potential. Estimates of intracellular $[Cl^-]$ from the distribution of $^{36}Cl^-$ in ob/ob mouse islets suggest that this is indeed the case, with apparent values for E_{Cl} of –18 to +2.5 mV being reported (Sehlin 1978). More recent fluorimetric measurements of intracellular $[Cl^-]$ also indicated that Cl^- is actively accumulated by β-cells, although at lower concentrations than suggested by the radioisotopic estimates (Eberhardson et al. 2000). The most widespread cellular mechanism for Cl^- accumulation is the $Na^+/K^+/2Cl^-$ co-transporter NKCC. There is functional evidence for the activity of this co-transporter in β-cells (Lindstrom et al. 1988; Sandstrom and Sehlin 1993) together with more recent evidence for molecular expression of the NKCC1 isoform (Majid et al. 2001). Loop diuretics, which inhibit this co-transporter, also inhibit insulin release (Sandstrom & Sehlin 1988; Sandstrom 1990) and exert a diabetogenic action (Furman 1981). Thus, NKCC1 activity may be important in the β-cell to maintain raised intracellular [Cl^-] so that VSAC activation leads to Cl^- efflux and depolarization (Fig. 3).

The above evidence indicates that activation of the VSAC by glucose could be an important mechanism coupling metabolism of the sugar to depolarisation, electrical activity and insulin release. Thus, the pancreatic β-cell could be equipped with two glucose-sensing mechanisms, the K_{ATP} channel and the VSAC. The K_{ATP} channel is regulated by glucose within the approximate range 0-5 mM, suggesting that the primary function of this channel might be to hyperpolarise the β-cell, and thus prevent insulin release, during hypoglycaemia. In contrast, activity of the VSAC appears to be sensitive to glucose over the concentration range (0-20 mM) effective in regulating electrical activity and insulin release (Sehlin 1978; Best 2000, 2002b). Unlike the K_{ATP} channel, which has been cloned and characterised in great detail, virtually nothing is known about the molecular identity of the glucose-sensitive VSAC. As noted above, drugs that interact with the channel are likely to be poorly selective, so that identification of the channel protein(s) and subsequent manipulation of expression will be invaluable in assessing its role in glucose-sensing by the β-cell.

4.4.1.6 Glucose directly activates voltage-sensitive Ca^{2+} channels

In addition to modulating Ca^{2+} entry indirectly via effects on β-cell membrane potential, there is evidence that glucose can have direct effects on Ca^{2+} channel activity. Thus, glucose has been shown to activate voltage-sensitive Ca^{2+} channels (Smith et al. 1989) and a Ca^{2+}-permeable non-selective cation channel (Rojas et al. 1990) in mouse and human pancreatic β-cells respectively. In both cases, channel activation by glucose appeared to require metabolism of the sugar but the nature of the intracellular signal(s) involved was unknown.

4.4.1.7 'Distal' effects of glucose on insulin secretion

It is well established that, in addition its effects on β-cell electrical activity, glucose can influence insulin release via regulation at one or more distal sites (reviewed in Henquin 2000). A number of experimental systems have been devised to study this 'post-ionic' mechanism of β-cell regulation by glucose. An approach commonly employed has been to clamp the membrane potential by the use of diazoxide to open K_{ATP} channels in the presence of a depolarising concentration of K^+ to maintain elevated cytosolic [Ca^{2+}] levels. Under such conditions, glucose has been shown to augment insulin release in a concentration-dependent manner (Gembal et al. 1993; Straub et al. 1998). It has also been reported that, under certain conditions, glucose can augment insulin release in the complete absence of Ca^{2+} (Komatsu et al. 1997), although the importance of this effect has been questioned (Sato et al. 1998).

In general, the distal effect of glucose requires metabolism of the sugar, is independent of protein kinases A and C, and could serve a physiological role in sensitising the exocytotic machinery to raised levels of cytosolic Ca^{2+} (Henquin

2000). Activation of this exocytotic pathway has been demonstrated in response to a variety of metabolisable nutrients, suggesting that changes in β-cell nucleotide levels, including ATP/ADP ratio, could be involved (Henquin 2000). An additional suggestion is that oxidative metabolism of glucose results in the generation of glutamate by mitochondria, and that this metabolite can directly activate the exocytotic machinery, possibly involving glutamate uptake into secretory vesicles (Maechler and Wollheim 1999). However, separate reports have shown a dissociation between glutamate production and insulin release, indicating that the two events were not causally related (MacDonald and Fahien 2000; Bertrand et al. 2002). Thus, the exact site of the distal regulation of secretion by glucose remains to be established, but it seems likely that it functions as an amplification pathway for exocytosis which is normally 'silent' at fasting glucose levels and can only be unmasked by short-circuiting the normal electrical response of the β-cell to glucose (Henquin 2000).

4.4.2 The pancreatic α-cell

In contrast to the pancreatic β-cell, exocytosis in the glucagon-secreting α-cell is stimulated during hypoglycaemia, and is inhibited at normal 'fasting' blood glucose concentrations and higher (>4-5mM; reviewed in Unger and Orci 1981). Thus, at low glucose concentrations, the α-cell displays electrical and secretory activity and a rise in glucose levels inhibits electrical activity and hence glucagon release. How this regulation by glucose is achieved is unknown. It has been suggested that γ-aminobutyric acid (GABA), released from neighbouring islet β-cells in response to glucose, could inhibit glucagon release by activation of a $GABA_A$ receptor-operated Cl^- channel, thus hyperpolarising the α-cell (Rorsman et al. 1989). A recent variation of this concept is that zinc co-released with insulin from β-cells might be implicated in the paracrine suppression of α-cell function (Ishihara et al. 2003). However, such mechanisms would not explain why glucagon secretion is inhibited at glucose concentrations lower than those, which stimulate insulin release (Gylfe 1990). It has also been suggested that somatostatin, released from islet δ-cells by glucose, might inhibit glucagon release, but in this case, hyperpolarisation of the α-cell would involve activation of a K^+ channel (Yoshimoto et al. 1999).

Several studies using isolated pancreatic α-cells indicate that glucose inhibits electrical and secretory activity via a direct effect, which could be of greater physiological importance than the above paracrine effects (Pipeleers et al. 1985; Barg et al. 2000; Gopel et al. 2000). As in the β-cell, the effects of glucose on the α-cell require metabolism of the sugar via the high K_m enzyme glucokinase (Heimburg et al. 1996). In contrast to pancreatic β-cells, the high K_m glucose transporter GLUT-2 is not expressed in α-cells (at least in the rat), emphasising the importance of glucose phosphorylation rather than transport as the rate-limiting step in glucose sensing. The cellular mechanisms by which glucose metabolism regulates α-cell function are even less clear than in the β-cell. In recent years, conflicting reports have appeared on this subject. An early prediction that glucose should hyperpolarise the α-cell (Rorsman et al. 1989) appeared to be substantiated by reports that glucose inhibited electrical activity by hyperpolarising the α-cell membrane potential thus inactivating voltage-sensitive Ca^{2+} channels (Bokvist et al. 1999; Barg et al. 2000). How glucose hyperpolarised the cell was uncertain, though this was clearly a direct action of glucose on the α-cell. However, a more recent study from this laboratory suggested that glucose might inhibit electrical activity in the α-cell by a mechanism similar to the stimulation of electrical activity in the β-cell, that is by depolarising the cell by an inhibition of K_{ATP} channel activity (Gopel et al. 2000). In essence, it is envisaged that, at low glucose concentrations, the membrane potential of the α-cell is -40 to -60mV, the threshold potential 'window' for activation of pacemaker voltage-gated ion channels, possibly T-type Ca^{2+} channels which have a lower voltage threshold than L-type Ca^{2+} channels involved in β-cell electrical activity. This allows regenerative action potentials resulting in Ca^{2+} entry and glucagon release. A rise in glucose concentration depolarises the cell further via inhibition of K_{ATP} channels such that the pacemaker ion channels are inactivated, thus inhibiting electrical and secretory activity. This hypothesis appears to be inconsistent with a number of observations. For example, depolarising mouse α-cells by current injection was reported to enhance, rather than inhibit, electrical activity (Barg et al. 2000). The model is also inconsistent with the observations that sulphonylureas and other inhibitors of K_{ATP} channel activity induce electrical and secretory activity in the α-cell (Bokvist et al. 1999; Ronner et al. 1993). Furthermore, arginine and certain other cationic amino acids depolarise the α-cell via electrogenic transport of the amino acid and also induce electrical and secretory activity (Rorsman and Hellman 1988; Pipeleers et al. 1985). Finally, a major role for K_{ATP} channels in glucose-sensing by the α-cell is inconsistent with a report that glucose-inhibited glucagon secretion is observed in islets from K_{ATP} channel knockout mice (Miki et al. 2001).

4.4.2.1 Anion channel hypothesis: the α-cell version

As outlined above, there is considerable confusion and conflicting information regarding the mechanisms by which glucose regulates glucagon release. Indeed, it is even unclear whether a rise in glucose concentration depolarises or hyperpolarises the pancreatic α-cell. The weight of evidence falls arguably in favour of the latter possibility, raising the key question of how glucose metabolism in the α-cell might produce a hyperpolarisation of the plasma membrane potential. We can propose a possible mechanism based on our hypothetical model for the regulation of β-cell membrane potential by glucose. There is evidence that glucose can regulate electrical and secretory activity in the pancreatic β-cell by activating a volume-sensitive anion channel (VSAC). Since E_{Cl} in the β-cell appears to be positive with respect to resting membrane potential, VSAC activation leads to Cl^- exit, thus generating an inward (depolarising) current. Intracellular [Cl^-] is maintained above its electrochemical equilibrium by the $Na^+/K^+/2Cl^-$ co-transporter NKCC, and possibly by other inward Cl^- transport systems such as Na^+/Cl^- (NCC). In contrast to the β-cell, pancreatic α-cells do not appear to express NKCC (Majid et al. 2001). Instead, there is evidence in α-cells for activity and expression of the K^+/Cl^- (KCC) co-transporter (Davies et al. 2002b), which under physiological conditions maintains low intracellular $[Cl^-]$. In such a case, activation of the VSAC by glucose would result in Cl^- entry into the cell, thus generating an outward, hyperpolarising current (see Fig. 3). Rat pancreatic α-cells do have a VSAC (unpublished data), and whilst it is not yet known whether the channel can be activated by glucose metabolism, glucose does appear to cause α-cell swelling (Davies et al. 2002a). The inhibition of glucagon release by GABA, which involves activation of a $GABA_A$ receptor-operated Cl^- channel, causing hyperpolarisation of the membrane potential and inhibition of electrical and secretory activity (Rorsman et al. 1989; Gilon et al. 1991) would be consistent with our hypothesis.

Thus, it is possible that activation of the VSAC could be a common mechanism in glucose regulation of both pancreatic β- and α-cell function. The opposing responses of these cells to glucose would be determined essentially by the $[Cl^-]$ gradient across the plasma membrane, and hence by the differential expression of Cl^- transport systems in the two cell types. Incidentally, the ability of GABA to depolarise or hyperpolarise neurons is also thought to be dependent on such a differential expression of Cl^- co-transporters (Rivera et al. 1999; Vardi et al. 2000), emphasizing the probable importance of transmembrane $[Cl^-]$ gradients in regulating cell membrane potential and excitability. A clearer understanding of Cl^- handling in islet cells and its regulation by glucose could be invaluable in elucidating how glucose regulates pancreatic hormone secretion, and possibly the activity of other glucose-regulated secretory cells.

4.4.3 Enteroendocrine cells (EEC)

Surprisingly, the epithelium lining the gastrointestinal tract represents both the largest sensory and the largest endocrine organ system in the body, operating in close communication with a vast network of subepithelial sensory nerves. The gut epithelium is the first point of biological contact for basic nutrients after release from more complex macromolecules in the lumen. Unlike other discrete endocrine organs, EEC are scattered individually throughout the intestinal epithelium, and represent about 1% of its total cell population. There are multiple classes of EEC. In the foregut, these include cells that secrete 5-hydroxytryptamine/orexin (enterochromaffin, EC cells), cholecystokinin (CCK; I cells), secretin (S cells), glucagon-like peptide-1 (GLP-1; L cells), gastric inhibitory peptide (GIP; K cells) and somatostatin (D cells). In the stomach, gastrin release is well characterised, but the recent description of gastric ghrelin is radically changing our understanding of the nutrient sensing role of the stomach.

EEC operate as chemosensory conduits within the epithelium, mediating transepithelial signal transduction. The dominant enterocyte and goblet cell populations do not appear to contribute an important signalling role. When a luminal nutrient change occurs EEC respond by the basolateral secretion of biologically active peptides and/or amines, to relay with subepithelial nerve fibres as well as

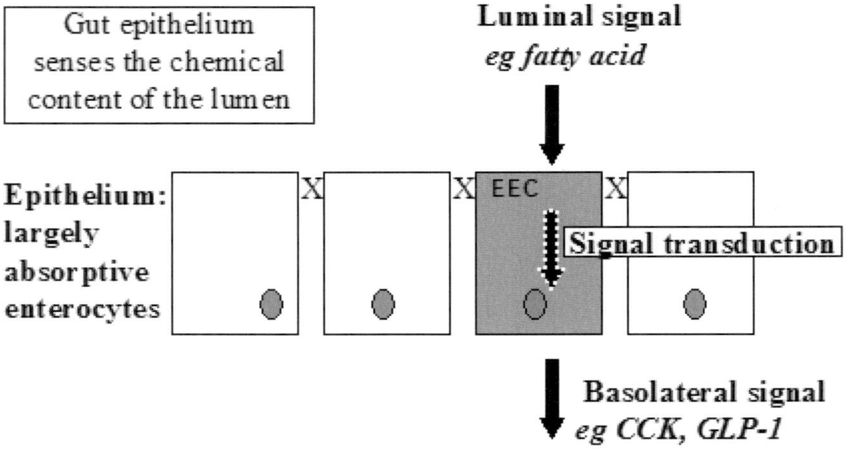

Fig. 4. Schematic diagram of Entreoendocrine cell (EEC) function. EEC are diffusely scattered within the gut epithelium, accounting for about 1% of the total cell population. Via apical microvilli open to the lumen, they are able to detect the chemical (nutrient) content of the gut. The molecular receptive mechanisms await elucidation. Peptides and amines are secreted basolaterally into the circulation to mediate appropriate physiological responses.

operating in a classical endocrine manner. These peptides control intrinsic gut functions, such as motility and secretion, in a co-ordinated manner maximising digestion and absorption. Indeed, this process achieves an efficiency far in excess of the needs of most non-food deprived subjects. Additionally, the signalling cascades relay more distantly to the central nervous system to regulate central responses, such as feeding behaviour and appetite, especially via the vagus nerves.

EEC extend long apical microvilli toward the gut lumen and these are believed to play a role in nutrient detection. However, the detailed cellular sensory physiology of EEC and interactions with other local cells have not been defined in any depth, and are largely confined to indirect observation such as plasma peptide assays and end organ responses. In particular, there are no primary cell preparation methodologies available for the study of EEC, and most cellular data are derived from immortal cell lines secreting gut hormones. Inevitably, these are limited in their physiological validity, and the detailed cellular physiology outlined above for islet cells is not yet available for EEC.

4.4.3.1 EEC responses to glucose

The presence of glucose in the upper gut has relatively little effect on local functions mediated by gut peptides, being a far weaker stimulus to most gut hormones than fat or protein. However, the 'incretin' effect, whereby peptides released from certain EEC in response to nutrients in the gut lumen stimulate insulin secretion, is induced by glucose in the upper small intestine. This entero-insular axis is evident since an oral glucose load induces greater insulin secretion than an isoglycaemic intravenous glucose infusion. Moreover, an identical amount of glucose produces a smaller plasma glucose response and greater insulin response when given orally rather than intravenously (Perley and Kipnis 1967). The two key gut peptides released are GLP-1 (Elliott et al. 1993), a cleavage product of proglucagon in intestinal L-cells (Drucker 2002), and GIP (Meier et al. 2002). The latter term was originally an abbreviation of gastric inhibitory polypeptide, but is now also referred as glucose-dependent insulinotropic polypeptide to reflect its putative role in the entero-insular axis. There is evidence for reduced incretin effects in type 2 diabetes mellitus and that GLP-1 analogues may improve postprandial glycaemia (Drucker 2002). Consistent with this, GLP-1 receptor knockout (Scrocchi et al. 1996) and GIP receptor knockout mice (Miyawaki et al. 1999) exhibit oral glucose intolerance and impaired insulinaemic responses.

Owing to the lack of primary models for EEC function, little is known of the stimulus-secretion coupling mechanisms involved in their nutrient responsiveness. Studies in an enriched isolated cell preparation suggest a direct effect of glucose on GIP-secreting cells, at least not requiring an intact innervation (Kieffer et al. 1994). However using the GLP-1 secreting cell line GLUTag, Reimann and Gribble (2002) recently demonstrated glucose and tolbutamide-induced action potentials and GLP-1 secretion, and detected expression of the K_{ATP} channel subunits $K_{IR6.2}$ and SUR1 and of glucokinase. Thus, L-cells may share cellular functions with pancreatic β-cells, perhaps not surprisingly since there are important commonalities in the genes encoding EEC and islet endocrine cell ontogeny (Rindi et al. 1999).

There is no good evidence that glucose is an important secretagogue for other key gastroduodenal peptides such as secretin, gastrin, or motilin, and it is at best a weak stimulus for CCK release. There are no data yet indicating which nutrient type, if any, mediates the fall in plasma ghrelin levels upon feeding, although a recent study suggests that ghrelin may have opposing actions on the β-cell to the incretins discussed above (Reimer et al. 2003).

The secretion of 5-hydroxytryptamine (serotonin) from EC cells has also been suggested as a mediator of some of the effects of glucose, such as delaying gastric emptying (Raybould and Zittel 1995) or inducing pancreatic exocrine secretion (Li et al. 2000). Raybould and Zittel further demonstrated that the gastric motility response to glucose was reproduced by its non-metabolisable analogue 3-O-methyl glucose, and was attenuated by perfusion of the intestine with phloridzin, a competitive blocker of the SGLT-1 co-transporter. A more recent report from the same laboratory demonstrated that glucose stimulated serotonin release from the human carcinoid-derived BON cell line, and proposed this as a useful experimental model (Kim et al. 2001).

4.4.4 Glucose regulation of hypothalamic neuronal activity

Most neurones utilise glucose as a major metabolic substrate. Indeed, neuronal activity throughout the brain requires glucose concentrations of at least 1 mM (Silver and Erecinska 1994), However, specialised neurones in the hypothalamus and certain other areas of the brain are able to detect and respond to changes in blood glucose concentrations within the normal physiological range (reviewed in Levin et al. 1999; Levin 2002). These neuronal responses to glucose, together with numerous other metabolic and hormonal signals, are important in regulating food intake and maintaining a balance with energy expenditure via the release of several neurotransmitters including neuropeptide Y, pro-opiomelanocortin, and GABA (Schwartz et al. 2000; Havel 2001). Indeed, alterations in neuronal glucose sensing are often associated with obesity and diabetes mellitus (Levin et al. 1999). The complex and heterogeneous nature of the hypothalamic neurones presents a similar problem to that in the gut in elucidating mechanisms of nutrient regulation. However, two basic types of glucose-regulated neurone have been identified. Glucose-responsive (GR) neurones, found predominantly in the ventromedial region of the hypothalamus (VMH), respond to a rise in glucose concentration by increasing their firing rate and are thought to be involved in the satiety response. Glucose-sensitive (GS) neurones, located mainly in the lateral hypothalamus (LH), increase their firing rate when glucose levels fall and are likely to play a role in the hunger response to fasting.

It is likely that the reciprocal responses of these neurones to glucose involve similar mechanisms to those found in pancreatic islet cells (Schuit et al. 2001). Although a number of substrates known to be ineffective in the β-cell were reported to activate hypothalamic neurones, these differences were accounted for by the ability of neurones to metabolise these substrates (Yang et al. 1999). As in the β-cell, regulation of hypothalamic neurones by glucose requires metabolism of the hexose (Ashford et al. 1990; Yang et al. 1999). There is evidence for expression of the high K_m glucose transporter GLUT-2, the pancreatic form of glucokinase and the K_{IR} and SUR1 subunits of the K_{ATP} channel in areas of the hypothalamus rich in GR neurones (Navarro et al. 1996; Lee et al. 1999; Lynch et al. 2000). In addition, inhibition of K_{ATP} channels has been demonstrated in GR neurones in response to addition of a high concentration of glucose (Ashford et al. 1990; Lee et al. 1999). K_{ATP} channel inhibition by sulphonylureas also results in electrical activity in these neurones (Ashford et al. 1990; Yang et al. 1999). In $K_{IR6.2}$ knockout mice, GR neurones were found to be spontaneously active, showing no further response to a rise in glucose concentration (Miki et al. 2001). This situation is analogous to that found in pancreatic β-cells from these mice (Miki et al. 1998) and suggests that K_{ATP} channels are necessary for the normal functioning of these glucose-regulated cell types.

It is important to emphasise that, as in the pancreatic β-cell, different cellular mechanisms may be responsible for detecting changes in glucose over low and high concentration ranges (Yang et al. 1999). In considering the glucose concentration-dependency of hypothalamic neuronal activity, an important consideration is the glucose concentrations to which the neurones are likely to be exposed. In particular, glucose levels in the cerebrospinal fluid are considerably lower than those in the blood (Segal 2001). Consequently, it is probably that GR neurones can detect changes in glucose within a rather narrow concentration range. To date, very few studies have addressed this point, and there have been no studies directly correlating K_{ATP} channel and electrical activities with respect to glucose concentration in GR neurones.

As with the pancreatic α-cell, the mechanisms underlying the inhibitory effect of glucose on firing in GS neurones is unknown. An early suggestion was activation by glucose of the Na^+/K^+-ATPase, producing a hyperpolarisation (Oomura et al. 1974). However, it was later shown that glucose increased membrane conductance in GS neurones (Mizuno and Oomura 1984), implying activation of one or more ion channels. A candidate for this effect is an ATP-activated K^+ channel, corresponding to the channel activated by a rise in glucose concentration and apparently specific to GS neurones (Rowe et al. 1996). The possible existence of a glucose-activated anion channel (VSAC) mechanism has not yet been explored in either GR or GS neurones. However, there was indirect evidence that the glucose-induced increase in membrane conductance recently reported in GS neurones could represent Cl^- channel activation (Song et al. 2001).

4.5 Regulation of secretion by amino acids

A wide range of amino acids can stimulate the release of hormones and neurotransmitters. On the one hand, substances such as γ-aminobutyric acid (GABA) and glutamate function essentially as locally-released neurotransmitter molecules and interact with cell surface receptors. This complex and extensive topic is beyond the scope of this article, and we shall focus primarily on amino acids that are likely to exert remote effects on cellular secretory activity via changes in their circulating concentrations. In this context, amino acids can influence secretion in two basic ways; via their cellular uptake, or as a result of their intracellular metabolism. As is the case with the regulation of exocytosis by glucose, pancreatic islet cells have provided a useful model for the study of amino acid regulation of secretory activity. Secretion of certain gut hormones is also regulated by amino acids, though rather less is understood regarding molecular mechanisms involved. There is some evidence that amino acids can regulate hypothalamic neuronal activity (Pavel 2001).

4.5.1 Amino acids and pancreatic islet cells

4.5.1.1 Activation by amino acid transport

Arginine and related cationic amino acids stimulate hormone release and electrical activity in both pancreatic β- and α-cells (Lambert et al. 1969; Henquin and Meissner 1981; Pipeleers et al. 1985). It is generally accepted that the major determinant of this stimulatory effect is uptake of the positively charged amino acid, resulting in depolarisation and electrical activity (Blachier et al. 1989; Thams and Capito 1999). The release of insulin and glucagon can also be enhanced by the neutral amino acids alanine and glycine (Lambert et al. 1969; Pipeleers et al.

1985; Tengholm et al. 1992). This effect is likely to be primarily the result of Na^+-coupled amino acid uptake (Lambert et al. 1969; Tengholm et al. 1992), causing a depolarisation of the cell membrane and increased $[Ca^{2+}]_i$ (Dunne et al. 1990). However, at least in the case of alanine, an additional metabolic signal has recently been suggested (Brennan et al. 2002).

4.5.1.2 Activation by amino acid metabolism

L-leucine is a neutral amino acid that causes a sustained stimulation of insulin release, an effect shared to a greater degree by its deaminated metabolite 2-ketoisocaproate (Panten et al. 1972). Both compounds are effectively metabolised by islet cells via the 2-keto acid dehydrogenase pathway, acetoacetate and CO_2 being the principal metabolites (Panten et al. 1972; Hutton et al. 1979). The effects of 2-ketoisocaproate on virtually all aspects of islet cell function closely resemble those of glucose, and have consequently been studied in considerable detail. Thus, the keto-acid has been reported to raise the ATP/ADP ratio in islets (Hutton et al. 1980), to inhibit K_{ATP} channel activity (Ashcroft et al. 1987), evoke electrical activity and increased cytosolic $[Ca^{2+}]$ (Hutton et al. 1980; Martin et al. 1995). In contrast to glucose, 2-ketoisocaproate is catabolised solely in mitochondria, thus reflecting the importance attributed to mitochondrial metabolism in regulating insulin release. However, studies with isolated islet cell mitochondria have shown that 2-ketoisocaproate is actually a poor mitochondrial substrate in terms of ATP generation, suggesting that a metabolic product, possibly acetoacetate, may provide a signal for insulin release (Lembert and Idahl 1998). Indeed, acetoacetate itself is an effective stimulus for β-cell electrical activity (Dean et al. 1975) and insulin release (Malaisse et al. 1990). Electrical activity evoked by 2-ketoisocaproate, as in response to glucose, is accompanied by activation of the VSAC (Best 1997, 1999, 2000). By analogy with the hypothetical mechanism outlined earlier for glucose activation of the VSAC (Fig. 3), it is possible that channel activation by 2-ketoisocaproate occurs in response to the generation of acetoacetate and its intracellular accumulation. Consistent with this idea, both electrical activity and insulin release in response to 2-ketoisocaproate are sensitive to inhibition by VSAC blockers (our unpublished findings).

In contrast to arginine and alanine, which stimulate glucagon release, 2-ketoisocaproate has been shown to inhibit glucagon secretion (Leclerq-Meyer et al. 1979). As in the β-cell, the response of the pancreatic α-cell to this keto-acid closely resembles the response to glucose, again suggesting common cellular mechanisms in the secretory responses to these two types of nutrient. The effects of leucine on the α-cell are more complex. In general, high concentrations of the amino acid are inhibitory (Leclerq-Meyer et al. 1985), presumably due to deamination to 2-ketoisocaproate. However, lower concentrations of both L- and D-leucine appear to enhance glucagon release by a mechanism that remains to be elucidated (Leclerq-Meyer et al. 1985). A potentially important action of leucine is its ability to stimulate the metabolism of endogenous substrates (notably glutamate) via an allosteric activation of glutamate dehydrogenase (GDH) activity, thereby increasing flux through the tricarboxylic acid cycle and hence generation of ATP (Sener and Malaisse 1980). This action of leucine is shared by its non-metabolisable analogue 2-aminobicyclo-[2,2,1]heptane-2-carboxylic acid (BCH) that is also a potent stimulus for β-cell electrical and secretory activity (Sener and Malaisse 1980; Henquin and Meissner 1981). The importance of glutamate dehydrogenase in regulating β-cell function is illustrated by cases of congenital hyperinsulinism and hypoglycaemia resulting from a GDH gene mutation causing increased enzymic activity (Stanley et al. 1998).

L-glutamine is efficiently metabolised by pancreatic islet cells, but does not enhance insulin release unless in the presence of another stimulus. This property is thought to reflect a role of glutamine as an energy source through its conversion to glutamate and subsequent entry into the citric acid cycle via glutamate dehydrogenase (Malaisse et al. 1980). Glutamine is, however, an effective stimulus for glucagon release (Leclerq-Meyer et al. 1985), although the cellular mechanism of this action is essentially unknown.

In common with glucose, a number of metabolisable amino acids including leucine (and 2-ketoisocaproate), glutamine and glutamate, have been reported to activate the 'distal' or amplifying pathway of exocytosis in the pancreatic β-cell described earlier (reviewed in Henquin 2000). Whether these effects involve increased ATP/ADP ratio, amino acid uptake into secretory vesicles or other unidentified mechanisms remains to be established.

4.5.2 Amino acids and enteroendocrine cells

The most detailed evidence concerning amino acid-induced gut hormone secretion relates to gastrin secretion from gastroduodenal G-cells. Taylor and colleagues (1982) undertook elaborate studies in humans, infusing individual amino acids and assaying plasma gastrin responses and gastric acid secretion. The aromatic amino acids L-phenylalanine and L-tryptophan were found to be the most potent stimuli for gastrin release and also that of pancreatic polypeptide. This is likely to be a direct, local action since amino acids are effective on isolated preparations enriched for G-cells (DelValle and Yamada 1990). The mechanism of amino acid activation of gastrin release has not been characterised. However, it is noteworthy that activity of the extracellular calcium-sensing receptor (CaR), which is activated selectively by aromatic amino acids (Conigrave et al. 2000) has been reported in G-cells (Buchan et al. 2001). In addition, aromatic amino acids have been shown to activate human embryonic kidney cells transfected with CaR (Conigrave et al. 2000).

An amino acid mixture containing L-phenylalanine, L-tryptophan, valine, and methionine has been shown to stimulate CCK *in vivo*, whilst a mixture of arginine, histidine, isoleucine, leucine, lysine, and threonine was ineffective (Colombel et al. 1988). More recently, L-phenylalanine alone was shown to have a similar effect, and it was suggested that CCK release could play a role in the satiety response to feeding (Ballinger and Clark 1994). The CCK secreting cell line STC-1 responds directly to L-phenylalanine (Mangel et al. 1995) via mechanism largely uncharacterised, but which does not appear to involve the CaR (J. McLaughlin, unpublished data). No direct evidence is available to link amino acid transport to EEC function although a neutral and basic amino acid transporter has been reported to be strongly expressed in rat EEC (Pickel et al. 1993).

4.6 Regulation of secretion by fatty acids

Free fatty acids (FFA) can exert profound effects on endocrine and neuroendocrine cells. However, acute effects of FFA can be different from long-term effects, whilst concentration, physical form of the fatty acid and degree of saturation can also be important determinants of activity.

4.6.1 Pancreatic islet cells

The effects of FFA on the endocrine pancreas are complex and the mechanisms poorly understood (reviewed in Roche et al. 2000; Haber et al. 2003). Several studies have demonstrated acute stimulation of insulin release by FFA, both *in vivo* (Balasse and Ooms 1973) and *in vitro* (Goberna et al. 1974; Warnotte et al. 1994; Stein et al. 1996). Palmitate has been shown to enhance β-cell electrical activity in the presence of glucose (Fernandez and Valdeolmillos 1998) and to preserve or restore the glucose responsiveness of β-cells following starvation (Stein et al. 1996). In general, enhancement of insulin release by FFA is especially pronounced with long chain saturated fatty acids (Stein et al. 1997). The underlying ionic mechanism of β-cell activation by FFA remains to be elucidated. It has been demonstrated that fatty acids can move rapidly across lipid bilayers (Kamp and Hamilton 1992) and can be metabolised by islet cells via mitochondrial oxidation and incorporation into various lipid classes (Sener et al. 1978). However, the secretory effect of palmitate does not appear to require its oxidation or to involve K_{ATP} channel inhibition (Warnotte et al. 1994). It has been suggested that fatty acyl CoA esters might be important in regulating exocytosis, possibly via protein acylation (Corkey et al. 2000) or protein kinase C activation (Thams and Capito 2001). However, it has recently been shown that FFA of C12 to C16 length raise cytosolic [Ca^{2+}] and enhance insulin release by interacting with GPR40, a G-protein coupled plasma membrane receptor strongly expressed in pancreatic β-cells (Itoh et al. 2003). No clear expression of GPR40 was detected in pancreatic α-cells, despite reports of inhibitory (Edwards et al. 1969) and stimulatory (Opara et al. 1990) effects of FFA on glucagon release. It will be of interest to establish whether GPR40 is expressed in other cell types where exocytosis can be regulated by FFA.

In contrast to the acute stimulatory effects of FFA on islet cell function, long-term exposure to FFA has been shown to impair glucose-induced insulin release, possibly due to impaired glucose metabolism resulting from fatty acid oxidation (Grill and Qvigstad 2000). Similar effects have been reported with prolonged exposure of islets to β-hydroxybutyrate (Zhou and Grill 1995), again contrasting with the acute stimulatory effects of ketones (Goberna et al. 1974; Malaisse et al. 1990). This detrimental effect on β-cell function of chronic exposure to FFA and ketones could represent a component of nutrient toxicity thought to play a role in the development of type 2 diabetes mellitus (Roche et al. 2000).

4.6.2 CCK cells

CCK is the key EEC peptide released in response to fat, mediating the greater part of the classical effects of increased gall bladder emptying, pancreatic exocrine secretion and delayed gastric emptying, and has recently been shown to play a key role in satiety. CCK is released *in vivo* in response to free fatty acids (Owyang et al. 1986), triglycerides being less effective (Guimbaud et al. 1997). Accordingly, a lipase inhibitor has been shown to prevent triglyceride-induced CCK release (Feinle et al. 2001), emphasising the importance of FFA in triggering this response. The stimulatory effect of FFA is exquisitely dependent on acyl chain length. Thus, saturated fatty acids of C12 or longer are effective whilst C11 and shorter are not (McLaughlin et al. 1999). Earlier suggestions that saturated fat was less potent than unsaturated fat (Beardshall et al. 1999) have not been confirmed, and probably reflect the need to overcome physicochemical differences in the mode of fat presentation. Fat also induces secretion of motilin, enteroglucagon, GIP, and gastrointestinal peptide YY (Spiller et al. 1988; Lardinois et al. 1988; Maas et al. 1998). As with CCK release, there is a general trend that long chain FFA are more potent than medium chain FFA in stimulating release of these gut peptides.

Studies using primary EEC are limited, although it has been reported that isolated gut mucosal cells in suspension release CCK in response to sodium oleate (Chang et al. 2000), suggesting a direct effect of FFA without need for luminal factors or intact innervation. The most detailed mechanistic data come from studies with STC-1 cells, which exhibit FFA chain-length dependency highly comparable to that observed *in vivo* (McLaughlin 1998, 1999). The stimulatory effect of FFA appears to be independent of fatty acid metabolism, but involves depolarization, activation of voltage-sensitive Ca^{2+} channels and a subsequent rise in $[Ca]_i$ and CCK secretion (McLaughlin 1998; Sidhu et al. 2000). Clearly, FFA are rapidly taken up into the cytoplasm of STC-1 cells (Sidhu et al. 2000) but whether transport is required to induce the above effects is unclear. One intriguing observation is that filtering FFA 'solutions' significantly reduces STC-1 cells' responses, raising the possibility that fatty acids may activate mechanical pathways by acting as insoluble aggregates, rather than as acyl monomers (Benson et al. 2002). However, this phenomenon could also reflect removal of the bulk of FFA from the system, and it is conceivable that FFA exerts its actions following dissociating from aggregates. The recent description of the FFA-activated, G-protein coupled receptor GPR40 in insulin-secreting cells now needs to be explored in the context of FFA regulation of CCK release.

4.6.3 Hypothalamic neurones

Application of a mixture of FFA has been shown to activate lateral hypothalamic GS neurones (Oomura et al. 1975). These authors suggested that this response could contribute to the hunger response to fasting, which is associated with raised FFA levels, and also reported that FFA inhibited GR cells in the ventromedial hypothalamus. The mechanisms of action of FFA on hypothalamic neurones is unclear, though a recent study (Loftus et al. 2000) has provided evidence that intracellular malonyl CoA could be a potential signalling molecule by analogy with the hypothetical model proposed for pancreatic β-cells (Corkey et al. 2000).

4.7 Concluding remarks

Nutrients can influence hormone and neurotransmitter release at a number of different cellular levels. There is evidence for common molecular mechanisms in different cells types for the detection of changes in levels of specific nutrients. In broad terms, two distinct sites of action can be identified: plasma membrane events such as interaction with a receptor or electrogenic transport and intracellular events involving metabolism of the nutrient itself or activation of endogenous nutrient metabolism. Whilst the processes of receptor binding and transport might be considered relatively straightforward (at least conceptually), the fundamental intracellular mechanisms coupling nutrient metabolism with exocytosis have remained highly elusive. Since increased nutrient metabolism can often be shown to coincide with changes in intracellular nucleotide levels (notably raised ATP/ADP ratio), much attention has focussed on the idea that such changes provide, directly or indirectly, a signal for exocytosis. However, it is also conceivable that changes in nucleotide levels occur in parallel to an increased rate of exocytosis, possibly due to raised cytosolic (and thus mitochondrial) [Ca^{2+}] and a consequent activation of mitochondrial dehydrogenases (Pralong et al. 1994). In this context, a rise in ATP/ADP ratio might be considered to fulfil a supporting role in augmented secretory activity, rather than providing a primary intracellular signal *per se*. This question should be resolved to a large extent by identifying additional cellular mechanisms coupling nutrient metabolism with exocytosis.

References

Ashcroft FM, Ashcroft SJ, Harrison DE (1987) Effects of 2-ketoisocaproate on insulin release and single potassium channel activity in dispersed rat pancreatic beta-cells. J Physiol 385:517-529

Ashcroft FM, Harrison DE, Ashcroft SJ (1984) Glucose induces closure of single potassium channels in isolated rat pancreatic beta-cells. Nature 312:446-448

Ashcroft FM, Ashcroft SJH, Harrison DE (1988) Properties of single potassium channels modulated by glucose in rat pancreatic β-cells. J Physiol 400:501-527

Ashcroft FM, Rorsman P (1989) Electrophysiology of the pancreatic β-cell. Prog Biophys Molec Biol 54:87-143

Ashcroft SJ (2000) The beta-cell K_{ATP} channel. J Membr Biol 176:187-206

Ashford ML, Boden PR, Treherne JM (1990) Glucose-induced excitation of hypothalamic neurones is mediated by ATP-sensitive K^+ channels. Pflugers Arch 415:479-483

Balasse EO, Ooms HA (1973) Role of plasma free fatty acids in the control of insulin secretion in man. Diabetologia 9:145-151

Ballinger AB, Clark ML (1994) L-phenylalanine releases cholecystokinin (CCK) and is associated with reduced food intake in humans: evidence for a physiological role of CCK in control of eating. Metabolism 43:735-738

Barg S, Galvanovskis J, Gopel SO, Rorsman P, Eliasson L (2000) Tight coupling between electrical activity and exocytosis in mouse glucagon-secreting α-cells. Diabetes 49:1500-1510

Beardshall K, Frost G, Morarji Y, Domin J, Bloom SR, Calam J (1989) Saturation of fat and cholecystokinin release: implications for pancreatic carcinogenesis. Lancet 2:1008-1010

Benson RS, Sidhu S, Jones MN, Case RM, Thompson DG (2002) Fatty acid signalling in a mouse enteroendocrine cell line involves fatty acid aggregates rather than free fatty acids. J Physiol 538:121-131

Bertrand G, Ishiyama N, Nenquin M, Ravier MA, Henquin JC (2002) The elevation of glutamate content and the amplification of insulin secretion in glucose-stimulated pancreatic islets are not causally related. J Biol Chem 277:32883-32891

Best L (1997) Glucose and a-ketoisocaproate induce transient inward currents in rat pancreatic beta cells. Diabetologia 40:1-6

Best L (1999) Cell-attached recordings of the volume-sensitive anion channel in rat pancreatic B-cells. Biochim Biophys Acta 1419:248-256

Best L (2000) Glucose-sensitive conductances in rat pancreatic beta-cells: contribution to electrical activity. Biochim Biophys Acta 1468:311-319

Best L (2002a) Evidence that glucose-induced electrical activity in rat pancreatic β-cells does not require K_{ATP} channel inhibition. J Membr Biol 185:193-200

Best L (2002b) Study of a glucose-activated anion-selective channel in rat pancreatic beta-cells. Pflugers Arch 445:97-104

Best L (2002c) Inhibition of glucose-induced electrical activity by 4-hydroxytamoxifen in rat pancreatic β-cells. Cell Signal 14:69-73

Best L, Brown PD, Sheader EA, Yates AP (2000) Selective inhibition of glucose-stimulated beta-cell activity by an anion channel inhibitor. J Membr Biol 177:169-175

Best L, Brown PD, Tomlinson S (1997) Anion fluxes, volume regulation and electrical activity in the mammalian pancreatic β-cell. Exp Physiol 82:957-966

Best L, Miley HE, Yates AP (1996a) Activation of an anion conductance and B-cell depolarization during hypotonically-induced insulin release. Exp Physiol 81:927-933

Best L, Sheader EA, Brown PD (1996b) A volume-activated anion conductance in insulin-secreting cells. Pflugers Arch 431:363-370

Best L, Speake T, Brown PD (2001) Characterization of the volume-sensitive anion channel in rat pancreatic beta-cells. Exp Physiol 86:145-150

Best L, Trebilcock R, Tomlinson S (1992a) Lactate transport in insulin-secreting β-cells: contrast between rat islets and HIT-T15 insulinoma cells. Mol Cell Endocr 86:49-56

Best L, Yates AP, Meats JE, Tomlinson S (1989) Effects of lactate on pancreatic islets: lactate efflux as a possible determinant of islet cell depolarization by glucose. Biochem J 259:507-511

Best L, Yates AP, Tomlinson S (1992b) Stimulation of insulin secretion by glucose in the absence of diminished potassium ($^{86}Rb^{+}$) permeability. Biochem Pharmacol 43:2483-2485

Blachier F, Mourtada A, Sener A, Malaisse WJ (1989) Stimulus-secretion coupling of arginine-induced insulin release. Uptake of metabolized and non-metabolized cationic amino acids by pancreatic islets. Endocrinology 124:134-141

Bokvist K, Olsen HL, Hoy M, Godfredsen CF, Holmes WF, Buschard K, Rorsman P, Gromada J (1999) Characterization of sulphonylurea and ATP-regulated K^{+} channels in rat pancreatic A-cells. Pflugers Arch 438:428-436

Brennan L, Shine A, Hewage C, Malthouse JP, Brindle KM, McClenaghan N, Flatt PR, Newsholme P (2002) A nuclear magnetic resonance-based demonstration of substantial oxidative L-alanine metabolism and L-alanine-enhanced glucose metabolism in a clonal pancreatic beta-cell line: metabolism of L-alanine is important to the regulation of insulin secretion. Diabetes 51:1714-1721

Buchan AM, Squires PE, Ring M, Meloche RM (2001) Mechanism of action of the calcium-sensing receptor in human antral gastrin cells. Gastroenterology 120:1128-1139

Carpinelli AR, Malaisse WJ (1981) Regulation of ^{86}Rb outflow from pancreatic islets: the dual effect of nutrient secretagogues. J Physiol 315:143-156

Chang CH, Chey WY, Chang TM (2000) Cellular mechanism of sodium oleate-stimulated secretion of cholecystokinin and secretin. Am J Physiol 279:G295-303

Colombel JF, Sutton A, Chayvialle JA, Modigliani R (1988) Cholecystokinin release and biliopancreatic secretion in response to selective perfusion of the duodenal loop with aminoacids in man. Gut 29:1158-1166

Conigrave AD, Quinn SJ, Brown EM (2000) L-amino acid sensing by the extracellular Ca^{2+}-sensing receptor. Proc Natl Acad Sci USA 97:4814-4819

Cook DL, Hales CN (1984) Intracellular ATP directly blocks K^{+} channels in pancreatic B-cells. Nature 311:271-273

Corkey BE, Deeney JT, Yaney GC, Tornheim K, Prentki M (2000) The role of long-chain fatty acyl-CoA esters in beta-cell signal transduction. J Nutr 130(2S suppl):299S-304S

Davies SL, Best L, Brown PD (2002a) Glucose-induced volume changes in α-cells isolated from rat pancreas. J Physiol 544P:91P

Davies SL, Williams K, Syer EG, Best L, Brown PD (2002b) Regulatory volume decrease in α-cells isolated from the rat pancreas involves K^{+}-Cl^{-} cotransporters. J Physiol 544P:91P

Dean PM, Matthews EK, Sakamoto Y (1975) Pancreatic islet cells: effects of monosaccharides, glycolytic intermediates and metabolic inhibitors on membrane potential and electrical activity. J Physiol 246:459-478

DelValle J, Yamada T (1990) Amino acids and amines stimulate gastrin release from canine antral G-cells via different pathways. J Clin Invest 85:139-143

Drews G, Zempel G, Krippeit-Drews P, Britsch S, Busch GL, Kaba NK, Lang F (1998) Ion channels involved in insulin release are activated by osmotic swelling of pancreatic B-cells. Biochim Biophys Acta 1370:8-16

Drucker DJ (2002) Biological actions and therapeutic potential of the glucagon-like peptides. Gastroenterology 122:531-544

Dunne MJ, Petersen OH (1986) Intracellular ADP activates K^+ channels that are inhibited by ATP in an insulin-secreting cell line. FEBS Lett 208:59-62
Dunne MJ, Yule DI, Gallacher DV, Petersen OH (1990) Effects of alanine on insulin-secreting cells: patch-clamp and single cell intracellular Ca^{2+} measurements. Biochim Biophys Acta 1055:157-164
Eberhardson M, Patterson S, Grappengiesser E (2000) Microfluorometric analysis of Cl^- permeability and its relation to oscillatory Ca^{2+} signalling in glucose-stimulated pancreatic β-cells. Cell Signal 12:781-786
Edwards JC, Howell SL, Taylor KW (1969) Fatty acids as regulators of glucagon secretion. Nature 224:808-809
Eggermont J, Trouet D, Carton I, Nilius B (2001) Cellular control of volume-regulated anion channels. Cell Biochem Biophys 35:263-274
Elliott RM, Morgan LM, Tredger JA, Deacon S, Wright J, Marks V (1993) Glucagon-like peptide-1 (7-36) amide and glucose-dependent insulinotropic polypeptide secretion in response to nutrient ingestion in man: acute post-prandial and 24-h secretion patterns. J Endocrinol 138:159-166
Feinle C, Rades T, Otto B, Fried M (2001) Fat digestion modulates gastrointestinal sensations induced by gastric distention and duodenal lipid in humans. Gastroenterology 120:1100-1107
Fernandez J, Valdeolmillos M (1998) Increased levels of free fatty acids in fasted mice stimulate *in vivo* β-cell electrical activity. Diabetes 47:1707-1712
Fon EA, Edwards RH (2001) Molecular mechanisms of neurotransmitter release. Muscle Nerve 24:581-601
Furman BL (1981) Impairment of glucose tolerance produced by diuretics and other drugs. Pharmacol Ther 12:613-649
Gembal M, Detimary P, Gilon P, Gao ZY, Henquin JC (1993) Mechanisms by which glucose can control insulin release independently from its action on adenosine triphosphate-sensitive K^+ channels in mouse B-cells. J Clin Invest 91:871-880
German MS (1993) Glucose sensing in pancreatic islet beta cells: The key role of glucokinase and the glycolytic intermediates. Proc Natl Acad Sci USA 90:1781-1785
Ghosh A, Ronner P, Cheong E, Khalid P, Matschinsky FM (1991) The role of ATP and free ADP in metabolic coupling during fuel-stimulated insulin release from islet beta-cells in the isolated perfused rat pancreas. J Biol Chem 266:22887-22892
Gilon P, Bertrand G, Loubatieres-Mariani MM, Remacle C, Henquin JC (1991) The influence of gamma-aminobutyric acid on hormone release by the mouse and rat endocrine pancreas. Endocrinology 129:2521-2529
Goberna R, Tamarit J Jr, Osorio J, Fussganger R, Tamarit J, Pfeiffer EF (1974) Action of β-hydroxybutyrate, acetoacetate and palmitate on insulin secretion by the perfused rat pancreas. Horm Metab Res 6:256-260
Gopel SO, Kanno T, Barg S, Weng X, Gromada J, Rorsman P (2000) Regulation of glucagon release in mouse α-cells by K_{ATP} channels and inactivation of TTX-sensitive Na^+ channels. J Physiol 528:509-520
Grill V, Qvigstad E (2000) Fatty acids and insulin secretion. Br J Nutr 83(suppl 1):S79-S84
Guimbaud R, Moreau JA, Bouisson M, Durand S, Escourrou J, Vaysse N, Frexinos J (1997) Intraduodenal free fatty acids rather than triglycerides are responsible for the release of CCK in humans. Pancreas 14:76-82
Gylfe E (1990) How secretion is inhibited. Nature 344:300

Havel PJ (2001) Peripheral signals conveying metabolic information to the brain: short-term and long-term regulation of food intake and energy homeostasis. Exp Biol Med 226:963-977

Heimberg H, De Vos A, Moens K, Quartier E, Bouwens L, Pipeleers D, Van Schaftingen E, Madsen O, Schuit F (1996) The glucose sensor protein glucokinase is expressed in glucagon-producing alpha-cells. Proc Natl Acad Sci 93:7036-7041

Henquin JC (1978) D-glucose inhibits potassium efflux from pancreatic islet cells. Nature 271:271-273

Henquin JC (1992) ATP-sensitive K^+ channels may not be the sole regulators of glucose-induced electrical activity in pancreatic B-cells. Endocrinology 131:127-131

Henquin JC (1998) A minimum of fuel is necessary for tolbutamide to mimic the effects of glucose on electrical activity in pancreatic β-cells. Endocrinology 139:993-998

Henquin JC (2000) Triggering and amplifying pathways of regulation of insulin secretion by glucose. Diabetes 49:1751-1760

Henquin JC, Meissner HP (1981) Effects of amino acids on membrane potential and $^{86}Rb^+$ fluxes in pancreatic beta-cells. Am J Physiol 240:E245-E252

Hutton JC, Sener A, Herchuelz A, Atwater I, Kawazu S, Boschero AC, Somers G, Devis G, Malaisse WJ (1980) Similarities in the stimulus-secretion coupling mechanism of glucose- and 2-keto acid-induced insulin release. Endocrinology 106:203-219

Hutton JC, Sener A, Malaisse WJ (1979) The metabolism of 4-methyl-2-oxopentanoate in rat pancreatic islets. Biochem J 184:291-301

Ishihara H, Maechler P, Gjinovci A, Herrera PL, Wollheim CB (2003) Islet beta-cell secretion determines glucagon release from neighbouring alpha-cells. Nature Cell Biol 5:330-335

Itoh Y, Kawamata Y, Harada M, Kobayashi M, Fujii R, Fukusumi S, Ogi K, Hosoya M, Tanaka Y, Uejima H, Tanaka H, Maruyama M, Satoh R, Okubo S, Kizawa H, Komatsu H, Matsumura F, Noguchi Y, Shinohara T, Hinuma S, Fujisawa Y, Fujino M (2003) Free fatty acids regulate insulin secretion from pancreatic β cells through GPR40. Nature 422:173-176

Kamp F, Hamilton JA (1992) pH gradients across phospholipid membranes caused by fast flip-flop of un-ionized fatty acids. Proc Natl Acad Sci USA 89:11367-11370

Kieffer TJ, Buchan AM, Barker H, Brown JC, Pederson RA (1994) Release of gastric inhibitory polypeptide from cultured canine endocrine cells. Am J Physiol 267:E489-E496

Kim M, Cooke HJ, Javed NH, Carey HV, Christofi F, Raybould HE (2001) D-glucose releases 5-hydroxytryptamine from human BON cells as a model of enterochromaffin cells. Gastroenterology 121:1400-1406

Kinard TA, Satin LS (1995) An ATP-sensitive Cl^- channel current that is activated by cell swelling, cAMP and glyburide in insulin-secreting cells. Diabetes 44:1461-1466

Komatsu M, Schermerhorn T, Noda M, Straub SG, Aizawa T, Sharp GW (1997) Augmentation of insulin release by glucose in the absence of extracellular Ca^{2+}: new insights into stimulus-secretion coupling. Diabetes 46:1928-1938

Lambert AE, Jeanrenaud B, Junod A, Renold AE (1969) Organ culture of fetal rat pancreas. II. Insulin release induced by amino and organic acids, by hormonal peptides, by cationic alterations of the medium and by other agents. Biochim Biophys Acta 184:540-553

Lang J (1999) Molecular mechanisms and regulation of insulin exocytosis as a paradigm of endocrine secretion. Eur J Biochem 259:3-17

Lardinois CK, Starich GH, Mazzaferri EL (1988) The postprandial response of gastric inhibitory polypeptide to various dietary fats in man. J Am Coll Nutr 7:241-247
Leclerq-Meyer V, Marchand J, Leclerq R, Malaisse WJ (1979) Interactions of alpha-ketoisocaproate, glucose and arginine in the secretion of glucagon and insulin from the perfused rat pancreas. Diabetologia 17:121-126
Leclerq-Meyer V, Marchand J, Woussen-Colle MC, Giroix MH, Malaisse WJ (1985) Multiple effects of leucine on glucagon, insulin and somatostatin secretion from the perfused rat pancreas. Endocrinology 116:1168-1174
Lee K, Dixon AK, Richardson PJ, Pinnock RD (1999) Glucose-receptive neurones in the rat ventromedial hypothalamus express K_{ATP} channels composed of Kir6.1 and SUR1 subunits. J Physiol 515:439-452
Lembert N, Idahl L-A (1998) Alpha-ketoisocaproate is not a true substrate for ATP production by pancreatic beta-cell mitochondria. Diabetes 47:339-344
Levin BE (2002) Metabolic sensors: viewing glucosensing neurons from a broader perspective. Physiol Behav 76:397-401
Levin BE, Dunn-Meynell AA, Routh VH (1999) Brain glucose sensing and body energy homeostasis: role in obesity and diabetes. Am J Physiol 276:R1223-R1231
Li Y, Hao Y, Zhu J, Owyang C (2000) Serotonin released from intestinal enterochromaffin cells mediates luminal non-cholecystokinin-stimulated pancreatic secretion in rats. Gastroenterology 118:1197-1207
Lindstrom P, Norlund L, Sandstrom P-E, Sehlin J (1988) Evidence for co-transport of sodium, potassium and chloride in mouse pancreatic islets. J Physiol 400:223-236
Lledo P-M (1997) Exocytosis in excitable cells: a conserved molecular machinery from yeast to neuron. Eur J Endocr 137:1-9
Loftus TM, Jaworsky DE, Frehywot GL, Townsend CA, Ronnett G, Lane MD, Kuhajda FP (2000) Reduced food intake and body weight in mice treated with fatty acid synthase inhibitors. Science 288:2299-2300
Lynch RM, Tompkins LS, Brooks HL, Dunn-Meynell AA, Levin BE (2000) Localization of glucokinase gene expression in the rat brain. Diabetes 49:693-700
Maas MI, Hopman WP, Katan MB, Jansen JB (1998) Release of peptide YY and inhibition of gastric acid secretion by long-chain and medium-chain triglycerides but not by sucrose polyester in men. Eur J Clin Invest 28:123-130
MacDonald MJ, Fahien LA (2000) Glutamate is not a messenger in insulin secretion. J Biol Chem 275:34025-34027
Maechler P, Wollheim CB (1999) Mitochondrial glutamate acts as a messenger in glucose-induced insulin exocytosis. Nature 402:685-689
Maechler P, Wollheim CB (2000) Mitochondrial signals in glucose-stimulated insulin secretion. J Physiol 529:49-56
Majid A, Speake T, Best L, Brown PD (2001) Expression of the Na-K-2Cl cotransporter in α and β cells isolated from the rat pancreas. Pflugers Arch 442:570-576
Malaisse WJ (1992) Glucose-sensing by the pancreatic B-cell: the mitochondrial part. Int J Biochem 24:693-701
Malaisse WJ (1995) The esters of carboxylic nutrients as insulinotropic tools in non-insulin-dependent diabetes mellitus. Gen Pharmacol 26:1133-1141
Malaisse WJ, Lea MA, Malaisse-Lagae F (1968) The effect of mannoheptulose on the phosphorylation of glucose and the secretion of insulin by islets of Langerhans. Metabolism 17:126-132

Malaisse WJ, Lebrun P, Yaylali B, Camara J, Valverde I, Sener A (1990) Ketone bodies and islet function:^{45}Ca handling, insulin synthesis, and release. Am J Physiol 259:E117-E122

Malaisse WJ, Malaisse-Lagae F (1968) Stimulation of insulin secretion by noncarbohydrate metabolites. J Lab Clin Med 72:438-448

Malaisse WJ, Sener A (1987) Glucose-induced changes in cytosolic ATP content in pancreatic islets. Biochim Biophys Acta 927:190-195

Malaisse WJ, Sener A, Herchuelz A, Hutton JC (1979) Insulin release: the fuel hypothesis. Metabolism 28:373-386

Malaisse WJ, Sener A, Carpinelli AR, Anjaneyulu K, Lebrun P, Herchuelz A, Christophe J (1980) The stimulus-secretion coupling of glucose-induced insulin release. XLVI Physiological role of L-glutamine as a fuel for pancreatic islets. Mol Cell Endocrinol 20:171-189

Mangel AW, Prpic V, Wong H, Basavappa S, Hurst LJ, Scott L, Garman RL, Hayes JS, Sharara AI, Snow ND (1995) Phenylalanine-stimulated secretion of cholecystokinin is calcium dependent. Am J Physiol 268:G90-G94

Martin F, Sanchez-Andres JV, Soria B (1995) Slow $[Ca^{2+}]_i$ oscillations induced by ketoisocaproate in single mouse pancreatic islets. Diabetes 44:300-305

Matschinsky FM, Glaser B, Magnuson MA (1998) Pancreatic β-cell glucokinse. Closing the gap between theoretical concepts and experimental realities. Diabetes 47:307-315

Matschinsky FM, Landgraf R, Ellerman J, Kotler-Brajtburg J (1972) Glucoreceptor mechanisms in islets of Langerhans. Diabetes 21(suppl 2):555-569

May AP, Whitehart SW, Weis WI (2001) Unravelling the mechanism of the vesicle transport ATPase NSF, the N-ethylmaleimide-sensitive factor. J Biol Chem 276:21991-21994

McLaughlin J, Grazia Luca M, Jones MN, D'Amato M, Dockray GJ, Thompson DG (1999) Fatty acid chain length determines cholecystokinin secretion and effect on human gastric motility. Gastroenterology 116:46-53

McLaughlin JT, Lomax RB, Hall L, Dockray GJ, Thompson DG, Warhurst G (1998) Fatty acids stimulate cholecystokinin secretion via an acyl chain length-specific, Ca^{2+}-dependent mechanism in the enteroendocrine cell line STC-1. J Physiol 513:11-18

Meier JJ, Nauck MA, Schmidt WE, Gallwitz B (2002) Gastric inhibitory polypeptide: the neglected incretin revisited. Regul Pept 107:1-13

Miki T, Liss B, Minami K, Shiuchi T, Saraya A, Kashima Y, Horiuchi M, Ashcroft FM, Minokoshi Y, Roeper J, Seino S (2001) ATP-sensitive K^+ channels in the hypothalamus are essential for the maintenance of glucose homeostasis. Nat Neurosci 4:507-512

Miki T, Nagashima K, Tashiro F, Kotake K, Yoshimoto H, Tamamoto A, Gonoi T, Iwanaga T, Miyazaki J-I, Seino S (1998) Defective insulin secretion and enhanced insulin action in K_{ATP} channel-deficient mice. Proc Natl Acad Sci USA 95:10402-10406

Miley HE, Sheader EA, Brown PD, Best L (1997) Glucose-induced swelling in rat pancreatic beta-cells. J Physiol 504:191-198

Misler S, Barnett DW, Gillis KD, Pressel DM (1992) Electrophysiology of stimulus-secretion coupling in human β-cells. Diabetes 41:1221-1228

Miyawaki K, Yamada Y, Yano H, Niwa H, Ban N, Ihara Y, Kubota A, Fujimoto S, Kajikawa M, Kuroe A, Tsuda K, Hashimoto H, Yamashita T, Jomori T, Tashiro F, Miyazaki J, Seino Y (1999) Glucose intolerance caused by a defect in the entero-insular axis: A study in gastric inhibitory polypeptide receptor knockout mice. Proc Natl Acad Sci USA 96:14843-14847

Mizuno Y, Oomura Y (1984) Glucose responding neurones in the nucleus tractus solitarius of the rat: *In vitro* study. Brain Res 307:109-116

Navarro M, Rodriguez de Fonseca F, Alvarez E, Chowen JA, Zueco JA, Gomez R, Eng J, Blazquez E (1996) Co-localisation of glucagon-like peptide-1 (GLP-1) receptors, glucose transporter GLUT-2 and glucokinase mRNAs in rat hypothalamic cells: evidence for a role of GLP-1 receptor agonists as an inhibitory signal for food and water intake. J Neurochem 67:1982-1991

Opara EC, Burch WM, Hubbard VS, Akwari OE (1990) Enhancement of endocrine pancreatic secretions by essential fatty acids. J Surg Res 48:329-332

Oomura Y, Nakamura T, Sugimori M, Yamada Y (1975) Effect of free fatty acid on the rat lateral hypothalamic neurons. Physiol Behav 14:483-486

Oomura Y, Ooyama H, Sugimori M, Nakamura T, Yamada Y (1974) Glucose inhibition of the glucose-sensitive neurone in the rat lateral hypothalamus. Nature 247:284-286

Owyang C, Louie DS, Tatum D (1986) Feedback regulation of pancreatic enzyme secretion. Suppression of cholecystokinin release by trypsin. J Clin Invest 77:2042-2047

Panten U, Schwanstecher M, Schwanstecher C (1996) Sulfonylurea receptors and mechanism of sulfonylurea action. Exp Clin Endocrinol Diabetes 104:1-9

Perley MJ, Kipnis DM (1967) Plasma insulin responses to oral and intravenous glucose: studies in normal and diabetic subjects. J Clin Invest 46:1954-1962

Pickel VM, Nirenberg MJ, Chan J, Mosckovitz R, Udenfriend S, Tate SS (1993) Ultrastructural localization of a neutral and basic amino acid transporter in rat kidney and intestine. Proc Natl Acad Sci USA 90:7779-7783

Pipeleers DG, Schuit FC, Van Schravendijk CF, Van de Winkel M (1985) Interplay of nutrients and hormones in the regulation of glucagon release. Endocrinology 117:817-823

Pralong WF, Spat A, Wollheim CB (1994) Dynamic pacing of cell metabolism by intracellular Ca^{2+} transients. J Biol Chem 269:27310-27314

Raybould HE, Zittel TT (1995) Inhibition of gastric motility induced by intestinal glucose in awake rats: role of Na^{+}-glucose co-transporter. Neurogastroenterol Motil 7:9-14

Reimann F, Gribble FM (2002) Glucose-sensing in glucagon-like peptide-1 secreting cells. Diabetes 51:2757-2763

Reimer MK, Pacini G, Ahren B (2003) Dose-dependent inhibition by ghrelin of insulin secretion in the mouse. Endocrinology 144:916-921

Rettig J, Neher E (2002) Emerging roles of presynaptic proteins in Ca^{++}-triggered exocytosis. Science 298:781-785

Rindi G, Ratineau C, Ronco A, Candusso ME, Tsai M, Leiter AB (1999) Targeted ablation of secretin-producing cells in transgenic mice reveals a common differentiation pathway with multiple enteroendocrine cell lineages in the small intestine. Development 126:4149-4156

Rivera C, Voipio J, Payne JA, Ruusuvuori E, Lahtinen H, Lamsa K, Pirvola U, Saarma M, Kaila K (1999) The K^{+}/Cl^{-} co-transporter KCC2 renders GABA hyperpolarizing during neuronal maturation. Nature 397:251-255

Roche E, Maestre I, Martin F, Fuentes E, Casero J, Reig JA, Soria B (2000) Nutrient toxicity in pancreatic beta-cell dysfunction. J Physiol Biochem 56:119-128

Rojas E, Hidalgo J, Carroll PB, Li MX, Atwater I (1990) A new class of calcium channels activated by glucose in human pancreatic beta-cells. FEBS Lett 261:265-270

Ronner P, Matschinsky FM, Hang TL, Epstien AJ, Buettger C (1993) Sulfonylurea-binding sites and ATP-sensitive K^+ channels in alpha-TC glucagonoma and beta-TC insulinoma cells. Diabetes 42:1760-1772

Ronner P, Naumann CM, Friel E (2001) Effects of glucose and amino acids on free ADP in betaHC9 insulin-secreting cells. Diabetes 50:291-300

Rorsman P, Berggren P-O, Bokvist K, Ericson H, Mohler H, Ostenson CG, Smith PA (1989) Glucose-inhibition of glucagon secretion involves activation of $GABA_A$-receptor chloride channels. Nature 341:233-236

Rorsman P, Hellman B (1988) Voltage-activated currents in guinea pig pancreatic alpha 2 cells. Evidence for Ca^{2+}-dependent action potentials. J Gen Physiol 91:223-242

Rowe IC, Treherne JM, Ashford M (1996) Activation by intracellular ATP of a potassium channel in neurones from rat basomedial hypothalamus. J Physiol 490, 97-113

Sandstrom P-E (1990) Bumetanide reduces insulin release by a direct effect on the pancreatic β-cell. Eur J Pharmacol 187:377-383

Sandstrom P-E, Sehlin J (1988) Furosemide reduces insulin release by inhibition of Cl and Ca fluxes in β-cells. Am J Physiol 255:E591-E596

Sandstrom P-E, Sehlin J (1993) Evidence for separate Na+, K+, Cl- and K+, Cl- cotransport systems in mouse pancreatic β-cells. Eur J Pharmacol 238:403-405

Sato Y, Nenquin M, Henquin J-C (1998) Relative contribution of Ca^{2+}-dependent and Ca^{2+}-independent mechanisms to the regulation of insulin secretion by glucose. FEBS Lett 421:115-119

Schuit FC, Huypens P, Heimberg H, Pipeleers DG (2001) Glucose sensing in pancreatic β-cells. A model for the study of other glucose-regulated cells in gut, pancreas and hypothalamus. Diabetes 50:1-11

Schwartz MW, Woods SC, Porte D, Seeley RJ, Baskin DG (2000) Central nervous system control of food intake. Nature 404:661-671

Scrocchi LA, Brown TJ, MaClusky N, Brubaker PL, Auerbach AB, Joyner AL, Drucker DJ (1996) Glucose intolerance but normal satiety in mice with a null mutation in the glucagon-like peptide 1 receptor gene. Nature Medicine 2:1254-1258

Segal MB (2001) Transport of nutrients across the choroids plexus. Microsc Res Tech 52:38-48

Seghers V, Nakasaki M, DeMayo F, Aguilar-Bryan L, Bryan J (2000) SUR1 knockout mice. J Biol Chem 275:9270-9277

Sehlin J (1978) Interrelationship between chloride fluxes in pancreatic islets and insulin release. Am J Physiol 235:E501-E508

Sehlin J (1987) Evidence for voltage-dependent Cl^- permeability in mouse pancreatic beta-cells. Biosci Rep 7:67-72

Sener A, Kawazu S, Hutton JC, Boschero AC, Devis G, Somers G, Herchuelz A, Malaisse WJ (1978) The stimulus-secretion coupling of glucose-induced insulin release. Effect of exogenous pyruvate on islet function. Biochem J 176:217-232

Sener A, Malaisse WJ (1976) Measurement of lactic acid in nanomolar amounts. Reliability of such a method as an index of glycolysis in pancreatic islets. Biochem Med 15:34-41

Sener A, Malaisse WJ (1980) L-leucine and a non-metabolized analogue activate pancreatic islet glutamate dehydrogenase. Nature 288:187-189

Sharma N, Crane A, Gonzalez G, Bryan J, Aguilar-Bryan L (2000) Familial hyperinsulinism and pancreatic beta-cell ATP-sensitive potassium channels. Kidney Int 57:803-808

Sidhu SS, Thompson DG, Warhurst G, Case RM, Benson RS (2000) Fatty acid-induced cholecystokinin secretion and changes in intracellular Ca^{2+} in two enteroendocrine cell lines, STC-1 and GLUTag. J Physiol 528:165-176

Silver IA, Erecinska M (1994) Extracellular glucose concentration in mammalian brain: continuous monitoring of changes during increased neuronal activity and upon limitation in oxygen supply in normo-, hypo- and hyperglycaemic animals. J Neurosci 14:5068-5076

Smith PA, Rorsman P, Ashcroft FM (1989) Modulation of dihydropyridine-sensitive Ca^{2+} channels by glucose metabolism in mouse pancreatic beta-cells. Nature 342:550-553

Spiller RC, Trotman IF, Adrian TE, Bloom SR, Misiewicz JJ, Silk DB (1988) Further characterisation of the 'ileal brake' reflex in man – effect of ileal infusion of partial digests of fat, protein, and starch on jejunal motility and release of neurotensin, enteroglucagon, and peptide YY. Gut 29:1042-1051

Stanley CA, Lieu YK, Hsu BY, Burlina AB, Greenberg CR, Hopwood NJ, Perlman K, Rich BH, Zammarchi E, Poncz M (1998) Hyperinsulinism and hyperammonemia in infants with regulatory mutations of the glutamate dehydrogenase gene. N Engl J Med 338:1352-1357

Stein DT, Esser V, Stevenson BE, Lane KE, Whiteside JH, Daniels MB, Chen S, McGarry JD (1996) Essentiality of circulating fatty acids for glucose-stimulated insulin secretion in the fasted rat. J Clin Invest 97:2728-2735

Stein DT, Stevenson BE, Chester MW, Basit M, Daniels MB, Turley SD (1997) The insulinotropic potency of fatty acids is influenced profoundly by their chain length and degree of saturation. J Clin Invest 100:398-403

Straub SG, James RF, Dunne MJ, Sharp GW (1998) Glucose activates both K(ATP) channel-dependent and K(ATP) channel-independent signalling pathways in human islets. Diabetes 47:758-763

Tal M, Liang Y, Najafi H, Lodish HF, Matschinsky FM (1992) Expression and function of GLUT-1 and GLUT-2 glucose transporter isoforms in cells of cultured rat pancreatic islets. J Biol Chem 267:17241-17247

Taylor IL, Byrne WJ, Christie DL, Ament ME, Walsh JH (1982) Effect of individual l-amino acids on gastric acid secretion and serum gastrin and pancreatic polypeptide release in humans. Gastroenterology 83:273-278

Tengholm A, McClenaghan N, Grappengiesser E, Gylfe E, Hellman B (1992) Glycine transformation of Ca^{2+} oscillations into a sustained increase parallels potentiation of insulin release. Biochim Biophys Acta 1137:243-247

Thams P, Capito K (1999) L-arginine stimulation of glucose-induced insulin secretion through membrane depolarization and independent of nitric oxide. Eur J Endocrinol 140:87-93

Thams P, Capito K (2001) Differential mechanisms of glucose and palmitate in augmentation of insulin secretion in mouse pancreatic islets. Diabetologia 44:738-746

Unger RH, Orci L (1981) Glucagon and the α-cell. N Engl J Med 304:1518-1524

Valentijn K, Valentijn JA, Jamieson JD (1999) Role of actin in regulated exocytosis and compensatory membrane retrieval: insights from an old acquaintance. Biochem Biophys Res Commun 266:652-661

Vardi N, Zhang LL, Payne JA, Sterling P (2000) Evidence that different cation chloride cotransporters in retinal neurons allow opposite responses to GABA. J Neurosci 20:7657-7663

Velho G, Blanche H, Vaxillaire M, Bellanne-Chantelot C, Pardinin VC, Timsit J, Passa P, Dechamps I, Robert JJ, Weber IT, Marotta D, Pilkis SJ, Lipkind GM, Bell GI, Froguel PH (1997) Identification of 14 new glucokinase mutations and description of the clinical profile of 42 MODY-2 families. Diabetologia 40:217-224

Warnotte C, Gilon P, Nenquin M, Henquin JC (1994) Mechanisms of the stimulation of insulin release by saturated fatty acids. A study of palmitate effects in mouse beta-cells. Diabetes 43:703-711

Yang X-j, Kow L-M, Funabashi T, Mobbs CV (1999) Hypothalamic glucose sensor. Similarities to and differences from pancreatic β-cell mechanisms. Diabetes 48:1763-1772

Yoshimoto Y, Fukuyama Y, Horio Y, Inanobe A, Gotoh M, Kurachi Y (1999) Somatostatin induces hyperpolarization in pancreatic islet α-cells by activating a G protein-gated K^+ channel. FEBS Lett 444:265-269

Zhao C, Wilson MC, Schuit F, Halestrap AP, Rutter GA (2001) Expression and distribution of lactate/monocarboxylate transporter isoforms in pancreatic islets and the exocrine pancreas. Diabetes 50:361-366

Zhou YP and Grill V (1995) Long term exposure to fatty acids and ketones inhibits B-cell functions in human pancreatic islets of Langerhans. J Clin Endocrinol Metab 80:1584-1590

Abbreviations

$[Ca^{2+}]_i$: cytosolic calcium concentration
CaR: calcium-sensing receptor
CCK: cholecystokinin
EC cells: enterochromaffin cells
E_{Cl}: chloride equilibrium potential
EEC: enteroendocrine cells
FFA: free fatty acids
GABA: γ-aminobutyric acid
GDH: glutamate dehydrogenase
GIP: gastric inhibitory peptide
GLP-1: glucagon-like peptide-1
GLUT-2: glucose transporter 2
GR: glucose responsive
GS: glucose sensitive
KCC: K^+/Cl^- co-transporter
LH: lateral hypothalamus
MCT: monocarboxylate transporter
NCC: Na^+/Cl^- co-transporter
NKCC: $Na^+/K^+/2Cl^-$ co-transporter
PHHI: persistent hyperinsulinaemic hypoglycaemia in infancy
PKA: cyclic AMP dependent protein kinase
SGLT-1: sodium-coupled glucose transporter 1
SNARE: soluble N-ethylmaleimide sensitive factor attachment protein receptor
SUR: sulphonylurea receptor

VMH: ventromedial hypothalamus
VSAC: volume-sensitive anion channel

5 Nutrient signaling through mammalian GCN2

Scot R. Kimball, Tracy G. Anthony, Douglas R. Cavener, and Leonard S. Jefferson

Abstract

mGCN2 is the mammalian ortholog of the protein kinase Gcn2p that is activated in *Saccharomyces cerevisiae* in response to nutrient starvation. In mammalian cells in culture, mGCN2 is also activated by deprivation of essential amino acids. However, in animals *in vivo*, mGCN2 is not activated by physiological changes in plasma amino acid concentrations such as occur in response to feeding. Instead, mGCN2 is activated in response to feeding a diet lacking single essential amino acids, suggesting that imbalanced plasma essential amino acid levels are involved in the response. The only known substrate for mGCN2 is the α-subunit of the translation initiation factor, eIF2. Hyperphosphorylation of eIF2α represses the translation of most mRNAs. However, in both yeast and mammals, amino acid deprivation results in only partial phosphorylation of eIF2α, such that the translation of most mRNAs is incompletely repressed. Moreover, the translation of a few mRNAs is enhanced by eIF2α phosphorylation, in particular translation of the mRNAs encoding the transcription factors Gcn4p and ATF4 is stimulated in yeast and mammals, respectively, in response to amino acid deprivation. In each case, the genes induced by the transcription factors provide a mechanism for relieving the stress induced by nutrient starvation.

5.1 Introduction

The protein kinase referred to as general control non-derepressing (GCN)2 was first identified in *Saccharomyces cerevisiae* (*S. cerevisiae*) as a gene product required for induction of the general control response that is initiated in cells starved for one or more amino acids (Hinnebusch 1985). During the general control response, translation of the mRNA encoding the transcription factor Gcn4p, which is normally repressed, is upregulated, resulting in increased production of the protein (reviewed in Hinnebusch and Natarajan 2002). Gcn4p enhances the transcription of a plethora of genes encoding proteins involved in amino acid biosynthesis, resulting in increased intracellular production of amino acids. In *S. cerevisiae* lacking GCN2, amino acid starvation has no effect on GCN4 mRNA translation and the general control response is not activated. Recently, orthologs of GCN2 were identified in higher eukaryotes including *Drosophila melanogaster* (Olsen et al. 1998; Santoyo et al. 1997) and mouse (Berlanga et al. 1999; Sood et al. 2000).

Topics in Current Genetics, Vol. 7
J. Winderickx, P.M. Taylor (Eds) Nutrient-Induced Responses in Eukaryotic Cells

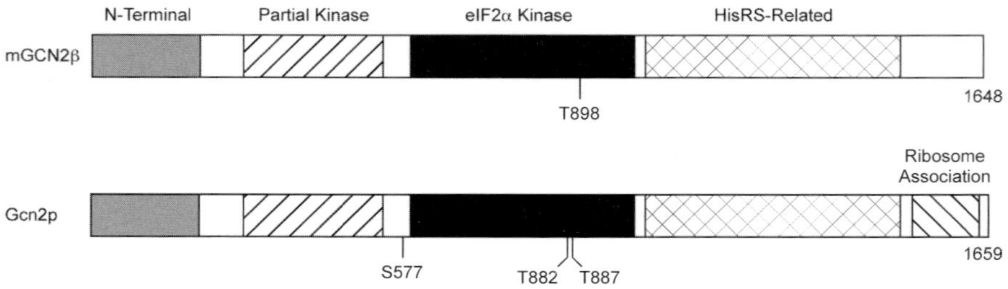

Fig. 1. Domain organization of mammalian and *S. cerevisiae* GCN2 protein kinases. The various domains of the β-isoform of mGCN2 and Gcn2p are denoted by shaded boxes and the number of amino acids encoded by the respective mRNAs are listed below and to the right. N-terminal - a domain exhibiting sequence homology between mGCN2β and Gcn2p, in *S. cerevisiae* the domain is involved in the association of Gcn2p with Gcn1p and Gcn20p; partial kinase – a domain that is required for eIF2α kinase activity, but is missing amino acid residues required for ATP binding and catalysis; eIF2α kinase – the authentic eIF2α protein kinase domain; HisRS-related – a domain exhibiting sequence homology to histidyl-tRNA synthetase; ribosome association – a domain in Gcn2p that is required for stable association of the protein with ribosomes. Phosphorylation sites described in the text are denoted as vertical lines below the box representing the protein.

The purpose of this article is to provide a summary of current knowledge on the mechanisms involved in regulating the activity of the mammalian ortholog of GCN2 (mGCN2) and the consequences of its activation.

5.2 Mechanism of GCN2 activation

5.2.1 Gcn2p

Based on the observation that a domain (HisRS domain) in the carboxy-terminal half of Gcn2p exhibits sequence homology to histidyl-tRNA synthetase (Fig. 1), it was proposed that binding of uncharged tRNA to Gcn2p rather than amino acid deprivation *per se* might activate Gcn2p (Wek et al. 1989). It is important to note that although the HisRS domain is most similar in sequence to histidyl-tRNA synthetase, Gcn2p binds several deacylated tRNAs with similar affinities (Dong et al. 2000), thus providing a mechanism through which deprivation of amino acids other than histidine might activate Gcn2p. Another study showing that Gcn2p is activated in cells containing a defective aminoacyl-tRNA synthetase, even in the presence of the cognate amino acid (Wek et al. 1995), supports the idea that uncharged tRNA activates Gcn2p. The same study provides evidence that uncharged tRNA binds to the HisRS domain of Gcn2p. However, the mechanism through which binding of uncharged tRNA to Gcn2p causes activation of the protein kinase domain has remained a mystery until recently. In a study by Dong and coworkers (Dong et al. 2000), it was shown that a C-terminal fragment of Gcn2p containing the HisRS domain binds to the isolated protein kinase domain *in vitro*.

Importantly, uncharged tRNA modulates the interaction between the two domains, suggesting that in the holoprotein, the HisRS domain binds to the protein kinase domain resulting in inhibition of protein kinase activity through steric hinderance. Based on these results, Dong and co-workers (Dong et al. 2000) proposed a model where upon binding of uncharged tRNA to the HisRS domain, the interaction between the HisRS and protein kinase domains is disrupted, freeing the protein kinase domain so that it can interact with and phosphorylate its substrate, the α-subunit of the translation initiation factor eukaryotic initiation factor (eIF)2 (Dever et al. 1992). Phosphorylation of the α-subunit of eIF2 (eIF2α) enhances the translation of the GCN4 mRNA resulting in induction of the general control response (reviewed in Hinnebusch 1997).

A more recent study (Qiu et al. 2002) describes a modified version of the activation model proposed by Dong et al. (Dong et al. 2000). In the study by Qiu et al. (Qiu et al. 2002), two point mutations in the protein kinase domain of Gcn2p were identified that bypass the requirement for uncharged tRNA in the activation of the protein. Based on the location of the residues in the predicted three-dimensional structure of the protein and the effect of the point mutations on the predicted structure, it was proposed that binding of uncharged tRNA to the HisRS domain results in a conformational change in the protein kinase domain, leading to its activation (Qiu et al. 2002). Thus, binding of uncharged tRNA to the HisRS domain may serve two functions: one function would be to alter the binding of the HisRS domain to the protein kinase domain and the second would be to cause a structural rearrangement in the protein kinase domain, leading to activation.

5.2.2 mGCN2

To date, no mechanistic studies have been performed using mGCN2. However, because like Gcn2p, mGCN2 has a domain in the carboxy-terminal portion of the protein that exhibits homology to histidyl-tRNA synthetase (Berlanga et al. 1999; Sood et al. 2000), it is likely that its kinase activity might also be stimulated through the association of uncharged tRNA with the HisRS domain. In this regard, in a variety of mammalian cell types, deprivation of single essential amino acids results in hyperphosphorylation of eIF2α (reviewed in Jefferson and Kimball 2003). Moreover, in response to leucine deprivation, mGCN2 is phosphorylated on Thr898, a residue located in the activation loop of the protein kinase domain (Harding et al. 2000). In Gcn2p, autophosphorylation of two residues in the activation loop, Thr882 and Thr887, is required for activation of protein kinase activity (Qiu et al. 2002), suggesting that phosphorylation of Thr898 on mGCN2, a residue that aligns with Thr882 in the Gcn2p sequence, likely reflects activation of its protein kinase domain. Interestingly, residue 903 (equivalent to Thr887 in Gcn2p) in mGCN2 is also a Thr; whether or not Thr903 is phosphorylated during activation of mGCN2 is unknown. Finally, phosphorylation of both eIF2α (Clemens et al. 1987) and mGCN2 at Thr898 (Harding et al. 2000) is increased in Chinese hamster ovary (CHO) cells containing a temperature-sensitive mutation in the leucyl-tRNA synthetase when the cells are incubated at the non-permissive temperature. In such cells, incubation at the non-permissive temperature inhibits leucyl-tRNA synthetase resulting in the accumulation of uncharged leucyl-tRNA which subsequently activates mGCN2. Overall, the available evidence supports a model whereby the binding of uncharged tRNA to the HisRS domain of mGCN2 results in activation of the protein kinase domain.

In contrast to yeast where Gcn2p is the only eIF2α kinase, mammalian cells have at least four protein kinases that phosphorylate the same residue, Ser51, on eIF2α (reviewed in Hinnebusch 2000). It is therefore relevant to ask whether mGCN2 is the only eIF2α kinase that is regulated by amino acid deprivation in mammalian cells. One approach that has been used to address this question involves the use of mouse embryonic stem (MES) cells containing a chromosomal disruption in the mGCN2 gene (MES-mGCN2-/- cells, Harding et al. 2000; Zhang et al. 2002). In such studies, eIF2α is hyperphosphorylated in wild type MES cells deprived of leucine for 15 min. In contrast, in MES-mGCN2-/- cells, eIF2α phosphorylation does not increase, even after incubation for three hours in medium lacking leucine (Zhang et al. 2002). Interestingly, during longer periods of leucine deprivation (e.g. 6-12 h) eIF2α is hyperphosphorylated even in MES-mGCN2-/- cells (Zhang et al. 2002), suggesting that a kinase other than mGCN2 is activated during prolonged amino acid deprivation. However, the kinase that is activated (or phosphatase that is repressed) during prolonged leucine deprivation has not been identified.

An increase in eIF2α phosphorylation is also observed in livers of rats (Kimball and Jefferson 1991) and mice (Zhang et al. 2002) perfused *in situ* with medium lacking single, essential amino acids. However, as reported for MES-mGCN2-/- cells, eIF2α phosphorylation does not increase in livers of mice with a chromosomal disruption of mGCN2 (mGCN2-/- mice) perfused with medium lacking histidine (Zhang et al. 2002). In fact, even in the presence of a complete amino acid mixture, eIF2α phosphorylation is lower in perfused livers from GCN2-/- mice than in livers from wild type mice, suggesting that under the conditions used for the perfusion, GCN2 accounts for a substantial proportion of basal eIF2α phosphorylation.

The results of the studies described above strongly suggest that the phosphorylation of eIF2α that occurs in response to amino acid deprivation of mammalian cells is mediated by activation of GCN2. However, in animals *in vivo*, the plasma concentration of amino acids is maintained at fairly constant values, and it is highly unlikely that cells would ever be exposed to a complete deprivation of an essential amino acid. In this regard, the increase in plasma amino acids that occurs in response to feeding a protein-containing meal to a fasted rat has no effect on eIF2α phosphorylation in liver or skeletal muscle (Yoshizawa et al. 1997). Thus, mGCN2 is not activated by physiological changes in plasma amino acid concentrations such as occurs in response to feeding. In contrast, feeding a complete meal lacking a single essential amino acid, e.g. leucine or tryptophan, significantly increases phosphorylation of eIF2α on Ser51 in rat liver (Anthony et al. 2001). An increase in eIF2α phosphorylation is not observed in livers of rats fed a complete meal lacking the nonessential amino acid, glycine, suggesting that it is an imbalance in the plasma concentration of essential amino acids that activates mGCN2. The mechanism through which imbalanced, but not balanced, changes in plasma essential amino acids promotes activation of mGCN2 is unknown. However, it is noteworthy that although the concentration of the amino acid that is lacking from the meal does not fall to zero, the concentration does decrease below the value observed in a fasted animal. Thus, it may be that simply decreasing the plasma concentration of one or more essential amino acids below a certain level results in activation of mGCN2.

To investigate the potential developmental and physiological functions of mGCN2 at the organismal level, a knockout mutation of mGCN2 (mGCN2 KO) was generated in mice (Zhang et al. 2002). Mice that are homozygous for the mGCN2 KO mutation (mGCN2-/-) in a mixed C57BL/6J and 129 SvEvTac are viable and fertile and do not display any overt morphological or physiological defects. The impact of amino acid limitation on the development of the mGCN2 KO mice were investigated in the context of a simple genetic scheme whereby heterozygous males (mGCN2+/-) were mated to homozygous mGCN2 mutant females (mGCN2-/-). Pregnant mGCN2-/- females bearing an expected 1:1 ratio of mGCN2-/- and mGCN2+/- pups were subjected to diets deficient in leucine, tryptophan, or glycine during the early phase of gestation and the differential effects on prenatal viability was assessed as deviations from this expected 1:1 ratio in live-born pups. Leucine deprivation exhibited the most profound effect as shown by a 63% reduction in the expected number of viable mGCN2-/- pups (Zhang et al. 2002). Although the developmental and molecular basis for this finding is unknown, these results are the first indication that mGCN2 is likely to play an important role in prenatal nutrition of mammals.

To establish whether or not mGCN2 is required for the increase in eIF2α phosphorylation that occurs in livers of rodents fed a complete meal lacking an essential amino acid, mGCN2-/- mice and their wild type litter mates (GCN2+/+ mice) were fed either a complete meal, a complete meal lacking glycine, or a complete meal lacking leucine. One hour after the commencement of feeding, eIF2α phosphorylation was measured in liver by Western blot analysis using an anti-phospho-eIF2α antibody that only recognizes eIF2α when it is phosphorylated on Ser51. As shown in Figure 2, feeding a complete meal or a meal lacking glycine had no effect on eIF2α phosphorylation in livers of fasted GCN2+/+ mice. However, feeding a complete meal lacking leucine increased eIF2α phosphorylation approximately 2.5-fold compared to the value observed in liver of fasted mice. In contrast, no change in eIF2α phosphorylation was observed in livers of GCN2-/- mice, regardless of the type of meal fed to the mice.

5.3 The mGCN2 substrate, eIF2α

mGCN2 is among three of the four known eIF2α kinases that have as their only known substrate the α-subunit of eIF2 (reviewed in Dever 1999). During the initiation phase of mRNA translation, eIF2 binds GTP and initiator methionyl-$tRNA_i$

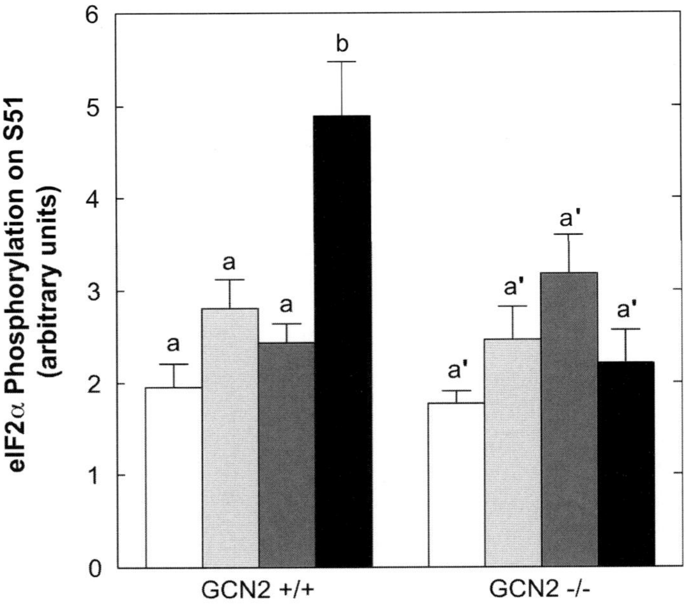

Fig. 2. Phosphorylation of eIF2α in livers of mice fed a diet lacking leucine is mediated by mGCN2. Mice were meal trained for 2 weeks and on the day of the study were either not fed (open bars), fed a complete meal (light gray bars), fed a complete meal lacking glycine (dark gray bars), or fed a complete meal lacking leucine (black bars) as described previously in a study using rats (Anthony et al. 2001). One hour after the commencement of feeding, livers were removed, homogenized, and subjected to Western blot analysis as previously described (Anthony et al. 2001). Phosphorylation of Ser51 on eIF2α was assessed using an anti-phospho-eIF2α antibody (BioSource International). Bars on the left represent results obtained using wild type (GCN2 +/+) mice and bars on the right represent results from mice containing a chromosomal disruption in the mGCN2 gene (GCN2 -/-). Bars not labeled with the same letter are significantly different from other results with the same mouse phenotype, $p<0.01$ by analysis of variance using the Tukey post-test.

(met-tRNA$_i$) to form a ternary complex which subsequently binds to the 40S ribosomal subunit to form a 43S preinitiation complex. During one of the last steps in translation initiation, the GTP bound to eIF2 is hydrolyzed to GDP and the eIF2•GDP binary complex is released from the 43S preinitiation complex. Prior to binding met-tRNA$_i$ and participating in another cycle of initiation, the GDP bound to eIF2 must be exchanged for GTP, a process mediated by eIF2B. However, when the α subunit of eIF2 is phosphorylated on Ser51, eIF2 is converted from a substrate into a competitive inhibitor of eIF2B (reviewed in Dever 1999). Thus, phosphorylation of eIF2α on Ser51 results in the effective sequestration of eIF2B into an inactive complex, reducing the translation of most mRNAs (Fig. 3).

Evidence that mGCN2 phosphorylates eIF2α on Ser51 is derived from studies using two different approaches. In the first approach, studies using anti-

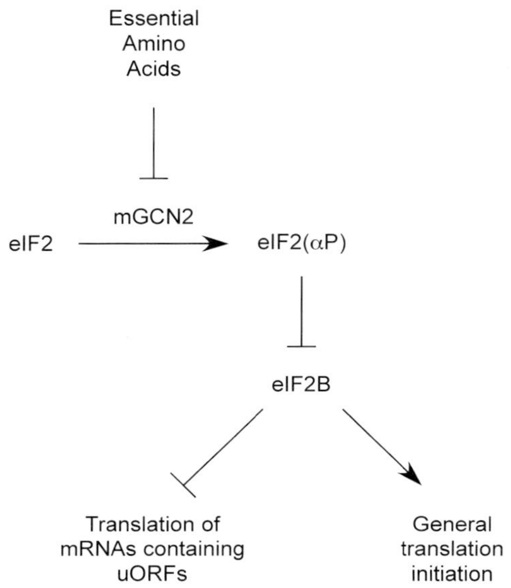

Fig. 3. Regulation of general and specific mRNA translation by activation of mGCN2. In mammals, essential amino acids maintain mGCN2 in an inactive state. When mGCN2 is activated by deprivation for one or more essential amino acids, it phosphorylates the α-subunit of eIF2 which forms a stable complex with the guanine nucleotide exchange factor eIF2B. Sequestration of eIF2B by phosphorylated eIF2 results in reduced translation of most mRNAs, but enhanced translation of specific mRNAs that have upstream open reading frames (uORFs) in their 5'-untranslated regions. Further details are provided in the text.

phospho-peptide antibodies that recognize eIF2α only when it is phosphorylated on Ser51 have shown that eIF2α(Ser51) is phosphorylated in wild type MES cells deprived of leucine but not MES-mGCN2-/- cells (Harding et al. 2000; Zhang et al. 2002). Second, expression of mGCN2 in *S. cerevisiae* lacking Gcn2p results in a slow-growth phenotype associated with phosphorylation of eIF2α on Ser 51 (Sood et al. 2000). Replacement of Ser51 in wild type eIF2α with alanine prevents both eIF2α phosphorylation and the slow-growth phenotype caused by expression of mGCN2. Results from the latter study (Sood et al. 2000) also provide evidence that phosphorylation of eIF2α(Ser51) induces the slow growth phenotype by inhibiting eIF2B activity. In that study, mGCN2 was expressed in a yeast strain that has a mutation in the δ-subunit of eIF2B that renders the guanine nucleotide exchange activity of eIF2B resistant to eIF2α phosphorylation. In contrast to a strain containing wild type eIF2Bδ, mGCN2 does not induce a slow growth phenotype in cells expressing mutant eIF2Bδ, suggesting that in such cells, phosphorylation of eIF2α does not inhibit the guanine nucleotide exchange activity of eIF2B.

The phosphorylation of eIF2α on Ser51 that occurs in livers of mice fed meals lacking leucine also results in inhibition of eIF2B activity. As shown in Fig. 4, feeding either a complete meal or a meal lacking glycine to GCN2+/+ mice has no

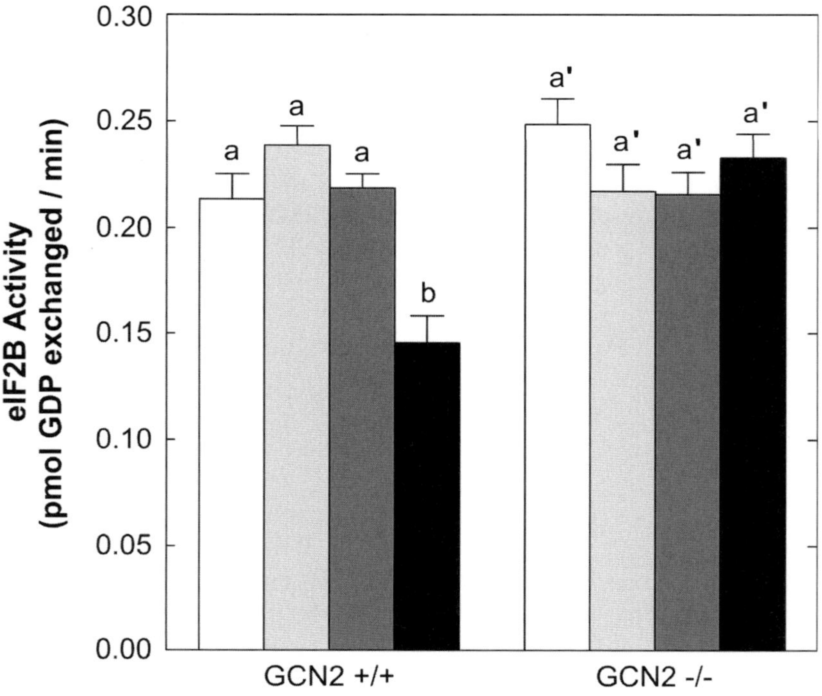

Fig. 4. Inhibition of eIF2B activity in livers of mice fed a diet lacking leucine is mediated by mGCN2. Mice were either not fed (open bars), fed a complete meal (light gray bars), fed a complete meal lacking glycine (dark gray bars), or fed a complete meal lacking leucine (black bars) as described in the legend to Fig. 2. Livers were homogenized and the guanine nucleotide exchange activity of eIF2B was measured using eIF2•[^{3}H]GDP as substrate as described previously (Kimball et al. 1989). Results are expressed as pmol GDP exchanged / min. Bars on the left represent results obtained using wild type (GCN2 +/+) mice and bars on the right represent results from mice containing a chromosomal disruption in the mGCN2 gene (GCN2 -/-). Bars not labeled with the same letter are significantly different from other results with the same mouse phenotype, $p<0.01$ by analysis of variance using the Tukey post-test.

effect on eIF2B activity, but feeding GCN2+/+ mice a meal lacking leucine results in a significant reduction in eIF2B activity in liver homogenates. In contrast, in liver homogenates from GCN2-/- mice, eIF2B activity is not affected by feeding any of the meals. Thus, eIF2B activity (Fig. 4) is directly proportional to changes in eIF2α phosphorylation (Fig. 2). Moreover, changes in eIF2α phosphorylation and eIF2B activity in response to feeding a meal lacking leucine is dependent upon mGCN2.

5.4 mGCN2 interacting proteins: GCN1 and GCN20

In *S. cerevisiae*, maximal activation of Gcn2p in response to amino acid deprivation requires two other proteins, Gcn1p and Gcn20p, that are found associated with 80S ribosomes and polysomes in growing cells (Marton et al. 1993; Vazquez de Aldana et al. 1995). Gcn1p and Gcn20p form a stable complex (Vazquez de Aldana et al. 1995) that binds to Gcn2p through the association of Gcn20p with an amino-terminal domain in Gcn2p (Garcia-Barrio et al. 2000). Interestingly, in mouse, three different mRNAs encoding mGCN2 have been identified that differ only in the region of their N-termini. (Sood et al. 2000). Although the functional significance is unknown, only one of the mGCN2 isoforms exhibits sequence homology to the domain in Gcn2p that binds to Gcn20p. This isoform, termed mGCN2β, is ubiquitously expressed in mouse tissues and will phosphorylate yeast eIF2α when expressed in *S. cerevisiae*. Whether or not mGCN2β, or either of the other two isoforms, binds to mGcn20p is unknown.

Although Gcn1p is not required for binding of Gcn20p to Gcn2p, the binding is enhanced in the presence of Gcn1p. In cells containing loss of function mutations in GCN1 (Marton et al. 1993) or GCN20 (Vazquez de Aldana et al. 1995) amino acid deprivation either does not stimulate eIF2α phosphorylation or eIF2α phosphorylation is greatly attenuated, respectively, without affecting Gcn2p expression. Mutations in GCN1 or GCN20 have no effect on Gcn2p activity *per se*, because the protein kinase activity of Gcn2p measured in Gcn2p immunoprecipitates is unaffected by such mutations. Thus, loss of function mutations in GCN1 and GCN20 do not directly inhibit mGcn2p, but instead prevent its activation during periods of amino acid starvation.

Both Gcn1p and Gcn20p have domains exhibiting sequence similarity to eukaryotic elongation factor (eEF)3. Because Gcn1p and Gcn20p bind to actively translating ribosomes and exhibit sequence homology with eEF3, it has been proposed that Gcn1p and Gcn20p function in Gcn2p activation by transferring uncharged tRNA from the acceptor (A) site on the ribosome to the HisRS domain of Gcn2p (reviewed in Hinnebusch 2000). In support of this model, point mutations in GCN2 that bypass the requirement for binding of uncharged tRNA in activation of Gcn2p also circumvent the need for association with Gcn1p and Gcn20p (Qiu et al. 2002).

In addition to the Gcn2p domain that binds to Gcn20p, a second domain referred to as the double-stranded RNA-binding domain (DRBD) is important in mediating the association of Gcn2p with ribosomes (Zhu and Wek 1998). Mutation of key lysine residues in this domain severely represses the association of Gcn2p with polysomes and free ribosomal subunits as well as impairs the enhanced Gcn4p mRNA translation normally induced by amino acid deprivation. Thus, in *S. cerevisiae*, the DRBD domain is critically important in activation of Gcn2p in response to nutrient deprivation. In mGCN2, there is little sequence homology with Gcn2p in the region of the DRBD (Sood et al. 2000). However, mGCN2 does contain clusters of basic residues in this region that may function in a manner similar to the lysine residues in the DRBD domain of Gcn2p.

Although little has been done to characterize the protein, a human ortholog of Gcn1p has been identified and is referred to as hsGCN1 (Marton et al. 1997). In addition to having sequence homology with Gcn1p, hsGCN1 also exhibits homology to eEF3. In fact, the region of greatest similarity between hsGCN1 and Gcn1p occurs over the same region where Gcn1p has highest similarity to eEF3. The mRNA encoding hsGCN1 is widely expressed in human tissues, and in particular is highly expressed in skeletal muscle, ovary, and testes. Interestingly, the phosphorylation of eIF2α that occurs in yeast expressing mGCN2 does not require Gcn1p (Sood et al. 2000), suggesting that, in contrast to Gcn2p, the mammalian kinase may be able to function independently of GCN1.

Positive identification of a mammalian equivalent of Gcn20p has not yet been accomplished, although a recent study (Tyzack et al. 2000) suggests that a protein termed ABC50 may be the elusive human ortholog. Thus, in addition to sequence homology to Gcn20p, ABC50 also exhibits homology to eEF3. Moreover, like Gcn20p, ABC50 binds to 40S and 60S ribosomal subunits and this association is enhanced in the presence of ATP (Tyzack et al. 2000). ABC50 also binds to eIF2 and stimulates the formation of the active eIF2•GTP•met-tRNA$_i$ ternary complex. However, ABC50 does not substitute for Gcn20p in a GCN20-deficient strain of *S. cerevisiae* and the N-termini of ABC50 and Gcn20p, the region of Gcn20p that binds to Gcn2p, are the regions of the proteins that exhibit the least similarity. Thus, it has been proposed (Tyzack et al. 2000) that in mammals, two or more proteins may perform the functions of Gcn20p and that ABC50 may work in conjunction with one or more other proteins to fulfill the role played by Gcn20p in *S. cerevisiae*.

5.5 Phosphorylation of eIF2α promotes specific alterations in mRNA translation

As noted above, the phosphorylation of eIF2α that occurs as a result of mGCN2 activation results in a repression of translation of most mRNAs. However, mRNAs encoding a few specific proteins continues to be translated under such conditions. The best characterized of these proteins is Gcn4p. Like most mRNAs, translation of the message encoding Gcn4p is thought to occur through the binding of the 43S preinitiation complex (i.e. the 40S ribosomal subunit associated with the eIF2-GTP-met-tRNA$_i$ ternary complex) to the m^{7}GTP cap structure at the 5'-end of the mRNA (reviewed in Hinnebusch and Natarajan 2002). The domain between the m7GTP cap and the AUG start codon of the main open reading frame is often referred to by the misnomer 5'-untranslated region (5'-UTR). Because of its widespread usage, the term 5'-UTR will be used herein to denote this region of the mRNA molecule. After binding to the 5'-end of the mRNA, the 43S preinitiation complex scans down the message until it reaches an AUG start codon. In most mRNAs, the first AUG represents the start site for the protein encoded by the mRNA. However, unlike most mRNAs, the message encoding Gcn4p contains four short open reading frames upstream of the authentic AUG start codon

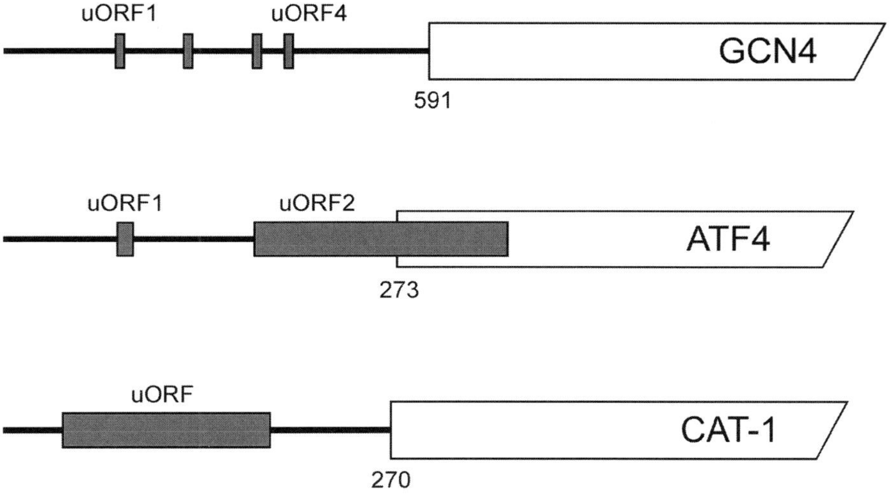

Fig. 5. Upstream open reading frames in the 5'-UTRs of the mRNAs encoding Gcn4p, ATF4 and CAT-1. The 5'-UTR of each mRNA is denoted by a black line and its length is listed below the mRNA. The partial coding regions for the individual proteins are shown as white boxes and uORFs are denoted as gray boxes. Note that for presentation purposes, the 5'-UTR of Gcn4p is not shown to scale.

(reviewed in Hinnebusch 1997; Hinnebusch and Natarajan 2002). Thus, in the Gcn4p mRNA, essentially all of the 40S ribosomal subunits initiate translation at upstream open reading frame (uORF)1 (Fig. 5). During the initiation process, the eIF2•GDP complex is released from the 40S ribosomal subunit and the 60S ribosomal subunit joins the 40S subunit to form the active 80S ribosome. After translating uORF1, approximately 50% of the ribosomes dissociate completely from the mRNA. The remainder of the ribosomes separate into 40S and 60S ribosomal subunits, but the 40S ribosomal subunit is thought to remain attached to the mRNA and continue to traverse the 5'-UTR. Under normal growth conditions, most of these 40S ribosomal subunits re-accumulate eIF2•GTP•met-tRNA$_i$ ternary complex, re-initiate at uORF4, and then dissociate from the mRNA. The few 40S ribosomal subunits that bypass uORF4 initiate at the AUG start codon of the open reading frame encoding Gcn4p, resulting in the very low basal level of expression observed in growing cells. In cells starved for amino acids, less eIF2•GTP•met-tRNA$_i$ ternary complex is available for translation initiation because, as described above, eIF2(αP) inhibits eIF2B, resulting in less eIF2•GTP being available to form the ternary complex. Thus, in amino acid starved cells, fewer 43S preinitiation complexes are available to bind to any mRNA, including the mGcn4p mRNA. However, it has been estimated that, in contrast to fed cells where most of the 40S ribosomal subunits re-accumulate ternary complex before reaching uORF4, under starvation conditions only about 50% of the 40S ribosomal subunits that are traversing the 5'-UTR of the Gcn4p mRNA are able to re-accumulate ternary complex before reaching uORF4 (Hinnebusch and Natarajan 2002). The 40S ribosomal subunits that do not acquire ternary complex before reaching uORF4 continue to move along the 5'-UTR and many of the subunits re-accumulate ternary complex before reaching the authentic Gcn4p start codon and initiate translation there. Thus, the reduced levels of eIF2•GTP•met-tRNA$_i$ ternary complex that result from activation of Gcn2p result in a 2 to 10-fold increase in translation of the GCN4 mRNA, depending upon the *S. cerevisiae* strain examined, compared to cells maintained in amino acid-replete medium (Hinnebusch and Natarajan 2002).

Although a mammalian ortholog of Gcn4p has not been identified, translation of mRNAs encoding other proteins have been shown to be regulated through a mechanism similar to that described for Gcn4p. For example, translation of the mRNA encoding activating transcription factor (ATF)4 (a.k.a. CREB-2) is upregulated in cells deprived of leucine, as assessed by an increase in the average number of ribosomes bound to ATF4 mRNAs (Harding et al. 2000). In contrast, enhanced ATF4 synthesis does not occur in MES-mGCN2-/- cells deprived of leucine, implicating activation of mGCN2 and phosphorylation of eIF2α in the response. Similar to the mRNA encoding Gcn4p, the mRNA encoding human ATF4 has multiple uORFs in its 5'UTR whereas the murine mRNA has two uORFs (Harding et al. 2000). The role of the uORFs in the regulation of ATF4 mRNA translation has been assessed by expressing in CHO-K1 cells a chimeric mRNA consisting of the murine ATF4 5'UTR and the coding region of luciferase (Harding et al. 2000). In that study, it was shown that the wild type ATF4 5'-UTR dramatically represses luciferase expression and that phosphorylation of eIF2α partially reverses the repression. Changing the AUG start codon of uORF2 to AUA prevents the repression of luciferase expression observed with the wild type uORF and eliminates translational regulation engendered by eIF2α phosphorylation. Deletion of the uORF has a similar, albeit less dramatic, effect. Although the mechanism for uORF-mediated translational control of ATF4 has not been elucidated, the results of the study by Harding and co-workers (Harding et al. 2000) strongly suggest that the uORFs in the 5'-UTR of the ATF4 mRNA are regulated by eIF2α phosphorylation in the same manner as is the mRNA encoding Gcn4p in *S. cerevisiae*.

Another example of an mRNA exhibiting enhanced translation in amino acid-deprived cells in a mGCN2-dependent manner is the mRNA encoding the cationic amino acid transporter (CAT)-1 (Fernandez et al. 2001; Fernandez et al. 2002; Fernandez et al. 2002a). Thus, CAT-1 expression is increased 58-fold in C6 rat glioma cells, a response that is obviated by expression of a kinase-dead variant of mGCN2 that acts as a dominant interfering kinase and prevents phosphorylation of eIF2α during starvation for amino acids (Fernandez et al. 2002). Similar to the control of Gcn4p and ATF4 expression, translation of the CAT-1 mRNA is regulated by regulatory elements in its 5'-UTR. However, unlike the mRNAs encoding Gcn4p and ATF4, the mRNA encoding CAT-1 has only a single uORF but also has a regulatory element referred to as an internal ribosome entry site (IRES), both of which are required for induction of CAT-1 expression by amino acid starvation (Fernandez et al. 2001; Fernandez et al. 2002; Fernandez et al. 2002a; Yaman et al. 2003). A model that has been proposed to describe the role of the uORF and IRES in regulating the translation of the CAT-1 mRNA (Yaman et al. 2003) posits that rather than directly modulating the translation of CAT-1 mRNA, eIF2 phosphorylation instead induces the synthesis of an IRES trans-acting factor (ITAF). In part, the requirement for an auxiliary factor is based on the observation that in C6 cells, phosphorylation of eIF2α is transient and returns to basal values within two hours of amino acid deprivation (Fernandez et al. 2001). In contrast, enhanced CAT-1 translation is not observed for at least six hours after removal of amino acids. The model also proposes that translation of the uORF permits ITAF to bind to the IRES leading to a change in its conformation which in turn allows binding of the ribosome to the 5'-UTR near the CAT-1 start codon and translate the message. In amino acid-replete cells, the uORF is translated, but because ITAF is not available, no binding of ribosomes to the IRES occurs. Thus, under non-starvation conditions, the uORF acts as a repressor of CAT-1 mRNA translation.

5.6 Other mechanisms for activating GCN2

In *S. cerevisiae*, Gcn4p expression is induced under conditions besides amino acid deprivation, including starvation for glucose (Yang et al. 2000) or purines (Rolfes and Hinnebusch 1993), or growth in medium containing ethanol (Rolfes and Hinnebusch 1993) or high salinity (Goossens et al. 2001). In each case, Gcn2p is required for Gcn4p induction. The mechanism through which purine limitation or growth on medium containing ethanol or high salt concentrations activate Gcn2p is unknown. However, a functional HisRS domain of Gcn2p is required for induction of Gcn4p synthesis, suggesting that uncharged tRNA is generated under each condition (reviewed in Hinnebusch and Natarajan 2002). An alternative explanation is that each of these conditions might promote dephosphorylation of Gcn2p on Ser577. Dephosphorylation of Ser577 increases the binding affinity of Gcn2p for uncharged tRNA, allowing for Gcn2p activation in the absence of increased uncharged tRNA content (Garcia-Barrio et al. 2002). In cells starved for glucose, cytoplasmic amino acid concentrations are reduced and the effects of glucose limitation on Gcn4p synthesis are attenuated by the addition of amino acids to the medium (Yang et al. 2000), suggesting that in part, glucose starvation is activating Gcn2p through the same mechanism as starvation for amino acids. However, unlike the response to amino acid deprivation, activation of Gcn2p by glucose starvation does not require Gcn20p and is largely independent of association of Gcn2p with ribosomes (Yang et al. 2000). Thus, glucose starvation may also cause activation of Gcn2p through mechanisms other than uncharged tRNA. Whether or not such conditions activate mGCN2 is largely unexplored.

Another condition that promotes the activation of both Gcn2p and mGCN2 is exposure of cells to irradiation with ultraviolet (UV) light. In mammalian cells exposed to UV light at 254 nm, enhanced eIF2α phosphorylation is observed as soon as one minute and reaches a maximum by 20 minutes post-exposure (Deng et al. 2002). UV-induced phosphorylation of eIF2α is observed in mouse embryo fibroblasts lacking either of the eIF2α kinases double-stranded RNA-activated protein kinase (PKR) or PKR-like endoplasmic reticulum-resident protein kinase

(PEK/PERK), but not in cells lacking mGCN2 (Deng et al. 2002). Moreover, UV irradiation induces phosphorylation of mGCN2 on Thr898 (Deng et al. 2002), a residue in the activation loop of the kinase that is phosphorylated upon activation of mGCN2 during leucine deprivation (Harding et al. 2000). These results suggest that exposure to UV irradiation activates mGCN2. However, it is unlikely that an increase in cellular content of uncharged tRNA is involved in the activation, because no change in tRNA acylation state is observed after UV irradiation (Deng et al. 2002). An alternative mechanism that has been proposed to explain the activation of mGCN2 is that UV irradiation promotes the crosslinking of mGCN2 to tRNA, which could result in activation of the kinase (Deng et al. 2002). It should be noted that another study has reached a different conclusion concerning the identity of the eIF2α kinase that is activated after exposure of cells to UV light. In the study by Wu et al. (Wu et al. 2002), expression of a dominant-interfering form of PERK blocked UV-induced phosphorylation of eIF2α, suggesting that PERK rather than mGCN2 is activated by UV irradiation. However, in the study by Wu et al. (Wu et al. 2002), different cell lines were used, and eIF2α phosphorylation was measured four hours after exposure to UV irradiation. Thus, mGCN2 may be activated shortly after exposure to UV irradiation and PERK may be activated later.

5.7 Summary

Many similarities exist between mGCN2 and Gcn2p in regards to their activation by amino acid deprivation and the ultimate response to their activation. However, many questions remain unanswered. For example, in rodents, feeding a meal lacking a single essential amino acid promotes eIF2α phosphorylation in an mGCN2-dependent manner. However, the mechanism through which imbalanced plasma levels of essential amino acids cause activation of mGCN2 is unknown. Moreover, although mGCN2 contains a domain that has sequence homology to histidyl-tRNA synthetase, it has not been shown that mGCN2 binds uncharged tRNA or that uncharged tRNA activates the kinase activity of the protein. In both yeast and mammals, a second protein kinase that responds to changes in amino acid availability is the target of rapamycin (referred to as TOR in yeast, mTOR in mammals). In yeast, TOR represses the activity of a type 2A protein phosphatase that normally dephosphorylates Ser577 on Gcn2p (Cherkasova and Hinnebusch 2003). Phosphorylation of Ser577 represses Gcn2p protein kinase activity and provides a possible mechanism for activation of Gcn2p independent of uncharged tRNA (Cherkasova and Hinnebusch 2003). Investigation of a potential link between mTOR and mGCN2 therefore seems worthy of investigation. Finally, although it is clear that amino acid availability modulates phosphorylation of Thr898 on mGCN2, evidence for phosphorylation of other sites, for example Thr903, is lacking.

Acknowledgements

The work described in this article that was performed in the author's laboratories was supported by grants DK15658 (LSJ), DK13499 (LSJ), and GM56957 (DRC) from the National Institutes of Health.

References

Anthony TG, Reiter AK, Anthony JC, Kimball SR, Jefferson LS (2001) Deficiency of essential dietary amino acids preferentially inhibits mRNA translation of ribosomal proteins in the liver of meal-fed rats. Am J Physiol 281:E430-E439

Berlanga JJ, Santoyo J, de Haro C (1999) Characterization of a mammalian homolog of GCN2 eukaryotic initiation factor 2α kinase. Eur J Biochem 265:754-762

Cherkasova VA, Hinnebusch AG (2003) Translational control by TOR and TAP42 through deposphorylation of eIF2α kinase GCN2. Genes Develop 17:859-872

Clemens MJ, Galpine A, Austin SA, Panniers R, Henshaw EC, Duncan R, Hershey JWB, Pollard JW (1987) Regulation of polypeptide chain initiation in Chinese hamster ovary cells with a temperature-sensitive leucyl-tRNA synthetase. Changes in phosphorylation of initiation factor eIF-2 and in the activity of the guanine nucleotide exchange factor GEF. J Biol Chem 262:767-771

Deng J, Harding H, Raught B, Gingras A, Berlanga J, Scheuner D, Kaufman R, Ron D, Sonenberg N (2002) Activation of GCN2 in UV-irradiated cells inhibits translation. Curr Biol 12:1279-1286

Dever TE, Feng L, Wek RC, Cigan AM, Donahue TF, Hinnebusch AG (1992) Phosphorylation of initiation factor 2α by protein kinase GCN2 mediates gene-specific translational control of *GCN4* in yeast. Cell 68:585-596

Dever TE (1999) Translation initiation: adept at adapting. Trends in biochemical sciences 24:398-403

Dong J, Qiu H, Garcia-Barrio M, Anderson J, Hinnebusch AG (2000) Uncharged tRNA activates GCN2 by displacing the protein kinase moiety from a bipartite tRNA-binding domain. Mol Cell 6:269-279

Fernandez J, Yaman I, Mishra R, Merrick WC, Snider MD, Lamers WH, Hatzoglou M (2001) Internal ribosome entry site-mediated translation of a mammalian mRNA is regulated by amino acid availability. J Biol Chem 276:12285-12291

Fernandez J, Yaman I, Merrick WC, Koromilas A, Wek RC, Sood R, Hensold J, Hatzoglou M (2002) Regulation of internal ribosome entry site-mediated translation by eukaryotic initiation factor-2α phosphorylation and translation of a small upstream open reading frame. J Biol Chem 277:2050-2058

Fernandez J, Yaman I, Sarnow P, Snider MD, Hatzoglou M (2002a) Regulation of internal ribosomal entry site-mediated translation by phosphorylation of the translation initiation factor eIF2alpha. J Biol Chem 277:19198-19205

Garcia-Barrio M, Dong J, Ufano S, Hinnebusch AG (2000) Association of GCN1-GCN20 regulatory complex with the N-terminus of eIF2α kinase GCN2 is required for GCN2 activation. EMBO J 19:1887-1899

Garcia-Barrio M, Dong J, Cherkasova VA, Zhang X, Zhang F, Ufano S, Lai R, Qin J, Hinnebusch AG (2002) Serine 577 is phosphorylated and negatively affects the tRNA binding and eIF2α kinase activities of GCN2. J Biol Chem 277:30675-30683

Goossens A, Dever TE, Pascual-Ahuir A, Serrano R (2001) The protein kinase Gcn2p mediates sodium toxicity in yeast. J Biol Chem 76:30753-30760

Harding HP, Novoa I, Zhang Y, Zeng H, Wek RC, Schapira M, Ron D (2000) Regulated translation initiation controls stress-induced gene expression in mammalian cells. Mol Cell 6:1099-1108

Hinnebusch AG (1985) A hierarchy of trans-acting factors modulates translation of an activator of amino acid biosynthetic genes in *Saccharomyces cerevisae*. Mol Cell Biol 5:2349-2360

Hinnebusch AG (1997) Translational Regulation of Yeast *GCN4* - a window on factors that control initiator-tRNA binding to the ribosome. J Biol Chem 272:21661-21664

Hinnebusch AG (2000) In: Sonenberg N, Hershey JWB, Mathews MB (eds) Translational Control of Gene Expression. Cold Spring Harbor Laboratory Press, Cold Spring Harbor, NY p 185-243

Hinnebusch AG, Natarajan K (2002) Gcn4p, a master regulator of gene expression, is controlled at multiple levels by diverse signals of starvation and stress. Eukaryotic Cell 1:22-32

Jefferson LS, Kimball SR (2003) Amino acids as regulators of gene expression at the level of mRNA translation. J Nutr 133:2046S-2051S

Kimball SR, Everson WV, Flaim KE, Jefferson LS (1989) Initiation of protein synthesis in a cell-free system prepared from rat hepatocytes. Am J Physiol 256:C28-C34

Kimball SR, Jefferson LS (1991) Mechanism of inhibition of peptide chain initiation by amino acid deprivation in perfused rat liver. Regulation involving inhibition of eukaryotic Initiation Factor 2α phosphatase activity. J Biol Chem 266:1969-1976

Marton MJ, Crouch D, Hinnebusch AG (1993) GCN1, a translational activator of *GCN4* in *Saccharomyces cerevisiae*, is required for phosphorylation of eukaryotic translation initiation factor 2 by protein kinase GCN2. Mol Cell Biol 13:3541-3556

Marton MJ, Vazquez de Aldana CRV, Qiu HF, Chakraburtty K, Hinnebusch AG (1997) Evidence that GCN1 and GCN20, translational regulators of GCN4, function on elongating ribosomes in activation of eIF2α kinase GCN2. Mol Cell Biol 17:4474-4489

Olsen DS, Jordan B, Chen D, Wek RC, Cavener DR (1998) Isolation of the gene encoding the *Drosophila melanogaster* homolog of the *Saccharomyces cerevisiae* GCN2 eIF-2alpha kinase. Genetics 149:1495-1509

Qiu H, Dong J, Hinnebusch AG (2002) Mutations that bypass the tRNA binding activate the intrinsically defective kinase domain in GCN2. Genes Develop 16:1271-1280

Rolfes RJ, Hinnebusch AG (1993) Translation of the yeast transcriptional activator *GCN4* is stimulated by purine limitation: Implications for activation of the protein kinase GCN2. Mol Cell Biol 13:5099-5111

Santoyo J, Alcalde J, Mendez R, Pulido D, de Haro C (1997) Cloning and characterization of a cDNA encoding a protein synthesis initiation factor-2α (eIF-2α) kinase from Drosophila melanogaster. Homology to yeast GCN2 protein kinase. J Biol Chem 272:12544-12550

Sood R, Porter AC, Olsen D, Cavener DR, Wek RC (2000) A mammalian homologue of GCN2 protein kinase important for translational control by phosphorylation of eukaryotic initiation factor-2α. Genetics 154:787-801

Tyzack JK, Wang X, Belsham GJ, Proud CG (2000) ABC50 interacts with eukaryotic initiation factor 2 and associates with the ribosome in an ATP-dependent manner. J Biol Chem 275:34131-34139

Vazquez de Aldana CR, Marton MJ, Hinnebusch AG (1995) GCN20, a novel ATP binding cassette protein, and GCN1 reside in a complex that mediates activation of the eIF-2α kinase GCN2 in amino acid-starved cells. EMBO J 14:3184-3199

Wek RC, Jackson BM, Hinnebusch AG (1989) Juxtaposition of domains homologous to protein kinases and histidyl-tRNA synthetases in GCN2 protein suggests a mechanism for coupling *GCN4* expression to amino acid availability. Proc Natl Acad Sci USA 86:4579-4583

Wek SA, Zhu S, Wek RC (1995) The histidyl-tRNA synthetase-related sequence in the eIF-2α protein kinase GCN2 interacts with tRNA and is required for activation in response to starvation for different amino acids. Mol Cell Biol 15:4497-4506

Wu S, Hu Y, Wang J, Chatterjee M, Shi Y, Kaufman RJ (2002) Ultraviolet light inhibits translation through activation of the unfolded protein response kinase PERK in the lumen of the endoplasmic reticulum. J Biol Chem 277:18077-18083

Yaman I, Fernandez J, Liu H, Caprara M, Komar AA, Koromilas AE, Zhou L, Snider MD, Scheuner D, Kaufman RJ, Hatzoglou M (2003) The zipper model of translation control: A small upstream ORF is the switch that controls structural remodeling of an mRNA leader. Cell 113:519-531

Yang R, Wek SA, Wek RC (2000) Glucose limitation induces *GCN4* translation by activation of Gcn2 protein kinase. Mol Cell Biol 20:2706-2717

Yoshizawa F, Kimball SR, Jefferson LS (1997) Modulation of translation initiation in rat skeletal muscle and liver in response to food intake. Biochem Biophys Res Commun 240:825-831

Zhang P, McGrath BC, Reinert J, Olsen DS, Lei L, Gill S, Wek SA, Vattem KM, Wek RC, Kimball SR, Jefferson LS, Cavener DR (2002) The GCN2 eIF2α kinase is required for adaptation to amino acid deprivation in mice. Mol Cell Biol 22:6681-6688

Zhu SH, Wek RC (1998) Ribosome-binding domain of eukaryotic initiation factor-2 kinase GCN2 facilitates translation control. J Biol Chem 273:1808-1814

Abbreviations

GCN: general control non-derepressing
HisRS: histidyl-tRNA synthetase
eIF: eukaryotic initiation factor
mGCN2: mammalian ortholog of Gcn2p
CHO: Chinese hamster ovary
MES: mouse embryonic stem
eEF: eukaryotic elongation factor
UTR: untranslated region
ATF: activating transcription factor
CAT-1: cationic amino acid transporter 1
IRES: internal ribosome entry site
ITAF: IRES trans-acting factor

UV: ultraviolet
PKR: double-stranded RNA-activated protein kinase
PEK/PERK: PKR-like endoplasmic reticulum-resident protein kinase
TOR: target of rapamycin
In yeast nomenclature, genes are typically designated as three capital letters followed by a number, e.g. GCN2, and proteins are designated as three letters, only the first of which is capitalized, followed by a number and the lower case p, e.g. Gcn2p.

6 Nutrient control of dimorphic growth in *Saccharomyces cerevisiae*

Toshiaki Harashima and Joseph Heitman

Abstract

In response to an abundant fermentable carbon source and limiting nitrogen, diploid yeast cells differentiate to form pseudohyphae that consist of chains of elongated cells. The G-protein coupled receptor Gpr1 senses extracellular glucose and signals via the coupled Gα subunit Gpa2. Gpa2 then stimulates cAMP production by adenylyl cyclase and activates the cAMP signaling pathway to promote pseudohyphal differentiation in diploid cells. Recently, the kelch repeat proteins Gpb1 and Gpb2 were identified as effectors and Gβ subunit mimics for Gpa2. The Gpr1-Gpa2-Gpb1/2-cAMP signaling cascade also controls a related cellular differentiation process, involving invasive growth that occurs in haploid cells grown on rich medium. Multiple signaling pathways function coordinately with the cAMP signaling pathway to govern both diploid pseudohyphal differentiation and haploid invasive growth. In this chapter, we review the current state of knowledge about the signaling cascades that sense nutrients and effect these alternative developmental cell fates.

6.1 Introduction

The yeast *Saccharomyces cerevisiae* is not commonly found in a haploid state and exists almost exclusively in the diploid state in nature. Diploid yeast cells grow vegetatively under nutrient rich conditions, sporulate when limited for nitrogen and carbon sources in the presence of a nonfermentable carbon source (such as ethanol), or differentiate pseudohyphae in response to an abundant fermentable carbon source and nitrogen limitation. Therefore it is necessary for yeast cells to sense extracellular and intracellular nutrient conditions. Multiple signaling pathways enable yeast cells to undergo proper cellular differentiation in response to nutrients.

Cell surface receptors sense extracellular signals and transduce this information to intracellular machineries that control developmental decisions and fates. The G-protein coupled receptors (GPCR) are a broadly conserved family of receptors that span the cell membrane seven times and signal via their ability to engage coupled heterotrimeric G proteins comprised of α, β, and γ subunits (Gilman 1987; Sprang 1997; Lefkowitz 2000). These receptors are conserved from yeast to mammals,

Topics in Current Genetics, Vol. 7
J. Winderickx, P.M. Taylor (Eds) Nutrient-Induced Responses in Eukaryotic Cells

and sense ligands as diverse as odorants, pheromones and hormones, photons, and sugars. Ligand binding triggers conformational changes in the receptor that recruit the heterotrimeric G protein, promote GDP-GTP exchange reaction on the Gα subunit, and release the Gβγ complex. Depending on the signaling cascade, liberated Gα-GTP, Gβγ, or both transmit signals by controlling downstream effectors such as adenylyl cyclase, which synthesizes the second messenger cAMP (Gilman 1987). Signaling is attenuated by intrinsic Gα GTPase activity, and by GTPase activating RGS (Regulators of G-protein Signaling) proteins, which both promote reformation of the inactive Gα-GDP-βγ complex (Ross and Wilkie 2000). The intracellular cAMP level depends on the relative activities of adenylyl cyclase and cAMP degrading enzymes, the cAMP phosphodiesterases.

cAMP dependent protein kinase (PKA) mediates the physiological effects of cAMP. The PKA holoenzyme consists of two regulatory subunits and two catalytic subunits that exist as an inactive tetramer under noninducing conditions. Both subunits of PKA are highly conserved among fungi and other eukaryotes. When conditions increase cAMP levels, cAMP binds to the PKA regulatory subunits and induces a conformational change that promotes dissociation of the catalytic subunits from the inactive tetramer. The free catalytic subunits then phosphorylate target substrates including metabolic enzymes and transcription factors to regulate an array of cellular processes.

The yeast *S. cerevisiae* expresses three GPCRs. Ste2 and Ste3 are pheromone receptors that are coupled to the Gα subunit Gpa1 and which function to control mating (Dohlman and Thorner 2001). The third GPCR is the glucose receptor Gpr1, which, via its coupled Gα subunit Gpa2, regulates the cAMP signaling pathway and ultimately filamentous growth (Yun et al. 1997, 1998; Xue et al. 1998; Kraakman et al. 1999; Lorenz et al. 2000b; Tamaki et al. 2000).

In addition to these GPCR systems, several plasma membrane-localized sensors, permeases, and transporters also have important roles for transducing extracellular cues to trigger intracellular events. In *S. cerevisiae*, Mep2 functions not only as an ammonium permease but also as an ammonium sensor that is required for diploid pseudohyphal growth (Lorenz and Heitman 1998). Snf3 and Rgt2 are divergent members of the hexose transporter family which no longer transport glucose but instead function as glucose sensors to regulate expression of genes encoding hexose transporters (Özcan et al. 1996a, 1996b, 1998). Finally hexose transporters and hexokinases are required for cAMP production in response to glucose in conjunction with the Gpr1-Gpa2 pathway (Rolland et al. 2000, 2001).

The yeast *S. cerevisiae*, as well as a number of human pathogenic fungi including *Cryptococcus neoformans* and *Candida albicans*, and plant pathogenic fungi, such as *Ustilago maydis*, are all known to undergo a dimorphic transition in response to extracellular stimuli (Madhani and Fink 1998; Lengeler et al. 2000; Pan et al. 2000; D'Souza and Heitman 2001). While the dimorphic change has been implicated as a crucial event in the pathogenesis of those organisms, the molecular mechanisms controlling those dimorphic transitions are far from understood in detail in any pathogenic fungi.

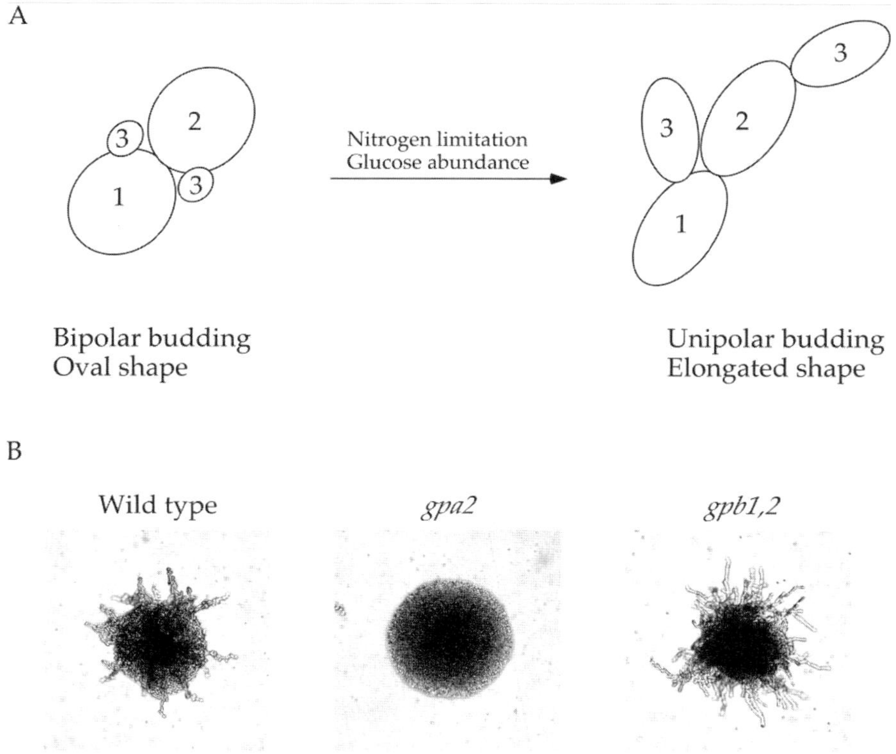

Fig. 1. Diploid pseudohyphal differentiation in *S. cerevisiae*. A) The morphological transition from unicellular yeast form to pseudohyphal form is accompanied by alteration of the budding pattern, changes in cell shape, and physical attachment between mother and daughter cells. B) Pseudohyphal differentiation of wild type cells, Gα-deficient *gpa2* mutant cells, and Gβ mimic-deficient *gbp1,2* cells is depicted

S. cerevisiae is an excellent model system with great potential to provide insights into molecular mechanisms by which the dimorphic transition is governed and ultimately to understand how virulence is produced in pathogenic fungi (Madhani and Fink 1998; Lengeler et al. 2000; D'Souza and Heitman 2001). In *S. cerevisiae*, diploid cells differentiate in response to nitrogen limitation and an abundant fermentable carbon source, such as glucose, to form pseudohyphae consisting of chains of elongated cells (Gimeno et al. 1992). Recent studies reveal that nutrient-sensing systems for glucose and nitrogen source play pivotal roles in pseudohyphal differentiation and are highly conserved among organisms. In this chapter, we review studies on nutrient sensing cascades that regulate diploid pseudohyphal growth and the related phenomenon of haploid invasive growth.

6.2 Signaling cascades controlling pseudohyphal growth in *Saccharomyces cerevisiae*

6.2.1 Diploid pseudohyphal differentiation in *S. cerevisiae*

In response to an abundant fermentable carbon source and limiting nitrogen, diploid *S. cerevisiae* cells form chains of elongated cells called pseudohyphae (Gimeno et al. 1992). This transition from the unicellular yeast form to the filamentous pseudohyphal form is accompanied by changes in several distinct cellular processes (Fig. 1A). First, the budding pattern of diploid cells is altered from a bipolar to a unipolar budding pattern to form filamentous pseudohyphae. Second, cell shape is altered from an oval yeast form to the elongated, thin hyphal form. Third, the mother and daughter cells remain physically attached. These morphological and physiological changes allow cells to form pseudohyphae and invade the agar and increase the cellular absorptive capacity by increasing the surface to volume ratio (Fig. 1B). This filamentous growth mode may enable this organism to forage for limiting nutrients.

Pseudohyphal growth is observed in strains of the ∑ background but not in conventional laboratory stains such as S288C. This is due to a nonsense mutation in the *FLO8* gene of S288C-derived strains (Liu et al. 1996). The *FLO8* gene encodes a transcriptional activator required for expression of the *FLO11* gene which encodes a glycosylphosphatidylinositol (GPI)-anchored cell surface protein essential for calcium-dependent cell-cell adhesion (also known as flocculation) (Lo and Dranginis 1996, 1998).

6.2.2 GPCR-G protein modules in *S. cerevisiae*

6.2.2.1 GPCR-G protein systems sense glucose and pheromones

The yeast *S. cerevisiae* expresses three GPCRs, two Gαs, and one set of Gβγ subunits. Two of the three GPCRs, Ste2 and Ste3, are pheromone receptors that function in mating. Ste2/3 are coupled to a heterotrimeric G protein comprised of the α, β, and γ subunits Gpa1, Ste4, and Ste18, respectively (Dohlman and Thorner 2001). The third GPCR in yeast is a novel receptor, Gpr1, which physically and functionally associates with the Gα subunit Gpa2.

Yeast cells occur in two haploid mating types, a and α, which communicate with each other via secreted peptide pheromones known as a-factor (produced by a cells) and α-factor (produced by α cells). a cells express the GPCR, Ste2, which is the receptor for α-factor and α cells express a different GPCR, Ste3, which is the receptor for a-factor. Both pheromone receptors are coupled to a heterotrimeric G protein comprised of the Gpa1 (Gα), Ste4 (Gβ), and Ste18 (Gγ) subunits. Pheromone binding to either receptor causes GDP-GTP exchange on the Gα subunit Gpa1 and release of the Ste4/Ste18 βγ complex. In conjunction with the Pak kinase Ste20, the free βγ complex activates the pheromone responsive MAP

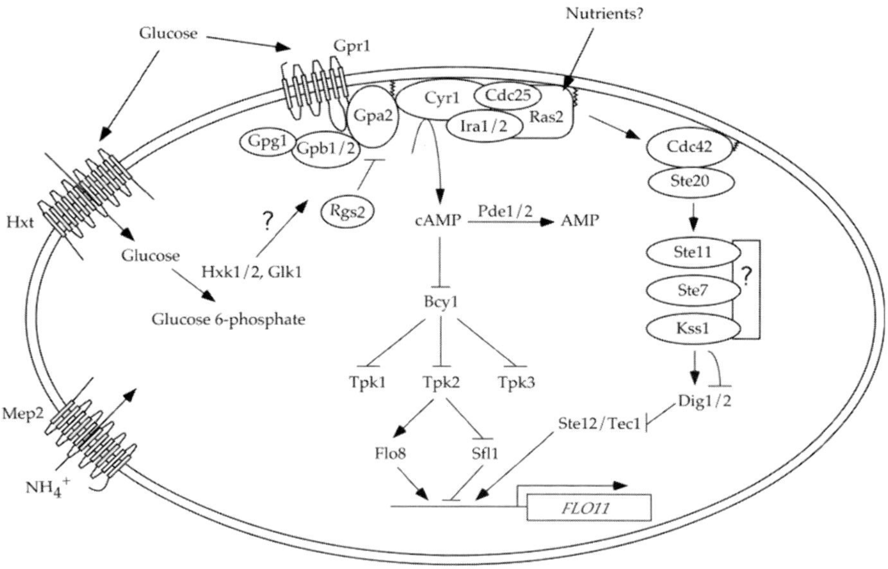

Fig. 2. Signaling pathways regulating pseudohyphal differentiation in *S. cerevisiae*. Multiple signaling cascades coordinately control pseudohyphal differentiation in *S. cerevisiae*, including elements of the Gpr1-Gpa2 nutrient sensing cAMP signaling pathway, the MAP kinase signaling pathway that also functions in mating of haploid cells, and the ammonium sensor Mep2.

kinase cascade by recruiting the Ste5 scaffold protein and associated kinases to the plasma membrane (Pryciak and Huntress 1998). Signaling is attenuated by the RGS protein, Sst2, which stimulates the GTPase activity of Gpa1 (Apanovitch et al. 1998).

Yeast cells also express a distinct GPCR receptor, Gpr1, which controls filamentous differentiation and invasive growth via the Gα subunit Gpa2 and its downstream effector adenylyl cyclase encoded by the *CYR1* gene (Fig. 2, Yun et al. 1997, 1998; Xue et al. 1998; Kraakman et al. 1999; Lorenz et al. 2000b; Tamaki et al. 2000). The Gα Gpa2 was discovered 15 years ago by low-stringency hybridization with a mammalian Gα subunit and early studies suggested that Gpa2 might play a role in regulating cAMP production in response to extracellular glucose (Nakafuku et al. 1988). Gpa2 was subsequently discovered to be required for yeast filamentous growth, and to signal in a pathway distinct from the MAP kinase cascade (Fig. 1B and 2, Kübler et al. 1997; Lorenz and Heitman 1997). The filamentous growth defect of *gpa2* mutant strains is suppressed by exogenous cAMP, providing evidence that Gpa2 regulates the cAMP signaling pathway to control filamentous growth.

The Gpr1 G protein-coupled receptor was subsequently identified by a two-hybrid screen with the Gα protein Gpa2 (Xue et al. 1998; Yun et al. 1997). This is one of only a few examples in which integral membrane proteins have been studied in the two-hybrid system. In this case, the soluble C-terminal tail of Gpr1 was found to interact with the coupled Gα protein Gpa2. In contrast to the pheromone GPCRs, which are expressed in haploid cells in a mating-type specific fashion, the Gpr1 receptor is expressed in both diploid and haploid cells of both mating types. Haploid cells lacking the Gpr1 receptor are defective in invasive growth and diploid cells lacking Gpr1 are defective in filamentous pseudohyphal differentiation (Lorenz et al. 2000b). As in the case of *gpa2* mutants, the pseudohyphal defect of *gpr1* mutant cells can also be suppressed by exogenous cAMP (Pan and Heitman 1999; Lorenz et al. 2000b). Furthermore, a dominant active Gpa2 allele also suppresses the pseudohyphal defect of mutant strains lacking the Gpr1 receptor, supporting the hypothesis that Gpa2 functions downstream of Gpr1 (Lorenz et al. 2000b). Recent evidence supports the hypothesis that Gpr1 is a receptor for glucose and structurally related sugars (Yun et al. 1998; Kraakman et al. 1999; Lorenz et al. 2000b; Rolland et al. 2000). When yeast cells are starved for glucose, readdition of glucose triggers a rapid and transient increase in intracellular cAMP levels, and importantly, both the Gpr1 receptor and the Gα protein Gpa2 are required for this glucose-induced event (Colombo et al. 1998; Yun et al. 1998; Kraakman et al. 1999; Lorenz et al. 2000b). Therefore Gpr1 regulates adenylyl cyclase activity via Gα Gpa2 in response to glucose. Consistent with this model, *gpr1* mutations, similar to *gpa2* mutations, are nearly synthetically lethal with mutations of the *RAS2* gene that controls adenylyl cyclase activity essential for cell viability (Kübler et al. 1997; Xue et al. 1998).

Similar to the role of the RGS protein Sst2 in pheromone signaling, the RGS protein Rgs2 accelerates the intrinsic GTPase activity of Gpa2 to downregulate glucose-induced cAMP signaling (Versele et al. 1999). *rgs2* mutations enhance cAMP production in response to glucose, whereas overexpression of Rgs2 attenuates this response. Since *rgs2* mutant cells form pseudohyphae like wild type cells (Harashima, unpublished data), a role for Rgs2 in pseudohyphal differentiation, if any, has not yet been established.

The Gpr1-Gpa2 signaling system also plays a role in sensing nitrogen starvation. The *GPR1* gene is transcriptionally induced in response to nitrogen limitation (Xue et al. 1998). Thus, nitrogen sources control the expression of the Gpr1 glucose receptor, increasing the sensitivity of this signaling cascade for the extracellular ligand. In this regard, the Gpr1-Gpa2 signaling pathway serves as a dual sensor of both carbon abundance and nitrogen limitation.

BLAST searches in NCBI (http://www.ncbi.nlm.nih.gov/blast) reveal fungal homologs of the *S. cerevisiae* glucose receptor Gpr1 in *Candida albicans* and *Neurospora crassa*. *Schizosaccharomyces pombe* also contains a glucose-sensing GPCR system that activates cAMP signaling analogous to the pathway in *S. cerevisiae*. In *S. pombe*, glucose-induced cAMP signaling inhibits gluconeogenesis by triggering the repression of the gene encoding fructose-1,6-bisphosphatase (*fbp1*). This requires the Gα protein Gpa2 and the Gpr1 glucose receptor homolog Git3 (Hoffman and Winston 1990; Nocero et al. 1994; Welton and Hoffman 2000). In *C. albicans* the cAMP signaling pathway is involved in the dimorphic transition from yeast to hyphal growth, similar to the induction of pseudohyphal differentiation in *S. cerevisiae* (Lengeler et al. 2000; Whiteway 2000). Although no experimental data as yet connect the *C. albicans* Gpr1 receptor homolog to glucose sensing or to cAMP signaling, the significant level of identity shared by these three GPCRs could be indicative of a new class of GPCRs involved in glucose sensing. Roles for the *N. crassa* Gpr1 homolog are also currently unknown.

6.2.2.2 Novel kelch proteins Gpb1 and Gpb2 as Gα protein partners

The yeast *S. cerevisiae* expresses two Gα subunits of heterotrimeric G proteins, namely Gpa1 and Gpa2. Gpa1 forms a canonical heterotrimeric G protein with the Ste4/Ste18 Gβγ subunits and specifically functions in the pheromone responsive pathway via its ability to couple to the pheromone receptors Ste2 and Ste3. On the other hand, Gpa2 acts in the cAMP signaling pathway. A critical question was whether Gpa2 functions with an unidentified βγ subunit complex, with a distinct type of partner subunit, or as a solo Gα protein.

Gpa2 is homologous not only to fungal Gα proteins but also to mammalian Gα subunits, and shares some 35-55% sequence identity with fungal and mammalian Gα subunits. Similar to the Giα1 G203A or Gs G226A mutant proteins that are catalytically active but unable to undergo the conformational change after GTP binding that is necessary for βγ subunit release, expression of a Gpa2 G299A mutant protein inhibits pseudohyphal development (Miller et al. 1988; Lee et al. 1992; Coleman et al. 1994; Lorenz and Heitman 1997). The structural features of a canonical Gβ subunit are seven WD-40 repeats, which form a β-propeller structure that is preceded by a helical N-terminal region. The yeast βγ subunits Ste4 and Ste18 function during mating, but are not expressed in diploid cells, and mutants lacking Ste4 or Ste18 exhibit no defects in invasion or filamentation (Liu et al. 1993). In addition, neither deletion nor overexpression of Ste4 affects cAMP signaling or PKA controlled phenotypes and, simultaneous overexpression of Ste4 and Ste18 also did not affect cAMP signaling, suggesting that the Gα subunit Gpa2 does not function with these canonical Gβγ subunits (Lorenz and Heitman 1997; Versele et al. 2001). Mutation of several candidate β and γ subunits had no effect on filamentous growth (Lorenz and Heitman 1997). Thus, it was unclear whether glucose binding to Gpr1 activated adenylyl cyclase by Gpa2 functioning alone or with atypical accessory subunits.

Recently, the kelch proteins Gpb1 and Gpb2 were identified and demonstrated to function as novel Gβ subunit structural mimics (Fig. 3, Harashima and Heitman 2002). Gpb2 was originally identified in yeast two-hybrid screens for Gpa2-interacting proteins that also identified the tail of the GPCR Gpr1, the meiotic kinase Ime2, and an unknown protein YGL121c (Yun et al. 1997; Xue et al. 1998; Kraakman et al. 1999; Lorenz et al. 2000b). Yeast cells also express a homolog of Gpb2, Gpb1, which shares ~35% identity and is encoded in a region of synteny (Harashima and Heitman 2002). Genetic analyses revealed that Gpb1 and Gpb2 are largely functionally redundant and loss of both Gpb1 and Gpb2 results in activation of the cAMP signaling pathway (Harashima and Heitman 2002). Namely, *gpb1,2* double mutant cells exhibit elevated expression of the *FLO11* gene, increased pseudohyphal formation (Fig. 1B) and invasive growth, sensitivity to

A

```
           289  KLPSLSY-HCSVELNDQLFIVGGLMACHRYDEEAPDLKDFYVDGIKNL
           375  PPPLVCT-QGCKLTERHIFLYGGFEIKS--ETQVDDKGRYFIRKRAFL
           435  KIELVAPPYQFAIYNNFSPRFGHMQASI-SNSNNNVSNENTTTSAKGR
Gpb1       529  KHCSTAT-HICSSVN-TILIFGGYSQTG--DDKYEAMNDMWKINIPVV
           700  TPPLNPS-KKCASIGRTIAFHGGSDGYDVCSDMWWFDFDSETWTKIDL
           759  PINLCMVGHSMTTVGHKVVLIGGLRQGD--VDRIYRDETLPEEVISGV
           846  PHQVLTVAGTIELVKGTMTLIGGVVAGREDISSLYLRGAVLQFILPSM
           276  KLPALTY-HCSVELNGNIYIFGGLMPCYSYEEDAPMLNDFFVDGIKNL
           362  PPPLLCV-QGSKLTDRHIFFYGGFEIRT--ETRGDENGKYHLKKRLYV
           422  KIDIIVQPSKYNAYPTMSSRFGHLQIS---IDNPNRRASVHSSSMNEI
Gpb2       487  KAAVLSS-LPHNTVH-TVIIFGGYRQTG--DDRYEAMNDLWKIEIPVI
           677  RQRLILS--QEKPVGKTVVLHGGSNGLNVLDDMWLMDLECETWTPIET
           738  SVNVGLVGHRMESIGRICVCIGGMVQED--VDQFYSENDDESPRKRKV
           829  SNIVVGVGGTSLQCDKSILIGGLISRRSNVKEIYLHGTITKSIFPSV
consensus                       hhhhGG
```

B

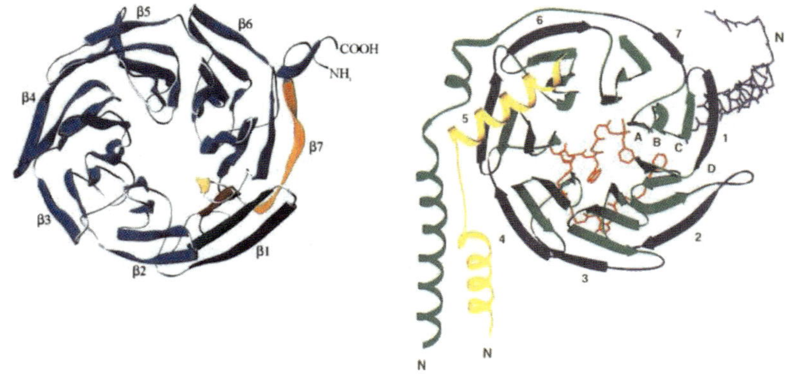

Fig. 3. Gpb1 and Gpb2 contain seven kelch repeats known to fold into a 7 bladed β-propeller. A) The 7 kelch repeats from both Gpb1 and Gpb2 were aligned and numbered according to the amino acid sequence of Gpb1/2. The general consensus is according to Adams et al. (2000) in which "h" indicates hydrophobic residues. Identical amino acids are shaded in black. Conservative substitutions in more than half the repeats are shaded in gray; amino acids were divided into hydrophobic (A, V, L, I, P, M, F, Y, W), hydrophilic (G, S, T, C, N, Q), acidic (D, E), and basic (K, R, H). Alignments were obtained using ClustalW with some manual adjustments (Thompson et al. 1994). B) Structural comparison between the conventional WD-40 based Gβ subunit and the seven kelch repeat protein galactose oxidase. Structures of the kelch protein galactose oxidase from *Hypomyces rosellus* (left) and the heterotrimeric G protein (right) were adapted from Adams et al. (2000), Ito et al. (1994), and Wall et al. (1995).

nitrogen starvation, loss of glycogen accumulation, heat shock sensitivity, and sporulation defects, all of which are characteristics of increased cAMP signaling (Harashima and Heitman 2002; Batlle et al. 2003). In addition, loss of Gpb1, Gpb2, or both results in perturbation of glucose-induced cAMP production (Harashima and Heitman 2002). These genetic studies revealed that Gpb1 and Gpb2 negatively regulate the cAMP signaling pathway by inhibiting both Gpa2 and an as yet unidentified target. Importantly, Gpa2 is still functional to some extent in the absence of either or both Gpb1 and Gpb2 (Harashima and Heitman 2002).

The Gpb1 and Gpb2 proteins lack the 7 signature WD-40 motifs characteristic of Gβ subunits. Instead, both have 7 kelch repeats, previously implicated in protein-protein interactions (Fig. 3, Adams et al. 2000). Both Gpb1 and Gpb2 preferentially interact with the GDP-bound form of Gpa2 *in vitro*, similar to Gβ subunits. Most interestingly, the crystal structure of one protein containing seven kelch repeats has been solved, namely that of the fungal enzyme galactose oxidase (Fig. 3, Ito et al. 1991, 1994). Remarkably, this protein folds into a seven bladed β-propeller structure that is strikingly similar to the fold adopted by the sequence unrelated G-protein β subunits in which the blades of the propeller are constructed from seven WD-40 repeats (Fig. 3, Wall et al. 1995; Lambright et al. 1996; Sondek et al. 1996). Based on these findings, Gpb1 and Gpb2 have been proposed to function as novel Gα Gpa2 protein partners, signaling effectors, and Gβ structural mimics (Harashima and Heitman 2002). These observations reveal a striking example of convergent evolution of divergent primary sequences to similar tertiary structures.

The identification of the protein partners of Gpa2 as novel Gβ mimics raises a question as to what determines the specificity of the protein-protein interactions between Gpa2 and Gpb1/2. Compared to known Gα subunits, the entire amino acid sequence of Gpa2 is very similar to that of known Gα subunits except for an amino terminal region of Gpa2 (Fig. 4A). Gpa2 has a unique N-terminal extension with short stretches of α-helical regions (residues 15 to 121). The amino-terminal α helical region of Gα subunits has been shown to interact with Gβγ subunits (Wall et al. 1998), and thus the amino-terminal extension of Gpa2 might confer specific protein-protein interactions with Gpb1/2 that specify a new class of unusual Gα subunit. Among fungal Gαs, the *Cryptococcus neoformans* Gpa1 and *Candida albicans* Gpa2 proteins also contain a unique amino-terminal extension (Fig. 4A, Tolkacheva et al. 1994; Sánchez-Martínez and Pérez-Martin 2002). Interestingly, homologs of the *S. cerevisiae* Gpa2 and Gpb1/2 proteins are conserved in *Saccharomyces* species including *Saccharomyces castellii* and *Saccharomyces kluyveri* (Fig. 4, http://genome-www4.stanford.edu/cgi-bin/SGD).

No conventional Gβ subunit for the *C. neoformans* Gpa1 or *C. albicans* Gpa2 protein has been identified. Interestingly, a protein that has significant homology with Gpb1/2 is present in the *C. albicans* genome based on the genome project at the Institut Pasteur (CandidaDB, http://genolist.pasteur.fr/CandidaDB). *C. albicans* Gpr1 and Gpa2 homologs have been isolated. These suggest that the Gpr1-Gpa2-Gpb1/2 regulatory system found in *S. cerevisiae* might be conserved in *C. albicans* as well. However, there are currently no experimental data in *C. albicans* that reveal the function for the Gpr1 and Gpb1/2 homologs and the Gpa2 protein has not been implicated in the nutrient-sensing cAMP signaling pathway. Rather, based on epistasis data, *C. albicans* Gpa2 has been proposed to function in the MAP kinase signaling cascade. However, these data can not establish a direct signaling link (Sánchez-Martínez and Pérez-Martin 2002). Given the dual, parallel

A

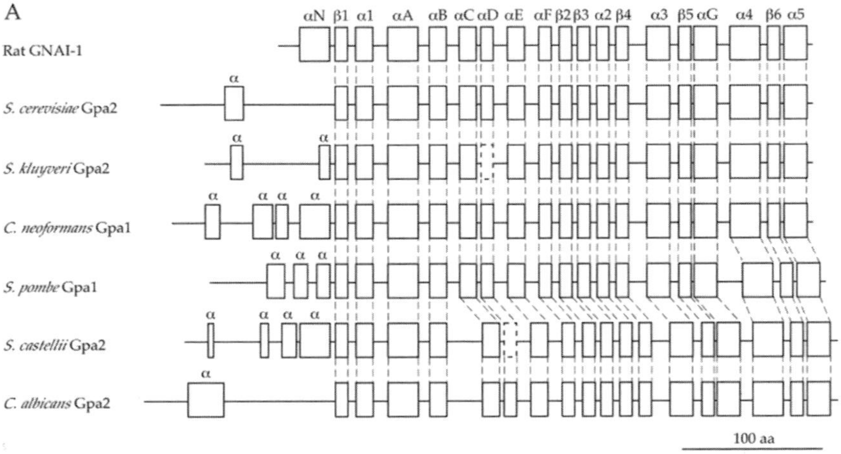

B

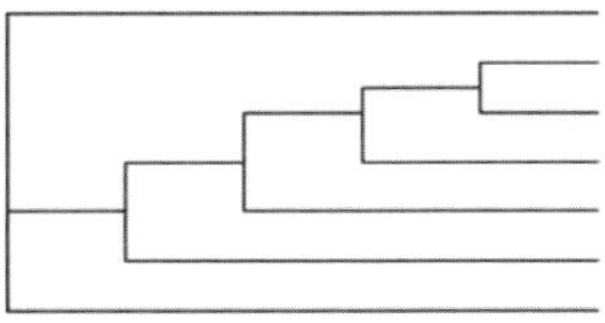

Gα	Organism	Length (aa)	Accession No	Partner
GNAI1	*Rattus norvegicus*	354	P10824	WD-40 Gβ
Gpa2	*Saccharomyces cerevisiae*	449	NP_010937	
Gpa2	*Saccharomyces castellii*	461		Kelch Gβ mimic (Gpb1/2)
Gpa2	*Saccharomyces kluyveri*	410		
Gpa2	*Candida albicans*	502		
Gpa1	*Cryptococcus neoformans*	433	P54853	Kelch Gβ mimic (Gpb1/2 or Ral2)
Gpa1	*Schizosaccharomyces pombe*	407	A41106	Kelch Gβ mimic (Ral2)

Fig. 4. Structural comparison of fungal Gα subunits that have a unique amino-terminal extension. A) Conserved secondary structures of Gα subunits from various fungi that have a unique amino-terminal extending region are shown schematically. Rat Gα subunit GNAI-1 is depicted as a representative that forms a canonical heterotrimeric G protein with a conventional WD-40 Gβ subunit. All Gα subunits are highly conserved over the entire molecule with some exceptions. Strikingly, the αN domain, which is located at the amino-terminal end of the rat Gα subunit GNAI-1 and contains residues required for Gβ binding, is highly diverse among fungal Gα subunits. There is no significant homology in the amino-terminal region among the Gα subunits presented here. Secondary structures for all Gα subunits were predicted by the PHD secondary structure prediction method (Rost and Sander 1993). The αD domain of *S. kluyveri* and *S. castellii* was not assigned by PHD but the corresponding regions are highly homologous to the αD domain from *S. cerevisiae* Gpa2. Secondary structure assignments were based on the ones corresponding to $G_{\alpha t/\alpha I}$ (Lambright et al. 1996). B) Phylogenetic tree for fungal Gα subunits with an extended unique amino terminal region. The phylogenetic tree was established by the Neighbor Joining method (Saitou and Nei 1987). Amino acid sequences of Gα subunits from *Saccharomyces castellii* Gpa2, *Saccharomyces kluyveri* Gpa2, and *Candida albicans* Gpa2 were obtained from the Saccharomyces Genome Database (SGD; http://genome-www4.stanford.edu/cgi-bin/SGD) and the Candida genome project at the Institut Pasteur (CandidaDB, http://genolist.pasteur.fr/CandidaDB), respectively. Defined or putative binding partners for Gα subunits are shown.

signaling nature of the MAP kinase cascade and the cAMP signaling pathway in *S. cerevisiae*, we propose that Gpa2 functions in a conserved nutrient sensing pathway in *C. albicans* that remains to be elucidated.

The *C. neoformans* Gα subunit Gpa1 was found to be homologous in sequence to the *S. cerevisiae* Gα subunit Gpa2 (Tolkacheva et al. 1994). Gpa1 plays a key role in virulence and nutrient sensing and activates cAMP production in response to extracellular signals including glucose (Alspaugh et al. 1997, 2002; D'Souza et al. 2001). Further studies focus on how the *C. neoformans* Gpa1 protein controls the cAMP signaling pathway in response to extracellular cues.

6.2.3 The cAMP signaling pathway

The target of cAMP in *S. cerevisiae* is the cAMP-dependent protein kinase (PKA). PKA is functionally and structurally conserved from yeast to mammals. In yeast cells, the PKA regulatory subunit is encoded by the *BCY1* gene and the three catalytic subunits are encoded by the *TPK1*, *TPK2*, and *TPK3* genes (Cannon and Tatchell 1987; Toda et al. 1987a, 1987b). In resting cells PKA is an inactive tetramer containing two regulatory subunits in complex with two catalytic subunits. In response to external signals that elevate intracellular cAMP levels, cAMP binds to the regulatory subunits and triggers conformational changes that release the active catalytic subunits. PKA activity is essential for vegetative growth in *S. cerevisiae*. Triple mutants lacking all three catalytic subunits, Tpk1, Tpk2, and Tpk3 are inviable, whereas mutant strains expressing any one of the three Tpk subunits are viable. These findings suggest that the three PKA catalytic subunits have redundant functions for cell viability.

Recent studies reveal that cAMP-dependent protein kinase plays a central role in regulating yeast pseudohyphal differentiation (Fig. 2, Robertson and Fink 1998; Pan and Heitman 1999). First, mutation of the PKA regulatory subunit Bcy1 dramatically enhances pseudohyphal growth (Pan and Heitman 1999). Second, the PKA catalytic subunits play distinct roles in regulating pseudohyphal growth. The Tpk2 subunit activates filamentous growth, whereas the Tpk1 and Tpk3 subunits primarily inhibit pseudohyphal differentiation (Robertson and Fink 1998; Pan and Heitman 1999). The unique activating function of the Tpk2 protein is attributable to structural differences in the kinase catalytic region and not to differences in gene regulation or the unique amino-terminal region of the protein, likely indicating differences in the substrate specificity of the divergent catalytic subunits (Pan and Heitman 1999). Based on epistasis analyses, Tpk2 functions downstream of the Gpr1 receptor and the coupled Gα protein Gpa2 (Pan and Heitman 1999; Lorenz et al. 2000b). Importantly, activation of PKA by mutation of the Bcy1 regulatory subunit restores pseudohyphal growth in mutants lacking components of the MAP kinase pathway, including *ste12*, *tec1*, and *ste12 tec1* mutants (Pan and Heitman 1999; Rupp et al. 1999). Thus, the MAP kinase and PKA pathways independently regulate filamentous growth.

Recent studies have defined a role for the PKA pathway in activating pseudohyphal growth via transcriptional regulation of the cell surface flocculin Flo11 by the Flo8 transcriptional activator and the transcriptional repressor Sfl1 (Fig. 2, Robertson and Fink 1998; Rupp et al. 1999; Conlan and Tzamarias 2001; Pan and Heitman 1999, 2002). Both Flo8 and Sfl1 are direct targets of Tpk2. Phosphorylation of Sfl1 by Tpk2 releases Sfl1 from the *FLO11* promoter by inhibiting dimerization, whereas phosphorylation of Flo8 by Tpk2 stimulates its binding to the *FLO11* promoter to activate Flo11 expression (Pan and Heitman 2002). Sfl1 and Flo8 bind to the same or adjacent regions of the *FLO11* promoter to control gene expression, supporting a model in which Flo8 antagonizes Sfl1 binding on the *FLO11* promoter (Pan and Heitman 2002). The *FLO11* gene has an extremely large promoter, ~3,000 bp, and is regulated by a complex set of transcription factors that includes Ste12, Tec1, Flo8, Sfl1, Msn1/Mss10/Phd2, and Mss11 (Robertson and Fink 1998; Gagiano et al. 1999; Rupp et al. 1999; Conlan and Tzamarias 2001; Pan and Heitman 1999, 2002). Two separate signaling pathways, the cAMP signaling pathway and the MAP kinase signaling pathway, converge on the *FLO11* promoter (Gagiano et al. 1999; Rupp et al. 1999). Taken together, these findings reveal an intimate role for the cAMP-dependent kinase in the regulation of yeast dimorphism.

Three other proteins have been shown to be components of the cAMP signaling pathway controlling pseudohyphal differentiation. One is a protein encoded by the *YGL121c* locus, which was originally isolated together with Gpr1, Gpb2, and Ime2 by yeast two-hybrid screens using the Gα subunit Gpa2 as the bait (Xue et al. 1998; Lorenz et al. 2000b). Subsequently, genetic studies revealed that YGL121c is involved in the cAMP signaling pathway. Cells lacking YGL121c have defects in pseudohyphal growth, invasive growth, and *FLO11* expression (Harashima and Heitman 2002). Yeast two-hybrid analyses showed that the protein encoded by *YGL121c* indirectly interacts with Gpa2 via association with Gpb1/2 (Harashima and Heitman 2002). *YGL121c* encodes a small, 126 residue protein, similar in size to known Gγ subunits. Secondary structure predictions suggest that YGL121c may fold into two linked α-helical domains participating in a coiled-coil structure, similar to known Gγ subunits. Based on these findings, YGL121c was named Gpg1 (G protein γ subunit mimic protein, Harashima and Heitman 2002). However Gpg1 has no C-terminal CaaX motif, which is required for prenylation and is a characteristic of heterotrimeric G protein γ subunits. Furthermore, unlike heterotrimeric G protein γ subunit mutants, the *gpg1* mutants in some cases exhibited phenotypes opposite to the *gpb1,2* double mutant cells (Harashima and Heitman 2002). Therefore, further analyses are required to understand how Gpg1 controls the cAMP signaling pathway.

The second class of proteins that function in cAMP signaling are cAMP phosphodiesterases. Yeast cells express two different cAMP phosphodiesterases: the high-affinity cAMP phosphodiesterase, Pde2, and the low-affinity cAMP phosphodiesterase, Pde1. *pde2* mutations enhance both diploid pseudohyphal differentiation and haploid invasive growth (Kübler et al. 1997; Lorenz and Heitman 1997; Pan and Heitman 1999). In contrast to the enhanced pseudohyphal phenotype of *pde2* mutant cells, *pde2* mutations have little effect on glucose induced cAMP production (Ma et al. 1999). The Pde1 phosphodiesterase has a single PKA consensus phosphorylation site and is a phosphoprotein *in vivo* (Ma et al. 1999). In addition, phosphorylation in crude extracts leads to modest increases in enzyme activity (Ma et al. 1999). These findings suggest that Pde1 could be part of the feedback loop that limits cAMP elevation in response to glucose signaling. Interestingly, *pde1* mutations dramatically enhance cAMP production in response to glucose but have no effect on pseudohyphal growth (our unpublished data).

The third protein that has been proposed to function in cAMP signaling is the phosphatidylinositol-specific phospholipase C (Plc1). Plc1 hydrolyzes phosphatidylinositol 4,5-bisphosphate (PIP_2) to produce inositol 1,4,5-triphosphate (IP_3) and diacylglycerol. In yeast, Plc1 has been reported to be in a physical complex with the Gpr1 receptor and the Gα subunit Gpa2 and to be necessary for pseudohyphal differentiation (Ansari et al. 1999). Although an IP_3 receptor has not yet been identified in *S. cerevisiae*, two inositol kinases, Ipk1 and Ipk2, have been identified (Saiardi et al. 1999, 2000; York et al. 1999; Odom et al. 2000). IP_3 is phosphorylated to IP_4 and IP_5 by the dual-specificity kinase Ipk2, which regulates transcription (Odom et al. 2000), and then IP_5 is phosphorylated to IP_6 by the Ipk1 kinase, which regulates mRNA export from the nucleus (York et al. 1999; Saiardi et al. 2000). Furthermore, recent studies have revealed that the polyinositol phosphates regulate chromatin remodeling (Shen et al. 2003; Steger et al. 2003). Therefore Plc1 might regulate gene expression required for pseudohyphal growth in response to nutrients.

Interestingly, Ras and Plc1 have been implicated in phosphatidylinositol metabolism in response to glucose (Kaibuchi et al. 1986) and Plc1 has also been found to be involved in turnover of phosphatidylinositol, activation of the plasma membrane H^+-ATPase, and calcium influx in response to glucose (Brandão et al. 1994; Coccetti et al. 1998; Tisi et al. 2002). Furthermore, addition of nitrogen to starved cells has been shown to induce IP_3 and diacylglycerol in a Cdc25-dependent manner (Schomerus and Küntzel 1992). However, inconsistently, it has also been reported that IP_3 production induced by the addition of nitrogen to nitrogen-starved cells is observed in *plc1* deleted cells (Bergsma et al. 2001). Further studies will be required to elucidate the precise role of Plc1 in nutrient responses and filamentous growth.

Yeast diploid cells undergo pseudohyphal growth under conditions of an abundant fermentable carbon source and nitrogen starvation, whereas under conditions limiting for both nitrogen and fermentable carbon sources, yeast cells sporulate. Recent studies indicate that the cAMP signaling pathway controls meiotic events. First, the meiosis-specific protein kinase Ime2, which is essential for meiosis and sporulation under starvation conditions, preferentially binds to the GTP-bound form of Gpa2, and this interaction inhibits Ime2 kinase activity (Yoshida et al. 1990; Donzeau and Bandlow 1999). Furthermore the presence of nitrogen induces the Gpa2-Ime2 interaction (Donzeau and Bandlow 1999). Second, Rim15, a protein kinase required for meiotic gene expression, has been identified as a target of PKA and phosphorylation of Rim15 by PKA inhibits the kinase activity of Rim15 (Vidan and Mitchell 1997; Reinders et al. 1998). These findings explain, at least in part, why hyperactivated cAMP signaling (in *gpb1,2* double mutants, a $RAS2^{G19V}$ mutant, and *bcy1* mutants) results in sporulation defects (Toda et al. 1985; Matsuura et al. 1990; Batlle et al. 2003). These findings also indicate that Gpa2 might receive input from other sources that sense nitrogen levels and that the Gpr1-Gpa2 pathway could function to coordinate and integrate sensing of both carbon and nitrogen sources.

6.2.4 Glucose uptake and glucose phosphorylation are required for cAMP production

In addition to the Gpr1-Gpa2 module, an intracellular glucose-sensing system is also required for glucose-induced cAMP production (Beullens et al. 1988). In *S. cerevisiae*, 18 putative hexose transporters have been identified (André 1995; Kruckeberg 1996; Boles and Hollenberg 1997). Because deletion of the *HXT1-7* genes abolishes growth on glucose media, these seven transporters seem to be the most physiologically important (Reifenberger et al. 1997). In *hxt1-7* mutants, glucose fails to induce a cAMP spike and expression of any one of the transporters overcomes this defect, indicating the glucose signal is not dependent on a specific hexose transporter (Rolland et al. 2000, 2001). Intracellular glucose is phosphorylated by three kinases in *S. cerevisiae*; two are hexokinases encoded by the *HXK1* and *HXK2* genes that can phosphorylate both glucose and fructose and the third is a glucokinase encoded by *GLK1*, which is specific for aldo-hexoses such as glucose (Entian et al. 1985; Fröhlich et al. 1985; Kopetzki et al. 1985; Albig and Entian 1988; Walsh et al. 1991). The presence of any one of the kinases is both necessary and sufficient for induction of cAMP in response to glucose (Beullens et al. 1988). However, evidence suggests that there is no correlation between the levels of glucose-induced cAMP and the levels of the product of the kinase reaction itself (Beullens et al. 1988; Rolland et al. 2001). Therefore, glucose kinases could themselves play dual functions in glucose metabolism and glucose sensing as recently found in the plant *Arabidopsis* (Moore et al. 2003). A role for the glucose transporters or kinases in pseudohyphal growth has not yet been elucidated.

Two hexose transporter-like glucose sensors have been found in *S. cerevisiae*, namely Snf3 and Rgt2 (Özcan et al. 1996a, 1998). Snf3 acts as a high-affinity glucose sensor, whereas Rgt2 acts as a low-affinity glucose sensor. Neither Snf1 nor Rgt2 is required for the glucose-induced cAMP signal (Rolland et al. 2001) and neither has been implicated in regulating dimorphic transitions of yeast cells (our unpublished data).

6.2.5 The Ras-mediated signaling pathway

6.2.5.1 Ras regulates adenylyl cyclase

Yeast cells express two Ras proteins, Ras1 and Ras2 (Powers et al. 1984). Ras activity is regulated by the guanine nucleotide exchange factors (GEF) Cdc25 and Scd25, and also by the GTPase activating proteins (GAP) Ira1 and Ira2 (Tanaka et al. 1990, 1991; Chevallier-Multon et al. 1993; Lai et al. 1993) (Fig. 2). Although neither Ras1 nor Ras2 is essential, *ras1 ras2* double mutant strains are inviable, indicating the two proteins are functionally redundant for cell viability (Kataoka et al. 1984). However, with respect to filamentous growth, Ras2 is required but Ras1 is not, and thus Ras2 has a specialized function in filamentous growth (Mösch et al. 1999). Ras2 has also been shown to control filamentous growth via both the cAMP and the MAP kinase signaling pathways (Mösch et al. 1996, 1999; Lorenz and Heitman 1997).

Evidence suggests that the target of Ras in the cAMP signaling pathway is adenylyl cyclase, which is encoded by the *CYR1* gene in *S. cerevisiae* (Fig. 2). Loss of Ras function causes the same lethal phenotype as loss of adenylyl cyclase (Kataoka et al. 1984; Broek et al. 1987). Viability of *ras1 ras2* double mutant cells can be restored by overexpression of the *CYR1* gene (Kataoka et al. 1985). Overexpression of a dominant active $RAS2^{G19V}$ allele results in increased levels of cAMP (Toda et al. 1985). Although both Ras1 and Ras2 control the activity of adenylyl cyclase and are in part functionally redundant, Ras2 seems to play a more critical role in cAMP production than Ras1 (Toda et al. 1985).

Both Ras2 and its guanine nucleotide exchange factor (GEF) Cdc25 appear to be involved in glucose-induced cAMP production. In the absence of functional Ras2, cells cannot stimulate cAMP production in response to glucose (Mbonyi et al. 1988). A mutant $RAS2^{C318S}$ allele, which abolishes palmitoylation required for membrane localization, prevents glucose-induced cAMP production (Bhattacharya et al. 1995; Jiang et al. 1998). The GEF protein Cdc25 is also required for the glucose induced cAMP response and physically interacts with adenylyl cyclase (Schomerus et al. 1990; Gross et al. 1999; Mintzer and Field 1999). Furthermore, mutations in the *IRA1* gene cause much higher levels of cAMP in response to glucose in a Ras2-dependent manner (Tanaka et al. 1989). These findings support a model in which Ras activity is required for both basal and glucose-induced adenylyl cyclase activities. However, in contrast to this hypothetical model, the ratio of GTP/GDP bound to Ras was not affected under conditions where glucose induces cAMP production although the ratio of GTP/GDP bound to Ras is altered by an intracellular acidification-inducing reagent (2,4-dinitrophenol) that induces a transient cAMP production (Colombo et al. 1998). Further experiments will be necessary to understand how Ras signals and plays a role in the stimulation of adenylyl cyclase by glucose.

6.2.5.2 Ras regulates the MAP kinase signaling pathway

Ras2 also controls the MAP kinase pathway via a different small G protein, Cdc42 (Fig. 2). The components of the MAP kinase pathway required for filamentous growth include the Ste20, Ste11, Ste7, and Kss1 kinases and the Ste12 and Tec1 transcription factors. A dominant active $Ras2^{G19V}$ mutant stimulates pseudohyphal growth and expression of the reporter gene, *FG(Ty)-lacZ*, whose expression is specifically controlled by components of the MAP kinase pathway required for filamentous growth (Mösch et al. 1996). Overexpression of dominant active alleles of *CDC42* including $CDC42^{G12V}$ and $CDC42^{Q61L}$, also highly stimulates pseudohyphal growth and *FG(Ty)-lacZ* expression in a Ste20-dependent manner.

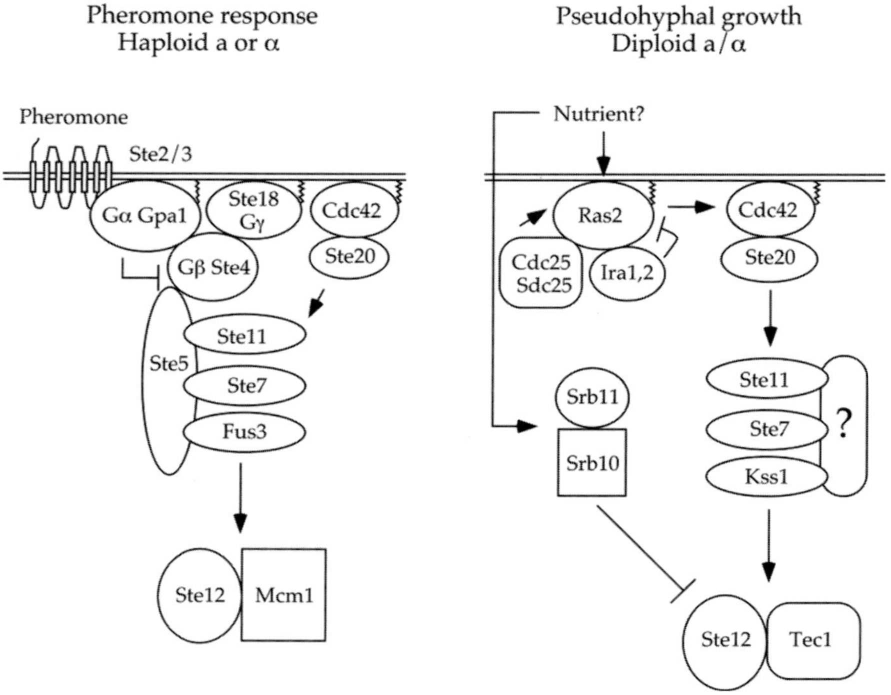

Fig. 5. Comparison of the components in the MAP kinase signaling pathway during haploid pheromone response and diploid pseudohyphal differentiation in *S. cerevisiae*. Signaling specificity is determined by unique signaling components in the pathways, including haploid specific pheromone receptors, the mating specific scaffold Ste5, and alternative Ste12 partner subunits, as described in the text.

In contrast, a dominant negative *CDC42*^{D118A} allele blocks the increased pseudohyphal growth and *FG(Ty)-lacZ* expression elicited by dominant active *RAS2*^{G19V} but does not inhibit the elevated expression of *FG(Ty)-lacZ* expression by *STE11-4*, an activated allele of *STE11* (Mösch et al. 1996). Therefore Ras2 controls filamentous growth via Cdc42 and its downstream MAP kinase cascade (Figs. 2, 5).

Yeast cells also require the MAP kinase signaling pathway for pheromone response and mating (Liu et al. 1993; Roberts and Fink 1994; Mösch et al. 1996). Thus, yeast cells can operate the same signaling cascade in response to different extracellular cues to adopt two different developmental fates: mating in haploid cells in response to pheromone, and diploid pseudohyphal growth in response to nutrient limitation and other environmental cues (Fig. 2 and 5). Specific signaling molecules in the cascade enable cells to accomplish distinct cellular developmental fates in response to particular extracellular signals. First, as mentioned above, the MAP kinase pathway is activated by a Ras2-Cdc42-mediated mechanism during filamentous growth. The 14-3-3 proteins Bmh1 and Bmh2, which physically interact with Ste20, are required for activation of the Ras2/MAP kinase pathway during filamentous growth but are dispensable for mating (Roberts et al. 1997).

For mating, the haploid specific pheromone receptors Ste2 and Ste3 and the coupled heterotrimeric G protein composed of Gpa1 (Gα), Ste4 (Gβ), and Ste18 (Gγ) are essential for activity of the MAP kinase pathway in response to pheromone (Liu et al. 1993). Second, the Ste5 protein is specifically involved in mating and functions as a kinase-scaffold protein that tethers the kinases Ste11, Ste7, and Fus3. Ste5, together with the kinases, is recruited to the plasma membrane via interactions with the Gβγ subunits Ste4/Ste18 (Pryciak and Huntress 1998). The Gβγ subunits Ste4/Ste18 facilitate signaling from the upstream pheromone receptor and Cdc42 to downstream transcription factors, including Ste12 and Mcm1. A scaffold protein implicated in pseudohyphal growth has not yet been identified. Third, the MAP kinases Fus3 and Kss1 establish two different developmental events, mating and filamentous growth. Fus3 is specialized to control mating and also inhibits haploid invasive growth (Madhani et al. 1997). On the other hand, Kss1 is specialized to promote haploid invasive and diploid pseudohyphal growth (Cook et al. 1997; Madhani et al. 1997). In the unphosphorylated form, Kss1 allows the transcription factor Ste12 to form an inactive complex with Dig1/2. Phosphorylation of Kss1 by Ste7 in response to nutrients releases the repression of Ste12 by Dig1/2 and then induces filamentous growth (Cook et al. 1996; Madhani and Fink 1997; Bardwell et al. 1998a, 1998b). Fourth, Ste12 forms a heterodimer with Tec1 and regulates expression of Tec1 itself and the cell surface flocculin Flo11, which is essential for agar invasion and filamentation during diploid pseudohyphal growth (Gavrias et al. 1996; Madhani and Fink 1997; Lo and Dranginis 1998; Rupp et al. 1999). On the other hand, Ste12 interacts with the Mcm1 protein to activate transcription of genes containing pheromone response elements during mating (Hwang-Shum et al. 1991; Oehlen et al. 1996).

The cyclin-dependent kinase Srb10/Cdk8/Ume5 functions with the C-type cyclin Srb11/Ume3 and regulates Ste12 protein stability during diploid pseudohyphal growth (Nelson et al. 2003). The Srb10-Srb11 complex phosphorylates Ste12 at Ser 261 and Ser 451, resulting in destabilization of Ste12. On the other hand, the S261A and S451A mutations stabilize Ste12 and promote pseudohyphal differentiation. Consistent with these observations, *srb10* mutant cells also form more pseudohyphae compared to wild type cells (Holstege et al. 1998; Nelson et al. 2003). In addition, *FLO11* expression is derepressed in *srb10*Δ mutant cells (Holstege et al. 1998). Therefore, the Srb10-Srb11 kinase inhibits pseudohyphal differentiation. Importantly, neither the S261A and S451A *ste12* mutations nor the *srb10*Δ mutation affect mating (Nelson et al. 2003). Under abundant glucose and nitrogen limitation conditions where pseudohyphal growth is induced, Srb10 protein levels are dramatically decreased, stabilizing Ste12 to induce pseudohyphal differentiation (Nelson et al. 2003). Interestingly Srb11 cyclin levels do not oscillate throughout the cell cycle, but are rather altered in response to nutrients and stress unlike cyclins that regulate mitotic cell division (Cooper et al. 1997). Therefore nutrients might control Srb10-Srb11 kinase activity by regulating protein levels of both Srb10 and Srb11 to induce pseudohyphal differentiation.

6.2.6 An ammonium sensor Mep2

Limiting extracellular nitrogen levels are crucial cues in pseudohyphal differentiation. Mep2, a high affinity ammonium permease, is required for pseudohyphal growth as an ammonium sensor (Lorenz and Heitman 1998). *S. cerevisiae* possesses 3 ammonium transporters, Mep1, Mep2, and Mep3, which belong to a Mep/Amt/Rh protein family that is highly conserved in bacteria, Archaea, fungi, plants, and vertebrates (Marini et al. 1994, 2000; Ninnemann et al. 1994; Montesinos et al. 1998; Marini and André 2000). Mep1 and Mep3 are highly homologous and share about 80% amino acid identity over their entire amino acid sequence, whereas Mep2 shares only 45% identity with either Mep1 or Mep3. All three transporters are predicted to contain 11 transmembrane domains and have distinct kinetic properties (Marini et al. 1997a; Marini and André 2000). Mep2 has the highest affinity for ammonium (*K*m 1-2 μM), Mep1 has a moderate affinity (*K*m 5-10 μM), and Mep3 shows a much lower affinity (*K*m, 1.4 to 2.1 mM).

Deletion of all three *MEP* genes abolishes cell growth on low ammonium medium as a sole nitrogen source, whereas single-deletion strains have no apparent growth defect (Lorenz and Heitman 1998). Interestingly, reintroduction of the *MEP2* gene on a 2μ plasmid into the *mep1,2,3Δ* cells enables cells to grow and differentiate pseudohyphae. However, neither the *MEP1* gene nor the *MEP3* gene was able to overcome the defect in filamentous growth, providing evidence that Mep2 acts not only as an ammonium transporter but also as an ammonium sensor.

Expression of the *MEP2* gene is regulated by the GATA transcription factor Gln3 and its associated cytoplasmic factor Ure2, both components of the nutrient sensing Tor pathway (Beck and Hall 1999; Cardenas et al. 1999; Rohde et al. 2001; Cooper 2002). Tor (Target of rapamycin) is a protein kinase that belongs to the phosphatidylinositol 3-kinase superfamily and is the target of the immunosuppressive FKBP12-rapamycin complex (Heitman et al. 1991a, 1991b; Cafferkey et al. 1993; Kunz et al. 1993; Helliwell et al. 1994). Under nutrient replete conditions, Tor is active and induces transcription of ribosomal genes, tRNA, and rRNA. Under nutrient starvation conditions or treatment with rapamycin, Tor is inactivated, resulting in induction of genes required for utilization of poor or limiting nitrogen sources, including the *MEP2* gene. Importantly, the Tor nutrient sensing pathway controls pseudohyphal differentiation (Cutler et al. 2001). Therefore yeast cells establish elaborate signaling networks to control cellular differentiation in response to nutrients.

Ammonium transport systems involving members of the Mep/Amt/Rh family are well conserved in organisms from yeast and bacteria to mammals (Marini et al. 2000; Javelle et al. 2001, 2003). In the ectomycorrhizal fungus *Hebeloma cylindrosporum*, three ammonium transporters, Amt1, Amt2, and Amt3, have been characterized (Javelle et al. 2001, 2003). Amt1 and Amt2 are high affinity transporters and Amt3 possesses a lower affinity. Both Amt1 and Amt2 can partially restore the pseudohyphal defect of *S. cerevisiae mep1,2,3Δ* mutant cells, whereas Amt3 cannot, indicating that Amt1 and Amt2 might function as ammonium sensors as well as ammonium transporters and that components of the Mep2-mediated signaling pathway regulating pseudohyphal differentiation in *S. cere-*

visiae might be conserved to control cell differentiation in *H. cylindrosporum*. Another example that demonstrates conservation among ammonium transport systems is presented in a mammalian homolog, RhAG, which is one of the Rh blood-group antigens (Marini et al. 2000). In spite of their biological importance, the physiological functions of these proteins were obscure until only recently. Marini *et al.* (2000) found that one of the Rh proteins, RhAG, can rescue the growth defect of *S. cerevisiae mep1,2,3*Δ mutant cells on media containing low ammonium. They also showed that RhAG is involved in excretion of ammonium. The RhAG homolog, RhGK, which is a kidney-specific protein that shares significant homology with Mep/Amt ammonium transporters (~24%), was also shown to play roles in ammonium transport and excretion. Therefore the Rh proteins may function to both import and export ammonium ions. Rh protein homologs have been reported in *Dictyostelium discoideum*, *Caenorhabditis elegans*, and *Drosophila melanogaster* (Marini et al. 1997b; Heitman and Agre 2000), but their functions have not yet been established in these organisms. Based on similarity of amino acid sequence, these proteins probably act as ammonium transporters and may also have a role in excretion of ammonia. In yeast, ammonia is used as a signaling molecule to communicate with neighboring colonies (Palková et al. 1997, 2002; Palková and Forstová 2000). Therefore the Rh proteins might be involved in a mammalian nutrient or toxin signaling pathway, analogous to the role of Mep2 in triggering yeast filamentous growth when ammonium ions are limiting.

6.3 Haploid invasive growth

Pseudohyphal differentiation occurs in diploid cells in response to the presence of an abundant fermentable carbon source and limiting nitrogen source. A related morphological process, termed haploid invasive growth, has also been described (Roberts and Fink 1994). In contrast to diploid pseudohyphal growth, haploid invasive growth normally occurs on rich media such as yeast extract-peptone-dextrose (YPD) medium during an extended period of incubation (Roberts and Fink 1994). Compared with haploid invasive growth, diploid cells do not penetrate and do not significantly extend filaments on the surface of the agar away from the colony under nutrient rich conditions. Many of the same signaling components that regulate diploid pseudohyphal growth are also required for haploid invasive growth, including components of the MAP kinase pathway (Ste20, Ste11, Ste7, Ste12, and Tec1) (Roberts and Fink 1994) and components of the cAMP signaling pathway (Gpr1, Gpa2, Ras2, Tpk2, Flo8, and Flo11) (Lo and Dranginis 1998; Mösch et al. 1999; Pan and Heitman 1999; Lorenz et al. 2000b). Taken together, haploid invasive growth shares many features with diploid pseudohyphal growth.

It has been suggested that nutritional limitation might occur beneath colonies and stimulate haploid invasive growth, even on rich medium (Roberts and Fink 1994). When yeast cells begin to consume nutrients in rich medium, cellular proteins and amino acids are catabolized to scavenge nitrogen, resulting in the production of short-chain alcohols called fusel alcohols. Recent studies reveal that several short-chain alcohols, including isoamyl alcohol and butanol, dramatically stimulate haploid yeast cells to differentiate into a filamentous form, which is similar to diploid pseudohyphal form (Dickinson 1996; Lorenz et al. 2000a). Under these conditions, Ste12 and Tec1, both components in the MAP kinase signaling pathway, are required for haploid filamentation but the Gpr1-Gpa2 system is not (Lorenz et al. 2000a). Fusel alcohols inhibit translation at the initiation step via Gcd1, which is a component of the eIF2B guanine nucleotide exchange complex that is responsible for recycling eIF2-GDP to eIF2-GTP (Ashe et al. 2001). Therefore, fusel alcohols may serve as metabolic byproducts that could serve as quorum sensing factors to induce signaling and differentiation. How these alcohols are sensed is not yet known, but these or other metabolic products could regulate differentiation under certain culture conditions.

Interestingly, haploid invasive growth was found to be induced by glucose depletion (Cullen and Sprague 2000). Under these conditions, haploid cells adopted an elongated morphology and a unipolar budding pattern and invaded the agar. These phenotypes were all suppressed by glucose addition. Signaling components required for glucose depletion-induced invasive growth are largely shared with those required for diploid pseudohyphal growth (Cullen and Sprague 2000, 2002; Kuchin et al. 2002; Palecek et al. 2002; Vyas et al. 2003). Importantly, the Ser/Thr protein kinase Snf1 is involved in multiple ways and one of the key roles of Snf1 is controlling Flo11 expression in response to glucose depletion (Kuchin et al. 2002; Vyas et al. 2003).

The Snf1 kinase belongs to a protein kinase family ubiquitously conserved from fungi to mammals that plays a key role in gene expression, metabolic events, and cellular development in response to glucose limitation (Gancedo 1998; Hardie et al. 1998). In *S. cerevisiae*, the Snf1 kinase forms three different complexes composed of the activate subunit Snf4 and one of three β subunits, Sip1, Sip2, or Gal83, which facilitate the involvement of the Snf1 kinase in diverse regulatory events in response to glucose limitation (Yang et al. 1994; Jiang and Carlson 1997). The active subunit Snf4 is required for Snf1 function (Celenza and Carlson 1989; Celenza et al. 1989; Woods et al. 1994). The three β subunits are in part functionally redundant but also have unique functions that confer substrate specificity of the Snf1 kinase (Yang et al. 1994; Schmidt and McCartney 2000). The three β subunits also control the subcellular localization of the Snf1 kinase, contributing to the specificity and the diverse functions of the kinase (Vincent et al. 2001).

During haploid invasive growth in response to glucose deprivation, filamentation and agar invasion are induced and the Snf1 kinase is essential for these events (Cullen and Sprague 2000). While the β subunit Gal83 is required for Flo11-dependent agar invasion, the β subunits Gal83 and Sip2 are functionally redundant with respect to filamentation (Vyas et al. 2003). In the presence of abundant glucose, Nrg1 and Nrg2, which are closely related repressors containing two C_2H_2 zinc-finger motifs, repress *FLO11* expression, whereas the Snf1-Gal83 kinase complex induces *FLO11* expression by antagonizing the Nrg1 and Nrg2 repressors in response to glucose limitation (Park et al. 1999; Kuchin et al. 2002; Vyas et al. 2003). In addition, *NRG1* expression itself is also repressed by glucose (Park et al.

1999). Nrg1 and Nrg2 also repress a subset of genes regulated by Snf1, including the *SUC2* and *DOG2* genes, and interact with Snf1 (Vyas et al. 2001). Although Nrg1 is implicated in repression of the *STA1* gene by recruiting the Ssn6-Tup1 repressor complex, it remains to be elucidated whether or not the Ssn6-Tup1 repressor complex is required for Nrg1/2-mediated repression of *FLO11* (Park et al. 1999). Importantly the cell surface glucose sensing Gpr1-Gpa2 system, which senses extracellular glucose and signals via the cAMP signaling pathway, and the Snf1-mediated signaling pathway, which senses intracellular glucose levels and stimulates *FLO11* expression in response to glucose starvation, both impinge on the *FLO11* promoter. Therefore yeast cells operate delicate signaling networks in response to nutrient availability to enable proper cellular differentiation under diverse environmental conditions.

Snf1 and Nrg1/2 are also involved in diploid pseudohyphal growth (Kuchin et al. 2002). *snf1* diploid cells exhibit a pseudohyphal defect. In contrast, *nrg1,2* diploid cells show an enhanced pseudohyphal phenotype. Interestingly, *nrg1,2* mutations are not able to restore the pseudohyphal defect of *snf1* cells completely, suggesting Snf1 controls diploid pseudohyphal growth via not only Nrg1/2 but also an as yet unidentified factor(s). The fact that Snf1 is required for diploid pseudohyphal growth supports a hypothesis in which the Snf1 kinase activity might also respond to nitrogen limitation and control expression of genes required for pseudohyphal growth, or expression of the *SNF1* gene might be regulated by nitrogen limitation and thus the Snf1 kinase might function as a dual sensor for glucose and nitrogen, similar to the glucose receptor Gpr1. Alternatively, the involvement of the Snf1 kinase in pseudohyphal growth might be indirect and function to enable metabolic adaptation under conditions where pseudohyphal differentiation is induced.

The human pathogen *C. albicans* possesses homologs of Snf1 and Nrg1/2 (Petter et al. 1997; Braun et al. 2001; Murad et al. 2001a, 2001b). As in *S. cerevisiae, C. albicans* Nrg1 functions as a repressor and inhibits filamentous growth by repressing hyphal-specific genes including *ECE1, HWP1, ALS3,* and *ALS8* (Braun et al. 2001; Murad et al. 2001a, 2001b). Genetic data suggest that the transcriptional repressor Tup1, which is a key regulator of filamentous growth in *C. albicans*, controls gene expression, at least in part, via Nrg1 (Braun et al. 2001; Murad et al. 2001a). Similar to *tup1* mutant cells, *nrg1* mutant cells are avirulent in the mouse tail-vein injection model (Braun et al. 2001; Murad et al. 2001b). Expression of the *NRG1* gene itself is down-regulated under conditions where filamentation is induced, consistent with its function (Braun et al. 2001; Murad et al. 2001b). This down-regulation is dependent on Efg1, a major activator of filamentous growth that is homologous to Phd1 of *S. cerevisiae*. The role of the Snf1 kinase in *C. albicans* remains to be elucidated since the *SNF1* gene is essential for cell growth (Petter et al. 1997). Taken together, components of the signaling pathways are well conserved between *S. cerevisiae* and *C. albicans* and control related developmental events in both organisms.

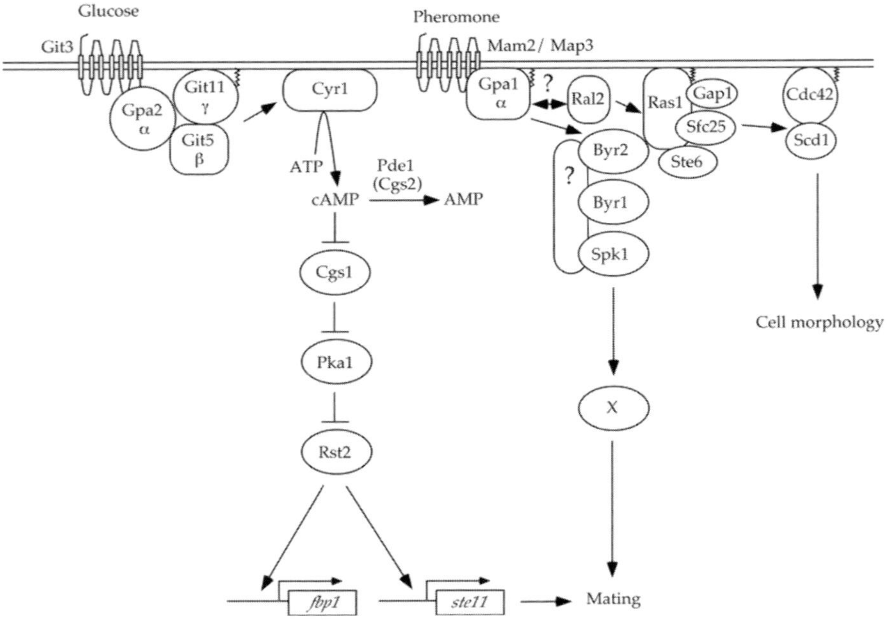

Fig. 6. Signaling pathways regulating mating and gene expression in *S. pombe*. Both the glucose sensing cAMP signaling pathway and the pheromone sensing MAP kinase signaling pathway coordinately control mating in *S. pombe*. This is a striking contrast to *S. cerevisiae* in which the glucose sensing cAMP signaling pathway plays no obligatory role in mating. In *S. pombe*, the glucose sensing cAMP signaling pathway also regulates expression of the *fbp1* gene, which encodes fructose-1,6-bisphosphatase involved in gluconeogenesis.

6.4 GPCRs in *Schizosaccharomyces pombe*

In contrast to *S. cerevisiae*, *S. pombe* cells exist predominantly as haploid cells in nature, the diploid state is relatively unstable, and meiosis and sporulation immediately occur after mating. In *S. pombe*, mating is controlled by both pheromones and nutrients. Starvation for carbon or nitrogen source induces mating and requires two signaling pathways, namely the nutrient-sensing cAMP signaling pathway and the pheromone responsive MAP kinase pathway (Fig. 6). Nutrient deprivation decreases intracellular cAMP levels by inhibiting adenylyl cyclase activity, and mating pheromones activate the pheromone responsive MAP kinase pathway. Similar to *S. cerevisiae*, *S. pombe* contains 3 GPCRs; two are pheromone receptors, Map3 and Mam2, and the third is the glucose sensing Git3 receptor that is homologous to *S. cerevisiae* Gpr1.

As in budding yeast, glucose stimulates cAMP production in *S. pombe* (Byrne and Hoffman 1993). Increased cAMP inhibits expression of both the *fbp1* and *ste11* genes, which encode the gluconeogenic enzyme fructose-1,6,-bisphosphatase and an HMG transcription factor required for mating and meiosis, respectively (Vassarotti and Friesen 1985; Hoffman and Winston 1990; Sugimoto et al. 1991; Mochizuki and Yamamoto 1992; Byrne and Hoffman 1993). In contrast, glucose limitation decreases intracellular cAMP levels and ultimately derepresses expression of both the *fbp1* and *ste11* genes. These observations illustrate a link between nutrients and cAMP production and mating and meiosis (Sugimoto et al. 1991; Mochizuki and Yamamoto 1992). Studies have revealed that the GPCR Git3, Gα subunit Gpa2, Gβ subunit Git5, Gγ subunit Git11, adenylyl cyclase Cyr1, the PKA regulatory subunit Cgs1, and the PKA catalytic subunit Pka1 are components of the cAMP signaling pathway that are quite similar to the elements in *S. cerevisiae* that regulate diploid pseudohyphal differentiation (Fig. 2 and 6, Yamawaki-Kataoka et al. 1989; Young et al. 1989; DeVoti et al. 1991; Hoffman and Winston 1991; Isshiki et al. 1992; Maeda et al. 1994; Landry et al. 2000; Welton and Hoffman 2000; Landry and Hoffman 2001). In *S. pombe*, the cAMP signaling pathway is not essential for cell viability, because neither adenylyl cyclase encoded by the *cyr1* gene nor the catalytic subunit of the cAMP-dependent protein kinase A encoded by the *pka1* gene is essential (Hoffman and Winston 1991; Maeda et al. 1994). This is in striking contrast to the essential role of the cAMP signaling pathway in *S. cerevisiae*.

The *S. pombe* Git3 protein has seven transmembrane domains characteristic of GPCRs and it functions in the nutrient sensing cAMP-signaling pathway (Welton and Hoffman 2000). Cells lacking the Gα subunit Gpa2 readily mate and sporulate even on rich media and do not respond to glucose to trigger cAMP production (Isshiki et al. 1992). In addition, *gpa2* mutants constitutively express the *fbp1* gene in the presence of glucose (Nocero et al. 1994). Similar to *gpa2* mutants, *git3* receptor deleted cells derepress *fbp1* gene expression and sporulate under rich conditions. The defects of *git3* mutant cells can be suppressed by exogenous cAMP or expression of an activated $gpa2^{R176H}$ allele. Conversely, overproduction of the Git3 receptor cannot suppress the adenylyl cyclase mutant phenotype (Welton and Hoffman 2000). Furthermore, *git3* mutant cells as well as *pka1* mutant cells show a delay in spore germination, and *git3 pka1* double mutant cells exhibited the same phenotype as *git3* or *pka1* single mutants (Welton and Hoffman 2000). These findings suggest that Git3 acts upstream of Gpa2 in the cAMP signaling pathway. The *git3* gene encodes a homolog of the *S. cerevisiae* GPCR Gpr1 glucose receptor (Welton and Hoffman 2000). Thus, *S. pombe* Git3 is thought to couple to the Gα subunit Gpa2, which is homologous to *S. cerevisiae* Gpa2. Interestingly, *git3 gpa2* double mutant cells showed a more severe defect than *gpa2* cells with respect to *fbp1* expression under rich conditions, indicating Git3 may also have functions independent of Gpa2 (Welton and Hoffman 2000).

Genetic and biochemical data revealed that *S. pombe* Gpa2 forms a heterotrimeric G protein with Gβ (Git5) and Gγ (Git11) subunits (Landry et al. 2000; Landry and Hoffman 2001). This is a striking contrast to the *S. cerevisiae* nutrient-sensing Gpr1-Gpa2 module that lacks canonical Gβγ subunits. Because both *git5* and *git11* mutant cells mate and derepress *fbp1* gene expression under nutrient-rich conditions, similar to *gpa2* mutant cells, the Git5 and Git11 Gβγ subunits cooperatively function with the Gα subunit Gpa2 (Landry et al. 2000; Landry and Hoffman 2001). The Gβ Git5 protein shares 43% identity with mammalian Gβ subunits and contains seven WD-40 repeats characteristic of canonical Gβ subunits. However, the Gβ Git5 subunit lacks the amino-terminal coiled-coil domain, which is typically essential for Gβ subunits to interact with Gγ subunits and to properly fold into 7 bladed β-propellers (Landry and Hoffman 2001). This atypical feature of the Git5 Gβ subunit may reflect unusual features of this signaling cascade (Landry and Hoffman 2001). Namely, the Gα subunit Gpa2 is partially functional in the absence of the Gβ subunit Git5 or the Gγ subunit Git11, suggesting Gβ Git5 can fold properly into a β-propeller and function without the Gγ subunit Git11. Although overproduction of the Gα subunit Gpa2 can suppress the defects of both Gβ *git5* and Gγ *git11* mutants, overproduction of either Gβ Git5 or Gγ Git11 cannot suppress the Gα *gpa1* mutant defects. Therefore, both Gβ Git5 and Gγ Git11 function exclusively via the Gα subunit Gpa2 (Landry and Hoffman 2001).

The fact that the *S. pombe* Gα subunit Gpa2 is still partially functional in the absence of Git5 Gβ, Git11 Gγ, or both subunits may imply that Gpa2 is anchored at the plasma membrane by itself and functionally interacts with the Git3 glucose receptor to signal adenylyl cyclase in response to glucose. This raises the question of how Gα Gpa2 is tethered at the plasma membrane to function. In canonical heterotrimeric G proteins, Gγ subunits are subject to prenylation at the cysteine residue of a carboxyl-terminal CaaX motif, which is essential for membrane localization and function of Gβγ subunits. Gβγ subunits are required for functional coupling between the receptor and the Gα subunit (Sternweis 1986; Blumer and Thorner 1990). Gα subunits are myristoylated, palmitoylated, or both and these lipid modifications are required for membrane localization and Gα function. Myristoylation occurs on members of the $G\alpha_i$ subfamily at the amino-terminal glycine (Gly2), which is the penultimate amino acid of mature Gα proteins after removal of the initiating methionine. On the other hand, palmitoylation occurs on all Gα subunits, except for members of the $G\alpha_t$ subfamily, at a cysteine residue (Cys) near the amino-terminus. Interestingly *S. pombe* Gα Gpa2 possesses neither the Gly2 nor an amino-terminal Cys; the most amino-terminal Cys is located at the 104th residue of the Gpa2 protein. Therefore, determining if Gpa2 is lipid modified and how it is recruited to the plasma membrane should provide new insights about mechanisms for lipid modification of novel Gα subunits.

Similar to *S. cerevisiae*, *S. pombe* has two mating types, h^+ and h^-. h^+ cells express the pheromone receptor Map3 whose ligand is the mating pheromone M factor secreted by h^- cells (Tanaka et al. 1993). h^- cells express the pheromone receptor Mam2 whose ligand is the mating pheromone P factor secreted by h^+ cells (Kitamura and Shimoda 1991). Both pheromone receptors have seven putative transmembrane domains, are homologous to the mating pheromone receptors in *S. cerevisiae*, and are required for mating in heterothallic haploid h^+ and h^- cells (Tanaka et al. 1993). Interestingly, homozygous *map3 mam2* double mutant diploid cells are unable to sporulate, suggesting that sensing of the mating pheromones is also required in diploid cells (Tanaka et al. 1993). This is in contrast to *S. cerevisiae* in which genes encoding mating pheromones and pheromone receptors are haploid specific and are repressed in diploid cells.

Pheromone signaling controls the MAP kinase pathway via the Gα subunit Gpa1, which has conserved domains found in heterotrimeric G protein α subunits (Fig. 6). Haploid cells lacking Gpa1 cannot mate and *gpa1* diploid cells fail to sporulate (Obara et al. 1991). Moreover overproduction of the dominant active allele *gpa1*^{Q244L} in either haploid mating type induces conjugation tube formation without a mating partner (Obara et al. 1991). Therefore Gpa1 positively regulates mating in contrast to *S. cerevisiae* Gα Gpa1, which plays a negative role in mating (Fig. 7). In the budding yeast, the Gα subunit Gpa1 forms a heterotrimeric G protein with the Gβγ subunits Ste4 and Ste18 and inhibits function of the Gβγ subunits, which activate the MAP kinase cascade in response to pheromone. As in *S. cerevisiae*, *S. pombe* expresses only one set of Gβγ subunits, namely the Gβ subunit Git5 and the Gγ subunit Git11. Several lines of evidence clearly indicate that the Gβγ subunits function in the nutrient sensing pathway together with the other Gα subunit Gpa2 (Fig. 6). Therefore, in this regard, the *S. pombe* Gpa1 Gα subunit is not a conventional Gα subunit. These observations illustrate interesting contrasts with respect to signaling systems (Fig. 7). In *S. cerevisiae*, the nutrient sensing GPCR Gpr1 is coupled to the unconventional Gα subunit Gpa2 to trigger cAMP signaling in response to glucose and the pheromone sensing GPCRs Ste2/3 are coupled to the conventional Gα subunit Gpa1 to activate the MAP kinase cascade. In *S. pombe*, the nutrient sensing GPCR Git3 is coupled to the conventional Gα subunit Gpa2 to trigger the cAMP signaling pathway in response to glucose and the pheromone sensing GPCRs Mam2/Map3 are coupled to the unconventional Gα subunit Gpa1 to activate the MAP kinase cascade.

As mentioned above, the novel kelch proteins Gpb1/2 have been identified as Gβ mimics for the Gα subunit Gpa2 in *S. cerevisiae*. We find no homolog of Gpb1/2 in *S. pombe*, based on results of a psi-BLAST search against the *S. pombe* genome sequence. Because kelch proteins are highly divergent, it might be difficult to identify a homolog of the Gpb1/2 kelch proteins. Alternatively, it is also possible that novel Gpa1 partners might be either a different type of kelch protein that does not have any significant homology with Gpb1/2 or an unrelated protein, since *S. pombe* Gpa1 does not have the long amino-terminal extension characteristic of *S. cerevisiae* Gpa2 although Gpa1 does have a more moderate amino-terminal extension (Obara et al. 1991).

Interestingly, a kelch protein called Ral2, which is required for mating, has been identified in *S. pombe* (Fukui and Yamamoto 1988; Fukui et al. 1989). Genetic data suggests that Ral2 functions with Ras1 to control mating (Fukui and Yamamoto 1988; Fukui et al. 1989). The Ral2 protein contains several kelch repeats containing significant homology with the p40 kelch protein that is proposed to be an effector of the small GTPase Rab9 involved in vesicle transport in human cells (Díaz et al. 1997). Moreover Ral2 protein also has significant homology with the kelch protein RasGEF in *Dictyostelium discoideum*. Thus a subgroup of kelch

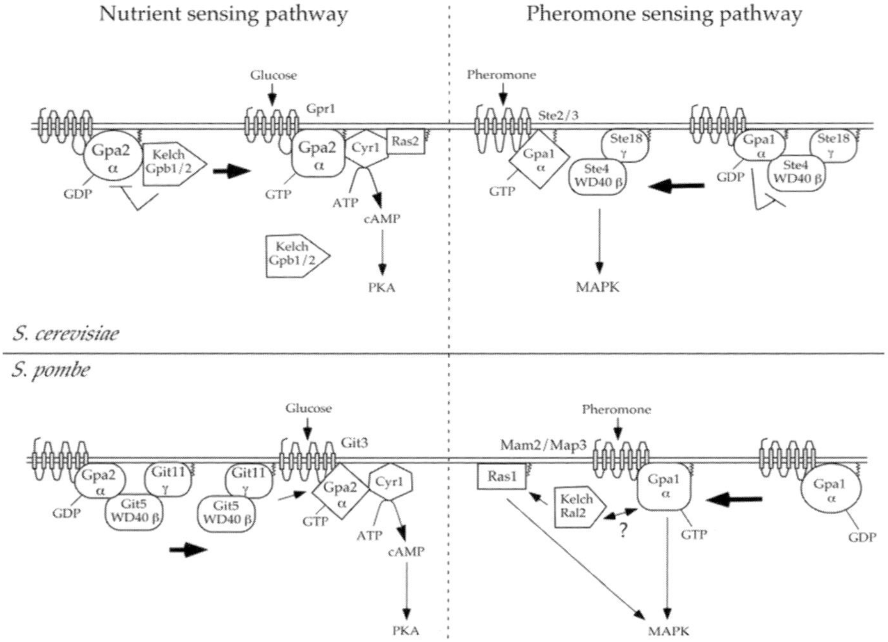

Fig. 7. Comparison of GPCR-heterotrimeric G protein systems in *S. cerevisiae* and *S. pombe*. Both yeasts have two GPCR-heterotrimeric G protein systems, one for glucose sensing and the other for pheromone sensing. Interestingly, the *S. cerevisiae* Gα subunit Gpa2 involved in glucose sensing does not form a conventional heterotrimeric G protein and instead forms a complex with the novel Gβ mimic proteins Gpb1/2, which contain seven kelch repeats and inhibit Gpa2 activity. On the other hand, the *S. pombe* Gα subunit Gpa1 required for pheromone sensing does not interact and function with the known Gβγ subunits Git5 and Git11. Instead the kelch protein Ral2 may function to control mating as a novel binding partner for the Gα subunit Gpa1, based on our observations in *S. cerevisiae.*

proteins seems to function in signaling together with G proteins, including small G proteins. In summary, the Gpb1/2 kelch protein Ral2 might act with the Gα subunit Gpa1 similar to the role of the kelch proteins in Gα signaling events in *S. cerevisiae* (Fig. 4B).

S. pombe contains a single ras homolog, Ras1 (Fig. 6 and 7, Fukui and Kaziro 1985). As mentioned above, Ras1 and Ras2 in *S. cerevisiae* activate adenylyl cyclase to stimulate cAMP production and Ras2 also positively controls the MAP kinase pathway during pseudohyphal growth. In contrast, Ras1 in *S. pombe* has a role in activating the MAP kinase signaling pathway to mate but does not play a clear role in activating adenylyl cyclase (Fukui et al. 1986). *ras1* mutant cells are sterile and a dominant active allele *ras1*^{G17V} exhibits a hypersporulation phenotype (Fukui et al. 1986; Neiman et al. 1993). Neither *ras1* null or *ras1*^{G17V} mutant cells show any defect in cAMP production (Fukui et al. 1986). Recent evidence suggests that *S. pombe* Ras1 functions in both activating the mating MAP kinase pathway via the MAPKKK Byr2 homologous to Ste11 of *S. cerevisiae* and morphogenesis via Scd1, a putative guanine nucleotide exchange factor for Cdc42 (Fig. 6, Chang et al. 1994; Barr et al. 1996; Chen et al. 1999). The functional specificity of Ras1 in the two signaling pathways depends on expression of two guanine nucleotide exchange factors, Ste6 and Efc25, for Ras1 (Papadaki et al. 2002). The components of the MAP kinase signaling pathway regulating mating in *S. pombe* have been identified (Fig. 6). Those include the MAPKKK Byr2, the MAPKK Byr1, and the MAPK Spk1 (Nadin-Davis and Nasim 1988; Toda et al. 1991; Wang et al. 1991; Gotoh et al. 1993; Neiman et al. 1993).

In summary, the function of both the heterotrimeric G-protein Gα subunits (Gpa1 and Gpa2) and small G-protein Ras (Ras1 and Ras2) have been specialized during evolution of budding and fission yeasts by the loss or acquisition of ability to partner with different signaling components, including βγ subunits, kelch based Gβ mimics, and adenylyl cyclase. Further studies in these two divergent yeasts promise to provide additional substantial insights into novel and conserved signaling paradigms.

6.5 Perspective

Studies on the role of nutrient sensing in the control of dimorphic transition in *S. cerevisiae* have revealed a novel GPCR signaling pathway that senses nutrients, defined a coupled Gα subunit that activates adenylyl cyclase bringing the yeast system in line with analogous signaling pathways in mammals, and resulted in the discovery of a novel class of heterodimer on heterotrimeric G proteins with canonical Gα subunits coupled to novel kelch repeat Gβ subunit mimics. These signaling paradigms established in budding yeast are conserved and elaborated in the fission yeast *S. pombe*. Analogous GPCR-Gα-kelch signaling pathways and a novel class of heterodimeric or heterotrimeric G proteins remain to be discovered in multicellular eukaryotes such as nematodes, flies, plants, mice, and humans. Continued studies on the yeast dimorphism-controlling signaling pathways promise to provide insights into how fungal pathogens sense and infect both plants and animals and how multicellular eukaryotic cells sense and respond to environmental cues.

Acknowledgments

We thank Stefan Hohmann and Joris Winderickxfor the invitation to prepare this review and Kirsten Nielsen, Julie Hicks, and John Rohde for critical reading. Joseph Heitman is an associate investigator of the Howard Hughes Medical Institute and a Burroughs Wellcome Scholar in molecular pathogenic mycology.

References

Adams J, Kelso R, Cooley L (2000) The kelch repeat superfamily of proteins: propellers of cell function. Trends Cell Biol 10:17-24

Albig W, Entian KD (1988) Structure of yeast glucokinase, a strongly diverged specific aldo-hexose-phosphorylating isoenzyme. Gene 73:141-152

Alspaugh JA, Perfect JR, Heitman J (1997) *Cryptococcus neoformans* mating and virulence are regulated by the G-protein α subunit GPA1 and cAMP. Genes Dev 11:3206-3217

Alspaugh JA, Pukkila-Worley R, Harashima T, Cavallo LM, Funnell D, Cox GM, Perfect JR, Kronstad JW, Heitman J (2002) Adenylyl cyclase functions downstream of the Gα protein Gpa1 and controls mating and pathogenicity of *Cryptococcus neoformans*. Eukaryot Cell 1:75-84

André B (1995) An overview of membrane transport proteins in *Saccharomyces cerevisiae*. Yeast 11:1575-1611

Ansari K, Martin S, Farkasovsky M, Ehbrecht IM, Küntzel H (1999) Phospholipase C binds to the receptor-like *GPR1* protein and controls pseudohyphal differentiation in *Saccharomyces cerevisiae*. J Biol Chem 274:30052-30058

Apanovitch DM, Slep KC, Sigler PB, Dohlman HG (1998) Sst2 is a GTPase-activating protein for Gpa1: Purification and characterization of a cognate RGS-Gα protein pair in yeast. Biochemistry 37:4815-4822

Ashe MP, Slaven JW, De Long SK, Ibrahimo S, Sachs AB (2001) A novel eIF2B-dependent mechanism of translational control in yeast as a response to fusel alcohols. EMBO J 20:6464-6474

Bardwell L, Cook JG, Voora D, Baggott DM, Martinez AR, Thorner J (1998a) Repression of yeast Ste12 transcription factor by direct binding of unphosphorylated Kss1 MAPK and its regulation by the Ste7 MEK. Genes Dev 12:2887-2898

Bardwell L, Cook JG, Zhu-Shimoni JX, Voora D, Thorner J (1998b) Differential regulation of transcription: Repression by unactivated mitogen-activated protein kinase Kss1 requires the Dig1 and Dig2 proteins. Proc Natl Acad Sci USA 95:15400-15405

Barr MM, Tu H, Van Aelst L, Wigler M (1996) Identification of Ste4 as a potential regulator of Byr2 in the sexual response pathway of *Schizosaccharomyces pombe*. Mol Cell Biol 16:5597-5603

Batlle M, Lu AL, Green DA, Xue Y, Hirsch JP (2003) Krh1p and Krh2p act downstream of the Gpa2p Gα subunit to negatively regulate haploid invasive growth. J Cell Sci 116:701-710

Beck T, Hall MN (1999) The TOR signalling pathway controls nuclear localization of nutrient-regulated transcription factors. Nature 402:689-692

Bergsma JC, Kasri NN, Donaton MC, De Wever V, Tisi R, de Winde JH, Martegani E, Thevelein JM, Wera S (2001) PtdIns(4,5)P_2 and phospholipase C-independent Ins(1,4,5)P_3 signals induced by a nitrogen source in nitrogen-starved yeast cells. Biochem J 359:517-523

Beullens M, Mbonyi K, Geerts L, Gladines D, Detremerie K, Jans AW, Thevelein JM (1988) Studies on the mechanism of the glucose-induced cAMP signal in glycolysis and glucose repression mutants of the yeast *Saccharomyces cerevisiae*. Eur J Biochem 172:227-231

Bhattacharya S, Chen L, Broach JR, Powers S (1995) Ras membrane targeting is essential for glucose signaling but not for viability in yeast. Proc Natl Acad Sci USA 92:2984-2988

Blumer KJ, Thorner J (1990) β and γ subunits of a yeast guanine nucleotide-binding protein are not essential for membrane association of the α subunit but are required for receptor coupling. Proc Natl Acad Sci USA 87:4363-4367

Boles E, Hollenberg CP (1997) The molecular genetics of hexose transport in yeasts. FEMS Microbiol Rev 21:85-111

Brandão RL, de Magalhães-Rocha NM, Alijo R, Ramos J, Thevelein JM (1994) Possible involvement of a phosphatidylinositol-type signaling pathway in glucose-induced activation of plasma membrane H^+-ATPase and cellular proton extrusion in the yeast *Saccharomyces cerevisiae*. Biochem Biophys Acta 1223:117-124

Braun BR, Kadosh D, Johnson AD (2001) *NRG1*, a repressor of filamentous growth in *C. albicans*, is down-regulated during filament induction. EMBO J 20:4753-4761

Broek D, Toda T, Michaeli T, Levin L, Birchmeier C, Zoller M, Powers S, Wigler M (1987) The *S. cerevisiae CDC25* gene product regulates the *RAS*/adenylate cyclase pathway. Cell 48:789-799

Byrne SM, Hoffman CS (1993) Six *git* genes encode a glucose-induced adenylate cyclase activation pathway in the fission yeast *Schizosaccharomyces pombe*. J Cell Sci 105:1095-1100

Cafferkey R, Young PR, McLaughlin MM, Bergsma DJ, Koltin Y, Sathe GM, Faucette L, Eng WK, Johnson RK, Livi GP (1993) Dominant missense mutations in a novel yeast protein related to mammalian phosphatidylinositol 3-kinase and VPS34 abrogate rapamycin cytotoxicity. Mol Cell Biol 13:6012-6023

Cannon JF, Tatchell K (1987) Characterization of *Saccharomyces cerevisiae* genes encoding subunits of cyclic AMP-dependent protein-kinase. Mol Cell Biol 7:2653-2663

Cardenas ME, Cutler NS, Lorenz MC, Di Como CJ, Heitman J (1999) The TOR signaling cascade regulates gene expression in response to nutrients. Genes Dev 13:3271-3279

Celenza JL, Carlson M (1989) Mutational analysis of the *Saccharomyces cerevisiae* SNF1 protein kinase and evidence for functional interaction with the SNF4 protein. Mol Cell Biol 9:5034-5044

Celenza JL, Eng FJ, Carlson M (1989) Molecular analysis of the *SNF4* gene of *Saccharomyces cerevisiae*: evidence for physical association of the SNF4 protein with the SNF1 protein kinase. Mol Cell Biol 9:5045-5054

Chang EC, Barr M, Wang Y, Jung V, Xu HP, Wigler MH (1994) Cooperative interaction of *S. pombe* proteins required for mating and morphogenesis. Cell 79:131-141

Chen CR, Li YC, Chen J, Hou MC, Papadaki P, Chang EC (1999) Moe1, a conserved protein in *Schizosaccharomyces pombe*, interacts with a Ras effector, Scd1, to affect proper spindle formation. Proc Natl Acad Sci USA 96:517-522

Chevallier-Multon MC, Schweighoffer F, Barlat I, Baudouy N, Fath I, Duchesne M, Tocqué B (1993) *Saccharomyces cerevisiae* CDC25 (1028-1589) is a guanine nucleotide releasing factor for mammalian ras proteins and is oncogenic in NIH3T3 cells. J Biol Chem 268:11113-11118

Coccetti P, Tisi R, Martegani E, Teixeira LS, Lopes Brandão R, de Miranda Castro I, Thevelein JM (1998) The *PLC1* encoded phospholipase C in the yeast *Saccharomyces cerevisiae* is essential for glucose-induced phosphatidylinositol turnover and activation of plasma membrane H^+-ATPase. Biochim Biophys Acta 1405:147-154

Coleman DE, Lee E, Mixon MB, Linder ME, Berghuis AM, Gilman AG, Sprang SR (1994) Crystallization and preliminary crystallographic studies of Giα1 and mutants of Giα1 in the GTP and GDP-bound states. J Mol Biol 238:630-634

Colombo S, Ma PS, Cauwenberg L, Winderickx J, Crauwels M, Teunissen A, Nauwelaers D, de Winde JH, Gorwa MF, Colavizza D, Thevelein JM (1998) Involvement of distinct G-proteins, Gpa2 and Ras, in glucose- and intracellular acidification-induced cAMP signalling in the yeast *Saccharomyces cerevisiae*. EMBO J 17:3326-3341

Conlan RS, Tzamarias D (2001) Sfl1 functions via the co-repressor Ssn6-Tup1 and the cAMP-dependent protein kinase Tpk2. J Mol Biol 309:1007-1015

Cook JG, Bardwell L, Kron SJ, Thorner J (1996) Two novel targets of the MAP kinase Kss1 are negative regulators of invasive growth in the yeast *Saccharomyces cerevisiae*. Genes Dev 10:2831-2848

Cook JG, Bardwell L, Thorner J (1997) Inhibitory and activating functions for MAPK Kss1 in the *S. cerevisiae* filamentous-growth signalling pathway. Nature 390:85-88

Cooper KF, Mallory MJ, Smith JB, Strich R (1997) Stress and developmental regulation of the yeast C-type cyclin Ume3p (Srb11p/Ssn8p). EMBO J 16:4665-4675

Cooper TG (2002) Transmitting the signal of excess nitrogen in *Saccharomyces cerevisiae* from the Tor proteins to the GATA factors: connecting the dots. FEMS Microbiol Rev 26:223-238

Cullen PJ, Sprague GF Jr (2000) Glucose depletion causes haploid invasive growth in yeast. Proc Natl Acad Sci USA 97:13619-13624

Cullen PJ, Sprague GF Jr (2002) The roles of bud-site-selection proteins during haploid invasive growth in yeast. Mol Biol Cell 13:2990-3004

Cutler NS, Pan X, Heitman J, Cardenas ME (2001) The TOR signal transduction cascade controls cellular differentiation in response to nutrients. Mol Biol Cell 12:4103-4113

D'Souza CA, Alspaugh JA, Yue CL, Harashima T, Cox GM, Perfect JR, Heitman J (2001) Cyclic AMP-dependent protein kinase controls virulence of the fungal pathogen *Cryptococcus neoformans*. Mol Cell Biol 21:3179-3191

D'Souza CA, Heitman J (2001) Conserved cAMP signaling cascades regulate fungal development and virulence. FEMS Microbiol Rev 25:349-364

DeVoti J, Seydoux G, Beach D, McLeod M (1991) Interaction between *ran1*$^{+}$ protein kinase and cAMP dependent protein kinase as negative regulators of fission yeast meiosis. EMBO J 10:3759-3768

Díaz E, Schimmöller F, Pfeffer SR (1997) A novel Rab9 effector required for endosome-to-TGN transport. J Cell Biol 138:283-290

Dickinson JR (1996) 'Fusel' alcohols induce hyphal-like extensions and pseudohyphal formation in yeasts. Microbiology 142:1391-1397

Dohlman HG, Thorner JW (2001) Regulation of G protein-initiated signal transduction in yeast: Paradigms and principles. Annu Rev Biochem 70:703-754

Donzeau M, Bandlow W (1999) The yeast trimeric guanine nucleotide-binding protein α subunit, Gpa2p, controls the meiosis-specific kinase Ime2p activity in response to nutrients. Mol Cell Biol 19:6110-6119

Entian KD, Hilberg F, Opitz H, Mecke D (1985) Cloning of hexokinase structural genes from *Saccharomyces cerevisiae* mutants with regulatory mutations responsible for glucose repression. Mol Cell Biol 5:3035-3040

Fröhlich KU, Entian KD, Mecke D (1985) The primary structure of the yeast hexokinase PII gene (*HXK2*) which is responsible for glucose repression. Gene 36:105-111

Fukui Y, Kaziro Y (1985) Molecular cloning and sequence analysis of a *ras* gene from *Schizosaccharomyces pombe*. EMBO J 4:687-691
Fukui Y, Kozasa T, Kaziro Y, Takeda T, Yamamoto M (1986) Role of a *ras* homolog in the life cycle of *Schizosaccharomyces pombe*. Cell 44:329-336
Fukui Y, Miyake S, Satoh M, Yamamoto M (1989) Characterization of the *Schizosaccharomyces pombe ral2* gene implicated in activation of the *ras1* gene product. Mol Cell Biol 9:5617-5622
Fukui Y, Yamamoto M (1988) Isolation and characterization of *Schizosaccharomyces pombe* mutants phenotypically similar to *ras1*$^-$. Mol Gen Genet 215:26-31
Gagiano M, van Dyk D, Bauer FF, Lambrechts MG, Pretorius IS (1999) Msn1p/Mss10p, Mss11p and Muc1p/Flo11p are part of a signal transduction pathway downstream of Mep2p regulating invasive growth and pseudohyphal differentiation in *Saccharomyces cerevisiae*. Mol Microbiol 31:103-116
Gancedo JM (1998) Yeast carbon catabolite repression. Microbiol Mol Biol Rev 62:334-361
Gavrias V, Andrianopoulos A, Gimeno CJ, Timberlake WE (1996) *Saccharomyces cerevisiae* TEC1 is required for pseudohyphal growth. Mol Microbiol 19:1255-1263
Gilman AG (1987) G Proteins: Transducers of receptor-generated signals. Annu Rev Biochem 56:615-649
Gimeno CJ, Ljungdahl PO, Styles CA, Fink GR (1992) Unipolar cell divisions in the yeast *S. cerevisiae* lead to filamentous growth: regulation by starvation and *RAS*. Cell 68:1077-1090
Gotoh Y, Nishida E, Shimanuki M, Toda T, Imai Y, Yamamoto M (1993) *Schizosaccharomyces pombe* Spk1 is a tyrosine-phosphorylated protein functionally related to *Xenopus* mitogen-activated protein kinase. Mol Cell Biol 13:6427-6434
Gross A, Winograd S, Marbach I, Levitzki A (1999) The N-terminal half of Cdc25 is essential for processing glucose signaling in *Saccharomyces cerevisiae*. Biochemistry 38:13252-13262
Harashima T, Heitman J (2002) The Gα protein Gpa2 controls yeast differentiation by interacting with kelch repeat proteins that mimic Gβ subunits. Mol Cell 10:163-173
Hardie DG, Carling D, Carlson M (1998) The AMP-activated/SNF1 protein kinase subfamily: metabolic sensors of the eukaryotic cell? Annu Rev Biochem 67:821-855
Heitman J, Agre P (2000) A new face of the Rhesus antigen. Nat Genet 26:258-259
Heitman J, Movva NR, Hall MN (1991a) Targets for cell cycle arrest by the immunosuppressant rapamycin in yeast. Science 253:905-909
Heitman J, Movva NR, Hiestand PC, Hall MN (1991b) FK 506-binding protein proline rotamase is a target for the immunosuppressive agent FK 506 in *Saccharomyces cerevisiae*. Proc Natl Acad Sci USA 88:1948-1952
Helliwell SB, Wagner P, Kunz J, Deuter-Reinhard M, Henriquez R, Hall MN (1994) TOR1 and TOR2 are structurally and functionally similar but not identical phosphatidylinositol kinase homologues in yeast. Mol Biol Cell 5:105-118
Holstege FC, Jennings EG, Wyrick JJ, Lee TI, Hengartner CJ, Green MR, Golub TR, Lander ES, Young RA (1998) Dissecting the regulatory circuitry of a eukaryotic genome. Cell 95:717-728
Hoffman CS, Winston F (1990) Isolation and characterization of mutants constitutive for expression of the *fbp1* gene of *Schizosaccharomyces pombe*. Genetics 124:807-816
Hoffman CS, Winston F (1991) Glucose repression of transcription of the *Schizosaccharomyces pombe fbp1* gene occurs by a cAMP signaling pathway. Genes Dev 5:561-571

Hwang-Shum JJ, Hagen DC, Jarvis EE, Westby CA, Sprague GF Jr (1991) Relative contributions of MCM1 and STE12 to transcriptional activation of a- and α-specific genes from *Saccharomyces cerevisiae*. Mol Gen Genet 227:197-204

Isshiki T, Mochizuki N, Maeda T, Yamamoto M (1992) Characterization of a fission yeast gene, *gpa2*, that encodes a Gα subunit involved in the monitoring of nutrition. Genes Dev 6:2455-2462

Ito N, Phillips SE, Stevens C, Ogel ZB, McPherson MJ, Keen JN, Yadav KD, Knowles PF (1991) Novel thioether bond revealed by a 1.7 Å crystal structure of galactose oxidase. Nature 350:87-90

Ito N, Phillips SEV, Yadav KDS, Knowles PF (1994) Crystal structure of a free radical enzyme, galactose oxidase. J Mol Biol 238:794-814

Javelle A, Morel M, Rodríguez-Pastrana BR, Botton B, André B, Marini AM, Brun A, Chalot M (2003) Molecular characterization, function and regulation of ammonium transporters (Amt) and ammonium-metabolizing enzymes (GS, NADP-GDH) in the ectomycorrhizal fungus *Hebeloma cylindrosporum*. Mol Microbiol 47:411-430

Javelle A, Rodríguez-Pastrana BR, Jacob C, Botton B, Brun A, Andre B, Marini AM, Chalot M (2001) Molecular characterization of two ammonium transporters from the ectomycorrhizal fungus *Hebeloma cylindrosporum*. FEBS Lett 505:393-398

Jiang R, Carlson M (1997) The Snf1 protein kinase and its activating subunit, Snf4, interact with distinct domains of the Sip1/Sip2/Gal83 component in the kinase complex. Mol Cell Biol 17:2099-2106

Jiang Y, Davis C, Broach JR (1998) Efficient transition to growth on fermentable carbon sources in *Saccharomyces cerevisiae* requires signaling through the Ras pathway. EMBO J 17:6942-6951

Kaibuchi K, Miyajima A, Arai K, Matsumoto K (1986) Possible involvement of *RAS*-encoded proteins in glucose-induced inositolphospholipid turnover in *Saccharomyces cerevisiae*. Proc Natl Acad Sci USA 83:8172-8176

Kataoka T, Broek D, Wigler M (1985) DNA sequence and characterization of the *S. cerevisiae* gene encoding adenylate cyclase. Cell 43:493-505

Kataoka T, Powers S, McGill C, Fasano O, Strathern J, Broach J, Wigler M (1984) Genetic analysis of yeast *RAS1* and *RAS2* genes. Cell 37:437-445

Kitamura K, Shimoda C (1991) The *Schizosaccharomyces pombe mam2* gene encodes a putative pheromone receptor which has a significant homology with the *Saccharomyces cerevisiae* Ste2 protein. EMBO J 10:3743-3751

Kopetzki E, Entian KD, Mecke D (1985) Complete nucleotide sequence of the hexokinase PI gene (*HXK1*) of *Saccharomyces cerevisiae*. Gene 39:95-101

Kraakman L, Lemaire K, Ma PS, Teunissen AWRH, Donaton MCV, Van Dijck P, Winderickx J, de Winde JH, Thevelein JM (1999) A *Saccharomyces cerevisiae* G-protein coupled receptor, Gpr1, is specifically required for glucose activation of the cAMP pathway during the transition to growth on glucose. Mol Microbiol 32:1002-1012

Kruckeberg AL (1996) The hexose transporter family of *Saccharomyces cerevisiae*. Arch Microbiol 166:283-292

Kübler E, Mösch HU, Rupp S, Lisanti MP (1997) Gpa2p, a G-protein α-subunit, regulates growth and pseudohyphal development in *Saccharomyces cerevisiae* via a cAMP-dependent mechanism. J Biol Chem 272:20321-20323

Kuchin S, Vyas VK, Carlson M (2002) Snf1 protein kinase and the repressors Nrg1 and Nrg2 regulate *FLO11*, haploid invasive growth, and diploid pseudohyphal differentiation. Mol Cell Biol 22:3994-4000

Kunz J, Henriquez R, Schneider U, Deuter-Reinhard M, Movva NR, Hall MN (1993) Target of rapamycin in yeast, TOR2, is an essential phosphatidylinositol kinase homolog required for G_1 progression. Cell 73:585-596

Lai CC, Boguski M, Broek D, Powers S (1993) Influence of guanine nucleotides on complex formation between Ras and CDC25 proteins. Mol Cell Biol 13:1345-1352

Lambright DG, Sondek J, Bohm A, Skiba NP, Hamm HE, Sigler PB (1996) The 2.0 Å crystal structure of a heterotrimeric G protein. Nature 379:311-319

Landry S, Hoffman CS (2001) The git5 Gβ and git11 Gγ form an atypical Gβγ dimer acting in the fission yeast glucose/cAMP pathway. Genetics 157:1159-1168

Landry S, Pettit MT, Apolinario E, Hoffman CS (2000) The fission yeast *git5* gene encodes a Gβ subunit required for glucose-triggered adenylate cyclase activation. Genetics 154:1463-1471

Lee E, Taussig R, Gilman AG (1992) The G226A mutant of Gsα highlights the requirement for dissociation of G protein subunits. J Biol Chem 267:1212-1218

Lefkowitz RJ (2000) The superfamily of heptahelical receptors. Nat Cell Biol 2:E133-E136

Lengeler KB, Davidson RC, D'Souza C, Harashima T, Shen WC, Wang P, Pan XW, Waugh M, Heitman J (2000) Signal transduction cascades regulating fungal development and virulence. Microbiol Mol Biol Rev 64:746-785

Liu HP, Styles CA, Fink GR (1993) Elements of the yeast pheromone response pathway required for filamentous growth of diploids. Science 262:1741-1744

Liu HP, Styles CA, Fink GR (1996) *Saccharomyces cerevisiae* S288C has a mutation in *FL08*, a gene required for filamentous growth. Genetics 144:967-978

Lo WS, Dranginis AM (1996) *FLO11*, a yeast gene related to the *STA* genes, encodes a novel cell surface flocculin. J Bacteriol 178:7144-7151

Lo WS, Dranginis AM (1998) The cell surface flocculin Flo11 is required for pseudohyphae formation and invasion by *Saccharomyces cerevisiae*. Mol Biol Cell 9:161-171

Lorenz MC, Cutler NS, Heitman J (2000a) Characterization of alcohol-induced filamentous growth in *Saccharomyces cerevisiae*. Mol Biol Cell 11:183-199

Lorenz MC, Heitman J (1997) Yeast pseudohyphal growth is regulated by GPA2, a G protein α homolog. EMBO J 16:7008-7018

Lorenz MC, Heitman J (1998) The MEP2 ammonium permease regulates pseudohyphal differentiation in *Saccharomyces cerevisiae*. EMBO J 17:1236-1247

Lorenz MC, Pan X, Harashima T, Cardenas ME, Xue Y, Hirsch JP, Heitman J (2000b) The G protein-coupled receptor Gpr1 is a nutrient sensor that regulates pseudohyphal differentiation in *Saccharomyces cerevisiae*. Genetics 154:609-622

Ma PS, Wera S, Van Dijck P, Thevelein JM (1999) The *PDE1*-encoded low-affinity phosphodiesterase in the yeast *Saccharomyces cerevisiae* has a specific function in controlling agonist-induced cAMP signaling. Mol Biol Cell 10:91-104

Madhani HD, Fink GR (1997) Combinatorial control required for the specificity of yeast MAPK signaling. Science 275:1314-1317

Madhani HD, Fink GR (1998) The control of filamentous differentiation and virulence in fungi. Trends Cell Biol 8:348-353

Madhani HD, Styles CA, Fink GR (1997) MAP kinases with distinct inhibitory functions impart signaling specificity during yeast differentiation. Cell 91:673-684

Maeda T, Watanabe Y, Kunitomo H, Yamamoto M (1994) Cloning of the *pka1* gene encoding the catalytic subunit of the cAMP-dependent protein kinase in *Schizosaccharomyces pombe*. J Biol Chem 269:9632-9637

Marini AM, Andre B (2000) *In vivo N*-glycosylation of the Mep2 high-affinity ammonium transporter of *Saccharomyces cerevisiae* reveals an extracytosolic N-terminus. Mol Microbiol 38:552-564

Marini AM, Matassi G, Raynal V, André B, Cartron JP, Chérif-Zahar B (2000) The human Rhesus-associated RhAG protein and a kidney homologue promote ammonium transport in yeast. Nat Genet 26:341-344

Marini AM, Soussi-Boudekou S, Vissers S, Andre B (1997a) A family of ammonium transporters in *Saccharomyces cerevisiae*. Mol Cell Biol 17:4282-4293

Marini AM, Urrestarazu A, Beauwens R, André B (1997b) The Rh (Rhesus) blood group polypeptides are related to NH_4^+ transporters. Trends Biochem Sci 22:460-461

Marini AM, Vissers S, Urrestarazu A, Andre B (1994) Cloning and expression of the *MEP1* gene encoding an ammonium transporter in *Saccharomyces cerevisiae*. EMBO J 13:3456-3463

Matsuura A, Treinin M, Mitsuzawa H, Kassir Y, Uno I, Simchen G (1990) The adenylate cyclase/protein kinase cascade regulates entry into meiosis in *Saccharomyces cerevisiae* through the gene *IME1*. EMBO J 9:3225-3232

Mbonyi K, Beullens M, Detremerie K, Geerts L, Thevelein JM (1988) Requirement of one functional *RAS* gene and inability of an oncogenic *ras* variant to mediate the glucose-induced cyclic AMP signal in the yeast *Saccharomyces cerevisiae*. Mol Cell Biol 8:3051-3057

Miller RT, Masters SB, Sullivan KA, Beiderman B, Bourne HR (1988) A mutation that prevents GTP-dependent activation of the α chain of Gs. Nature 334:712-715

Mintzer KA, Field J (1999) The SH3 domain of the *S. cerevisiae* Cdc25p binds adenylyl cyclase and facilitates Ras regulation of cAMP signalling. Cell Signal 11:127-135

Mochizuki N, Yamamoto M (1992) Reduction in the intracellular cAMP level triggers initiation of sexual development in fission yeast. Mol Gen Genet 233:17-24

Montesinos ML, Muro-Pastor AM, Herrero A, Flores E (1998) Ammonium/methylammonium permeases of a Cyanobacterium. Identification and analysis of three nitrogen-regulated *amt* genes in *Synechocystis* sp. PCC 6803. J Biol Chem 273:31463-31470

Moore B, Zhou L, Rolland F, Hall Q, Cheng WH, Liu YX, Hwang I, Jones T, Sheen J (2003) Role of the *Arabidopsis* glucose sensor HXK1 in nutrient, light, and hormonal signaling. Science 300:332-336

Mösch HU, Kübler E, Krappmann S, Fink GR, Braus GH (1999) Crosstalk between the Ras2p-controlled mitogen-activated protein kinase and cAMP pathways during invasive growth of *Saccharomyces cerevisiae*. Mol Biol Cell 10:1325-1335

Mösch HU, Roberts RL, Fink GR (1996) Ras2 signals via the Cdc42/Ste20/mitogen-activated protein kinase module to induce filamentous growth in *Saccharomyces cerevisiae*. Proc Natl Acad Sci USA 93:5352-5356

Murad AM, d'Enfert C, Gaillardin C, Tournu H, Tekaia F, Talibi D, Marechal D, Marchais V, Cottin J, Brown AJ (2001a) Transcript profiling in *Candida albicans* reveals new cellular functions for the transcriptional repressors CaTup1, CaMig1 and CaNrg1. Mol Microbiol 42:981-993

Murad AM, Leng P, Straffon M, Wishart J, Macaskill S, MacCallum D, Schnell N, Talibi D, Marechal D, Tekaia F, d'Enfert C, Gaillardin C, Odds FC, Brown AJ (2001b) *NRG1* represses yeast-hypha morphogenesis and hypha-specific gene expression in *Candida albicans*. EMBO J 20:4742-4752

Nadin-Davis SA, Nasim A (1988) A gene which encodes a predicted protein kinase can restore some functions of the *ras* gene in fission yeast. EMBO J 7:985-993

Nakafuku M, Obara T, Kaibuchi K, Miyajima I, Miyajima A, Itoh H, Nakamura S, Arai K, Matsumoto K, Kaziro Y (1988) Isolation of a second yeast *Saccharomyces cerevisiae* gene (*GPA2*) coding for guanine nucleotide-binding regulatory protein: Studies on its structure and possible functions. Proc Natl Acad Sci USA 85:1374-1378

Neiman AM, Stevenson BJ, Xu HP, Sprague GF Jr, Herskowitz I, Wigler M, Marcus S (1993) Functional homology of protein kinases required for sexual differentiation in *Schizosaccharomyces pombe* and *Saccharomyces cerevisiae* suggests a conserved signal transduction module in eukaryotic organisms. Mol Biol Cell 4:107-120

Nelson C, Goto S, Lund K, Hung W, Sadowski I (2003) Srb10/Cdk8 regulates yeast filamentous growth by phosphorylating the transcription factor Ste12. Nature 421:187-190

Ninnemann O, Jauniaux JC, Frommer WB (1994) Identification of a high affinity NH_4^+ transporter from plants. EMBO J 13:3464-3471

Nocero M, Isshiki T, Yamamoto M, Hoffman CS (1994) Glucose repression of *fbp1* transcription in *Schizosaccharomyces pombe* is partially regulated by adenylate cyclase activation by a G protein α subunit encoded by *gpa2* (*git8*). Genetics 138:39-45

Obara T, Nakafuku M, Yamamoto M, Kaziro Y (1991) Isolation and characterization of a gene encoding a G-protein α subunit from *Schizosaccharomyces pombe*: Involvement in mating and sporulation pathways. Proc Natl Acad Sci USA 88:5877-5881

Odom AR, Stahlberg A, Wente SR, York JD (2000) A role for nuclear inositol 1,4,5-trisphosphate kinase in transcriptional control. Science 287:2026-2029

Oehlen LJ, McKinney JD, Cross FR (1996) Ste12 and Mcm1 regulate cell cycle-dependent transcription of *FAR1*. Mol Cell Biol 16:2830-2837

Özcan S, Dover J, Johnston M (1998) Glucose sensing and signaling by two glucose receptors in the yeast *Saccharomyces cerevisiae*. EMBO J 17:2566-2573

Özcan S, Dover J, Rosenwald AG, Wölfl S, Johnston M (1996a) Two glucose transporters in *Saccharomyces cerevisiae* are glucose sensors that generate a signal for induction of gene expression. Proc Natl Acad Sci USA 93:12428-12432

Özcan S, Leong T, Johnston M (1996b) Rgt1p of *Saccharomyces cerevisiae*, a key regulator of glucose-induced genes, is both an activator and a repressor of transcription. Mol Cell Biol 16:6419-6426

Palecek SP, Parikh AS, Huh JH, Kron SJ (2002) Depression of *Saccharomyces cerevisiae* invasive growth on non-glucose carbon sources requires the Snf1 kinase. Mol Microbiol 45:453-469

Palková Z, Devaux F, Řičicová M, Mináriková L, Le Crom S, Jacq C (2002) Ammonia pulses and metabolic oscillations guide yeast colony development. Mol Biol Cell 13:3901-3914

Palková Z, Forstová J (2000) Yeast colonies synchronise their growth and development. J Cell Sci 113:1923-1928

Palková Z, Janderová B, Gabriel J, Zikánová B, Pospíšek M, Forstová J (1997) Ammonia mediates communication between yeast colonies. Nature 390:532-536

Pan X, Harashima T, Heitman J (2000) Signal transduction cascades regulating pseudohyphal differentiation of *Saccharomyces cerevisiae*. Curr Opin Microbiol 3:567-572

Pan X, Heitman J (1999) Cyclic AMP-dependent protein kinase regulates pseudohyphal differentiation in *Saccharomyces cerevisiae*. Mol Cell Biol 19:4874-4887

Pan X, Heitman J (2002) Protein kinase A operates a molecular switch that governs yeast pseudohyphal differentiation. Mol Cell Biol 22:3981-3993

Papadaki P, Pizon V, Onken B, Chang EC (2002) Two ras pathways in fission yeast are differentially regulated by two ras guanine nucleotide exchange factors. Mol Cell Biol 22:4598-4606

Park SH, Koh SS, Chun JH, Hwang HJ, Kang HS (1999) Nrg1 is a transcriptional repressor for glucose repression of *STA1* gene expression in *Saccharomyces cerevisiae*. Mol Cell Biol 19:2044-2050

Petter R, Chang YC, Kwon-Chung KJ (1997) A gene homologous to *Saccharomyces cerevisiae SNF1* appears to be essential for the viability of *Candida albicans*. Infect Immun 65:4909-4917

Powers S, Kataoka T, Fasano O, Goldfarb M, Strathern J, Broach J, Wigler M (1984) Genes in *S. cerevisiae* encoding proteins with domains homologous to the mammalian *ras* proteins. Cell 36:607-612

Pryciak PM, Huntress FA (1998) Membrane recruitment of the kinase cascade scaffold protein Ste5 by the Gβγ complex underlies activation of the yeast pheromone response pathway. Genes Dev 12:2684-2697

Reifenberger E, Boles E, Ciriacy M (1997) Kinetic characterization of individual hexose transporters of *Saccharomyces cerevisiae* and their relation to the triggering mechanisms of glucose repression. Eur J Biochem 245:324-333

Reinders A, Bürckert N, Boller T, Wiemken A, De Virgilio C (1998) *Saccharomyces cerevisiae* cAMP-dependent protein kinase controls entry into stationary phase through the Rim15p protein kinase. Genes Dev 12:2943-2955

Roberts RL, Fink GR (1994) Elements of a single MAP kinase cascade in *Saccharomyces cerevisiae* mediate two developmental programs in the same cell type: mating and invasive growth. Genes Dev 8:2974-2985

Roberts RL, Mösch HU, Fink GR (1997) 14-3-3 proteins are essential for RAS/MAPK cascade signaling during pseudohyphal development in *S. cerevisiae*. Cell 89:1055-1065

Robertson LS, Fink GR (1998) The three yeast A kinases have specific signaling functions in pseudohyphal growth. Proc Natl Acad Sci USA 95:13783-13787

Rohde J, Heitman J, Cardenas ME (2001) The TOR kinases link nutrient sensing to cell growth. J Biol Chem 276:9583-9586

Rolland F, de Winde JH, Lemaire K, Boles E, Thevelein JM, Winderickx J (2000) Glucose-induced cAMP signalling in yeast requires both a G-protein coupled receptor system for extracellular glucose detection and a separable hexose kinase-dependent sensing process. Mol Microbiol 38:348-358

Rolland F, Wanke V, Cauwenberg L, Ma P, Boles E, Vanoni M, de Winde JH, Thevelein JM, Winderickx J (2001) The role of hexose transport and phosphorylation in cAMP signalling in the yeast *Saccharomyces cerevisiae*. FEMS Yeast Res 1:33-45

Ross EM, Wilkie TM (2000) GTPase-activating proteins for heterotrimeric G proteins: Regulators of G protein signaling (RGS) and RGS-like proteins. Annu Rev Biochem 69:795-827

Rost B, Sander C (1993) Prediction of protein secondary structure at better than 70% accuracy. J Mol Biol 232:584-599

Rupp S, Summers E, Lo HJ, Madhani H, Fink G (1999) MAP kinase and cAMP filamentation signaling pathways converge on the unusually large promoter of the yeast *FLO11* gene. EMBO J 18:1257-1269

Saiardi A, Caffrey JJ, Snyder SH, Shears SB (2000) Inositol polyphosphate multikinase (ArgRIII) determines nuclear mRNA export in *Saccharomyces cerevisiae*. FEBS Lett 468:28-32

Saiardi A, Erdjument-Bromage H, Snowman AM, Tempst P, Snyder SH (1999) Synthesis of diphosphoinositol pentakisphosphate by a newly identified family of higher inositol polyphosphate kinases. Curr Biol 9:1323-1326

Saitou N, Nei M (1987) The neighbor-joining method: A new method for reconstructing phylogenetic trees. Mol Biol Evol 4:406-425

Sánchez-Martínez C, Pérez-Martín J (2002) Gpa2, a G-protein α subunit required for hyphal development in *Candida albicans*. Eukaryot Cell 1:865-874

Schmidt MC, McCartney RR (2000) β-subunits of Snf1 kinase are required for kinase function and substrate definition. EMBO J 19:4936-4943

Schomerus C, Küntzel H (1992) *CDC25*-dependent induction of inositol 1,4,5-trisphosphate and diacylglycerol in *Saccharomyces cerevisiae* by nitrogen. FEBS Lett 307:249-252

Schomerus C, Munder T, Küntzel H (1990) Site-directed mutagenesis of the *Saccharomyces cerevisiae CDC25* gene: effects on mitotic growth and cAMP signalling. Mol Gen Genet 223:426-432

Shen X, Xiao H, Ranallo R, Wu WH, Wu C (2003) Modulation of ATP-dependent chromatin-remodeling complexes by inositol polyphosphates. Science 299:112-114

Sondek J, Bohm A, Lambright DG, Hamm HE, Sigler PB (1996) Crystal structure of a G_A protein βγ dimer at 2.1 Å resolution. Nature 379:369-374

Sprang SR (1997) G protein mechanisms: Insights from structural analysis. Annu Rev Biochem 66:639-678

Steger DJ, Haswell ES, Miller AL, Wente SR, O'Shea EK (2003) Regulation of chromatin remodeling by inositol polyphosphates. Science 299:114-116

Sternweis PC (1986) The purified α subunits of G_o and G_i from bovine brain require βγ for association with phospholipid vesicles. J Biol Chem 261:631-637

Sugimoto A, Iino Y, Maeda T, Watanabe Y, Yamamoto M (1991) *Schizosaccharomyces pombe ste11*$^+$ encodes a transcription factor with an HMG motif that is a critical regulator of sexual development. Genes Dev 5:1990-1999

Tamaki H, Miwa T, Shinozaki M, Saito M, Yun CW, Yamamoto K, Kumagai H (2000) *GPR1* regulates filamentous growth through *FLO11* in yeast *Saccharomyces cerevisiae*. Biochem Biophys Res Commun 267:164-168

Tanaka K, Davey J, Imai Y, Yamamoto M (1993) *Schizosaccharomyces pombe map3*$^+$ encodes the putative M-factor receptor. Mol Cell Biol 13:80-88

Tanaka K, Lin BK, Wood DR, Tamanoi F (1991) IRA2, an upstream negative regulator of RAS in yeast, is a RAS GTPase-activating protein. Proc Natl Acad Sci USA 88:468-472

Tanaka K, Matsumoto K, Toh-e A (1989) *IRA1*, an inhibitory regulator of the *RAS*-cyclic AMP pathway in *Saccharomyces cerevisiae*. Mol Cell Biol 9:757-768

Tanaka K, Nakafuku M, Satoh T, Marshall MS, Gibbs JB, Matsumoto K, Kaziro Y, Toh-e A (1990) *S. cerevisiae* genes *IRA1* and *IRA2* encode proteins that may be functionally equivalent to mammalian *ras* GTPase activating protein. Cell 60:803-807

Thompson JD, Higgins DG, Gibson TJ (1994) CLUSTAL W: improving the sensitivity of progressive multiple sequence alignment through sequence weighting, position-specific gap penalties and weight matrix choice. Nucleic Acids Res 22:4673-4680

Tisi R, Baldassa S, Belotti F, Martegani E (2002) Phospholipase C is required for glucose-induced calcium influx in budding yeast. FEBS Lett 520:133-138

Toda T, Cameron S, Sass P, Zoller M, Scott JD, McMullen B, Hurwitz M, Krebs EG, Wigler M (1987a) Cloning and characterization of *BCY1*, a locus encoding a regula-

tory subunit of the cyclic AMP-dependent protein kinase in *Saccharomyces cerevisiae*. Mol Cell Biol 7:1371-1377
Toda T, Cameron S, Sass P, Zoller M, Wigler M (1987b) Three different genes in *S. cerevisiae* encode the catalytic subunits of the cAMP-dependent protein kinase. Cell 50:277-287
Toda T, Shimanuki M, Yanagida M (1991) Fission yeast genes that confer resistance to staurosporine encode an AP-1-like transcription factor and a protein kinase related to the mammalian ERK1/MAP2 and budding yeast *FUS3* and *KSS1* kinases. Genes Dev 5:60-73
Toda T, Uno I, Ishikawa T, Powers S, Kataoka T, Broek D, Cameron S, Broach J, Matsumoto K, Wigler M (1985) In yeast, *RAS* proteins are controlling elements of adenylate cyclase. Cell 40:27-36
Tolkacheva T, Mcnamara P, Piekarz E, Courchesne W (1994) Cloning of a *Cryptococcus neoformans* gene, *GPA1*, encoding a G-protein α-subunit homolog. Infect Immun 62:2849-2856
Vassarotti A, Friesen JD (1985) Isolation of the fructose-1,6-bisphosphatase gene of the yeast *Schizosaccharomyces pombe*. Evidence for transcriptional regulation. J Biol Chem 260:6348-6353
Versele M, de Winde JH, Thevelein JM (1999) A novel regulator of G protein signalling in yeast, Rgs2, downregulates glucose-activation of the cAMP pathway through direct inhibition of Gpa2. EMBO J 18:5577-5591
Versele M, Lemaire K, Thevelein JM (2001) Sex and sugar in yeast: two distinct GPCR systems. EMBO Reports 2:574-579
Vidan S, Mitchell AP (1997) Stimulation of yeast meiotic gene expression by the glucose-repressible protein kinase Rim15p. Mol Cell Biol 17:2688-2697
Vincent O, Townley R, Kuchin S, Carlson M (2001) Subcellular localization of the Snf1 kinase is regulated by specific β subunits and a novel glucose signaling mechanism. Genes Dev 15:1104-1114
Vyas VK, Kuchin S, Berkey CD, Carlson M (2003) Snf1 kinases with different β-subunit isoforms play distinct roles in regulating haploid invasive growth. Mol Cell Biol 23:1341-1348
Vyas VK, Kuchin S, Carlson M (2001) Interaction of the repressors Nrg1 and Nrg2 with the Snf1 protein kinase in *Saccharomyces cerevisiae*. Genetics 158:563-572
Wall MA, Coleman DE, Lee E, Iniguez-Lluhi JA, Posner BA, Gilman AG, Sprang SR (1995) The structure of the G protein heterotrimer $G_{i\alpha 1}\beta_1\gamma_2$. Cell 83:1047-1058
Wall MA, Posner BA, Sprang SR (1998) Structural basis of activity and subunit recognition in G protein heterotrimers. Structure 6:1169-1183
Walsh RB, Clifton D, Horak J, Fraenkel DG (1991) *Saccharomyces cerevisiae* null mutants in glucose phosphorylation: metabolism and invertase expression. Genetics 128:521-527
Wang Y, Xu HP, Riggs M, Rodgers L, Wigler M (1991) *byr2*, a *Schizosaccharomyces pombe* gene encoding a protein kinase capable of partial suppression of the *ras1* mutant phenotype. Mol Cell Biol 11:3554-3563
Welton RM, Hoffman CS (2000) Glucose monitoring in fission yeast via the gpa2 Gα, the git5 Gβ and the git3 putative glucose receptor. Genetics 156:513-521
Whiteway M (2000) Transcriptional control of cell type and morphogenesis in *Candida albicans*. Curr Opin Microbiol 3:582-588

Woods A, Munday MR, Scott J, Yang X, Carlson M, Carling D (1994) Yeast SNF1 is functionally related to mammalian AMP-activated protein kinase and regulates acetyl-CoA carboxylase *in vivo*. J Biol Chem 269:19509-19515

Xue Y, Batlle M, Hirsch JP (1998) *GPR1* encodes a putative G protein-coupled receptor that associates with the Gpa2p G_α subunit and functions in a Ras-independent pathway. EMBO J 17:1996-2007

Yamawaki-Kataoka Y, Tamaoki T, Choe HR, Tanaka H, Kataoka T (1989) Adenylate cyclases in yeast: A comparison of the genes from *Schizosaccharomyces pombe* and *Saccharomyces cerevisiae*. Proc Natl Acad Sci USA 86:5693-5697

Yang X, Jiang R, Carlson M (1994) A family of proteins containing a conserved domain that mediates interaction with the yeast SNF1 protein kinase complex. EMBO J 13:5878-5886

York JD, Odom AR, Murphy R, Ives EB, Wente SR (1999) A phospholipase C-dependent inositol polyphosphate kinase pathway required for efficient messenger RNA export. Science 285:96-100

Yoshida M, Kawaguchi H, Sakata Y, Kominami K, Hirano M, Shima H, Akada R, Yamashita I (1990) Initiation of meiosis and sporulation in *Saccharomyces cerevisiae* requires a novel protein kinase homologue. Mol Gen Genet 221:176-186

Young D, Riggs M, Field J, Vojtek A, Broek D, Wigler M (1989) The adenylyl cyclase gene from *Schizosaccharomyces pombe*. Proc Natl Acad Sci USA 86:7989-7993

Yun CW, Tamaki H, Nakayama R, Yamamoto K, Kumagai H (1997) G-protein coupled receptor from yeast *Saccharomyces cerevisiae*. Biochem Biophys Res Commun 240:287-292

Yun CW, Tamaki H, Nakayama R, Yamamoto K, Kumagai H (1998) Gpr1p, a putative G-protein coupled receptor, regulates glucose-dependent cellular cAMP level in yeast *Saccharomyces cerevisiae*. Biochem Biophys Res Commun 252:29-33

7 Regulation of the yeast general amino acid control pathway in response to nutrient stress

Ronald C. Wek, Kirk A. Staschke, and Jana Narasimhan

Abstract

In response to starvation for amino acids, yeast cells induce the expression of a collection of genes involved in amino acid metabolism. This pathway referred to as the general amino acid control is a true cross-pathway response that induces expression of genes contributing to the synthesis of all amino acids independent of which amino acid is limiting. In this review, we discuss the three basic parts of the general control pathway: 1) sensing of amino acid starvation by the protein kinase Gcn2p; 2) induction of the "master regulator" Gcn4p that involves a classic mechanism of translation control elicited by Gcn2p phosphorylation of eukaryotic initiation factor -2 (eIF2); and 3) the coordinate expression of stress remedy genes through Gcn4p-directed regulation of transcription. Recent studies suggest that Gcn4p can also interface with other stress response pathways, allowing the cell to direct the timing and content of stress gene expression in response to multiple stress inputs.

7.1 Major themes in the general amino acid control pathway

In response to environmental stresses, cells induce a program of gene expression designed to remedy the underlying cellular disturbance. A well characterized example of such a stress response is the general amino acid control pathway. In this pathway, starvation for amino acids induces a large number of genes predominantly involved in metabolism of amino acids. This is a true cross-pathway response that facilitates induction of genes important for the biosynthesis of virtually all amino acids independent of which amino acid is limiting. Because studies on the general control pathway have historically focused on *Saccharomyces cerevisiae* this review will highlight the key molecular and physiological features in this yeast, with some comparisons and discussions of this pathway in other eukaryotes as indicated.

The general control response can be divided into three basic parts. The first concerns the mechanism by which cells monitor amino acid levels. This sensing mechanism is carried out by the protein kinase Gcn2p and involves direct interaction between Gcn2p and uncharged tRNA that accumulates in cells severely limit-

Topics in Current Genetics, Vol. 7
J. Winderickx, P.M. Taylor (Eds) Nutrient-Induced Responses in Eukaryotic Cells

ing for amino acids. The second part involves elevated levels of the transcriptional activator Gcn4p in response to starvation for amino acids. A central feature of this induced expression involves increased translation of *GCN4* mRNA, a mechanism that has become a classic example of gene-specific translational control. The third part is the coordinate expression of hundreds of genes through Gcn4p-directed regulation of transcription. To mediate this regulation of mRNA synthesis, Gcn4p binds to a defined promoter sequence and enhances access for the RNA polymerase II transcriptional apparatus. This results in the activation of a collection of genes important for stress remedy and the salvaging of nutrients important for renewal. In addition to amino acid limitation, activation of the general control pathway occurs during other environmental stress conditions including other nutrient limitations such as carbohydrate or purine deprivation. The mechanistic details central to each of these parts of the general control pathway will be described in detail below. Importantly, many of these conceptual features are conserved not only in Gcn2p-mediated stress response pathways among other eukaryotic organisms, but also in other stress management pathways whereby complex stress conditions are recognized and processed to coordinate gene expression.

7.2 Recognition of amino acid starvation and activation of Gcn2 protein kinase

Starvation for any one of at least ten different amino acids studied induces expression of Gcn4p and its target genes. Mutations in aminoacyl-tRNA synthetase genes, such as *HTS1* important for charging of $tRNA^{His}$, elicit the general control response in yeast even in the presence of abundant cognate amino acid (Wek et al. 1995). Hence, elevated levels of uncharged tRNA that accumulate during amino acid starvation is thought to be the direct signal activating the general control pathway.

The sensor for uncharged tRNA levels in yeast is the multi-domain protein Gcn2p. The central domain of Gcn2p is directly involved in catalyzing the phosphorylation of eukaryotic initiation factor -2 (eIF2) in response to stress, an event that as described further below, modifies the activity of this translation factor and triggers increased Gcn4p synthesis (Dever et al. 1992). Recognition and activation of Gcn2p by elevated levels of uncharged tRNA involves a regulatory domain that has sequence homology with almost the entire length of the histidyl-tRNA synthetase (HisRS) enzymes (Fig. 1) (Wek et al. 1989). To address whether this so-called HisRS-related domain of Gcn2p participates in the monitoring of each of these starvation conditions, residue substitutions in the HisRS domain were analyzed for their effects on *GCN4* expression. In particular one mutant gcn2-m2p contains residue substitutions in motif 2 (Y1119L and R1120L), a conserved region among class II synthetases that directly interacts with tRNA substrates (Wek et al. 1995). The gcn2-m2p mutant was unable to phosphorylate eIF2 *in vivo* or *in vitro* and was blocked for induced expression of *GCN4* or its target genes in yeast cells starving for any one of at least six different amino acids

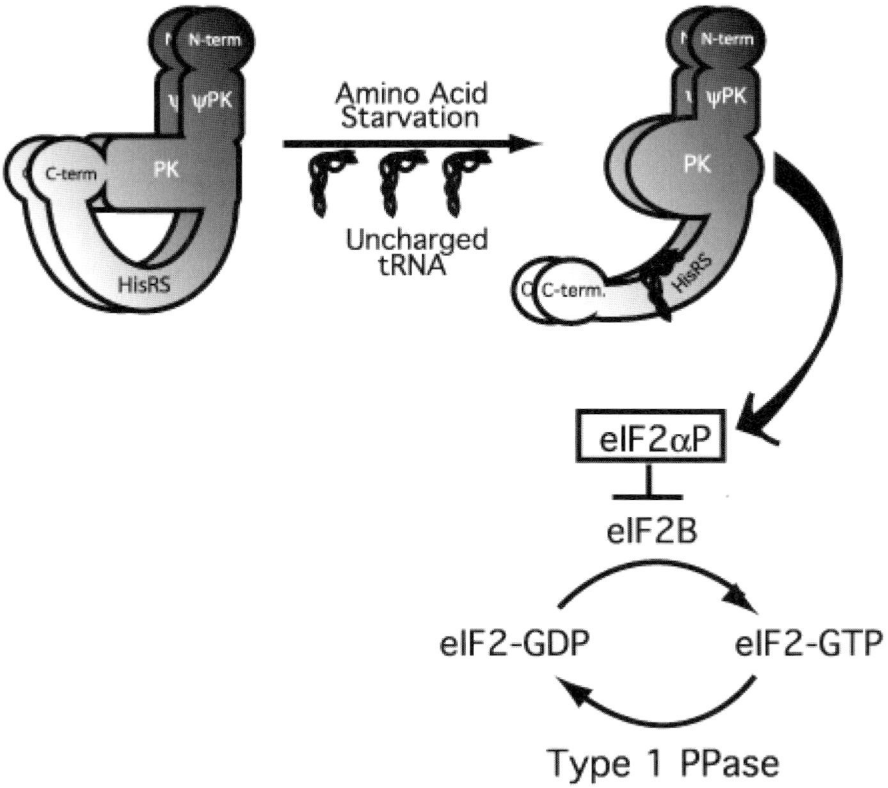

Fig. 1. Uncharged tRNA that accumulates during amino acid starvation binds and activates the eIF2 kinase activity of Gcn2p. Gcn2p is a multi-domain protein kinase (PK) that induces *GCN4* translation. Flanking this catalytic region is the HisRS-related domain that binds uncharged tRNA, leading to activation of Gcn2p through release of a proposed inhibitory interaction between the protein kinase domain and the C-terminus. The C-terminus of Gcn2p is also required for Gcn2p dimerization and association with the ribosomal 60S subunit. The N-terminus of Gcn2p binds with the Gcn1p/Gcn20p complex, and this interaction is suggested to facilitate passage of uncharged tRNA from the ribosomal A site to the HisRS-domain of Gcn2p. Flanking the N-terminus is the partial- or pseudo- protein kinase (ψPK) region that is required for Gcn2p function. Activation of Gcn2p leads to phosphorylation of the Ser51 residue in the α subunit of eIF2, altering this translation factor from a substrate to an inhibitor of the guaninine nucleotide exchange factor, eIF2B. Reduction in eIF2B activity lowers the levels of eIF2-GTP, thwarting association with initiator Met-$tRNA_i^{Met}$. Dephosphorylation of eIF2 is catalyzed by the Type 1 protein phosphatase Glc7p.

(Wek et al. 1995; Zhu et al. 1996). The motif 2 alterations also severely reduced binding of uncharged tRNA to the HisRS-related domain of GCN2 in *in vitro* experiments involving either northwestern binding or electromobility shift assays (EMSA) (Wek et al. 1995; Dong et al. 2000). The extreme carboxy terminus of

Gcn2p is multi-functional, and it has been suggested that its ability to dimerize is central to facilitate the HisRS-domain binding to uncharged tRNA (Dong et al. 2000). Gcn2p binding with uncharged tRNA does not appear to be restricted to uncharged $tRNA^{His}$; therefore, the HisRS-related domain has diverged sufficiently from the bona fide HisRS enzyme so that it can bind effectively to many different uncharged tRNA species that accumulate during amino acid starvation conditions. Furthermore, Gcn2p has reduced affinity for aminoacylated tRNA *in vitro*, consistent with the idea that it is activated by only uncharged tRNA (Dong et al. 2000).

Induction of Gcn2p by uncharged tRNA is proposed to involve a transition from an inhibited to catalytically active conformation that is signaled by direct contacts between the protein kinase domain, HisRS-regulatory region, and the extreme carboxy terminus of Gcn2p (Qiu et al. 1998, 2001). Biochemical and genetic studies examining the dynamic interactions between the domains of Gcn2p suggest that there is inhibitory contact between the protein kinase domain and the Gcn2p carboxy terminus that is relieved upon binding of uncharged tRNA to the HisRS-related domain (Dong et al. 2000; Qiu et al. 2001, 2002). However, release of this inhibitory contact does not appear to be sufficient for induced eIF2 kinase activity. Association of uncharged tRNA with the Gcn2p is also thought to contribute to a positive-acting contact between the amino terminal portion of the HisRS-region and the protein kinase domain (Qiu et al. 2001, 2002). Interaction between the HisRS and protein kinase regions is proposed to realign kinase subdomains V and VIb, including residues Arg794 and Phe842, opening the substrate binding cleft of the catalytic domain and allowing for eIF2 binding and phosphorylation. Gcn2p also contains a second region sharing homology with protein kinases. This so-called partial- or pseudo- kinase domain is located amino terminal to the bona fide Gcn2p kinase domain and is required for induction of eIF2 phosphorylation in response to amino acid limitation. Supporting the model that release of the autoinhibitory interaction between the protein kinase and extreme carboxy terminal regions of Gcn2p is not sufficient for activation is the observation that the Gcn2p kinase domain by itself does not appreciably phosphorylate eIF2 *in vivo* or *in vitro*. However, this isolated Gcn2p kinase domain becomes hyperactive after residue substitutions in the key subdomains V and VIb of Gcn2p (R794G and F842L, designated Gcn2p-Hyper) that are proposed to direct an active conformation independent of interaction with the HisRS-related or partial kinase regions (Qiu et al. 2002).

Accompanying this activated conformation of Gcn2p is autophosphorylation at threonine residues 882 and 887 in the so-called activation loop in subdomain VII of the kinase domain (Romano et al. 1998). This autophosphorylation may occur in trans between Gcn2p dimers. Dimerization of Gcn2p appears to occur independent of amino acid starvation, and involves predominantly the extreme carboxy terminus of Gcn2p, as well as weaker contributions between the HisRS-related and protein kinase domains. Upon self phosphorylation, Gcn2p is presumed to retain its induced eIF2 kinase activity until it is dephosphorylated by protein phosphatases. Dephosphorylation of eIF2 is thought to be mediated by a Type I protein phosphatase encoded by *GLC7* (Wek et al. 1992). The activity of the Glc7p is regulated by multiple regulatory proteins that associate with this phosphatase, enhancing its recognition for phosphorylated protein substrates. In mammalian cells, this regulatory protein is Gadd34, which itself is induced in response to eIF2 phosphorylation, indicating a feedback mechanism controlling stress gene expression (Connor et al. 2001; Novoa et al. 2001, 2003). It remains to be determined whether a Gadd34 orthologue functions in the regulation of *GCN4* expression in fungi.

There has also been reported examples of stress induction of *GCN4* expression that is independent of Gcn2p. Induction of Ras2p which leads to activation of protein kinase A in yeast is suggested to enhance Gcn4p synthesis (Engelberg et al. 1994). Furthermore, defects in tRNA processing or nuclear transport enhances *GCN4* translation independent of eIF2 phosphorylation (Vazquez de Aldana et al. 1994; Qiu et al. 2000). The mechanistic details of Gcn2p-independent induction of *GCN4* expression is not known, but it may involve direct or indirect reduction in eIF2 activity.

7.3 Ribosome association of Gcn2p is required for activation in response to amino acid starvation

The process by which Gcn2p monitors the charging levels of uncharged tRNA requires targeting of Gcn2p to the ribosomal machinery. This ribosome binding occurs through association with the extreme carboxy terminus of Gcn2p (Ramirez et al. 1991; Zhu and Wek 1998). Substitutions of three lysine residues in the carboxy terminal segment (designated Gcn2-605p) selectively blocked interaction between Gcn2p and 60S ribosomal subunits and reduced eIF2 phosphorylation in response to amino acid limitation. Ribosomal sites important for contact with the carboxy terminus of Gcn2p have not yet been delineated, but regions within rRNA may be critical given that this portion of Gcn2p has affinity for RNA (Zhu and Wek 1998).

A second interface between ribosomes and Gcn2p involves the amino terminus of this eIF2 kinase, from residues 1 to 272 (Garcia-Barrio et al. 2000; Kubota et al. 2000). This portion of Gcn2p has two features, an amino-terminal core conserved between Gcn2p orthologues (designated CNT) and a highly charged region, that together contribute to the physical interaction with a protein complex consisting of Gcn1p and Gcn20p which itself is associated with ribosomes (Marton et al. 1997; Garcia-Barrio et al. 2000). Gcn1p is a very large protein- almost 300 kD, that contains a central region with homology to fungal elongation factor 3 (EF3) (Marton et al. 1993). EF3 facilitates the release of uncharged tRNA from the exit (E) site of the ribosome, thereby promoting entry of charged tRNA into the acceptor (A) site during translation elongation (Chakraburtty 1999; Anand et al. 2001). The Gcn1p-Gcn20p complex is proposed to be in proximity to the ribosomal acceptor site, and a carboxy-terminal segment of Gcn1p from residues 2052 to 2428, including a critical Arg2259 residue, interacts directly with Gcn2p (Sattlegger and Hinnebusch 2000). However, ribosome association of either Gcn2p or the Gcn1p-Gcn20p complex is not dependent on this Gcn protein interaction.

How does ribosome targeting enhance activation of Gcn2p eIF2 kinase activity in response to amino acid starvation? Several models have been proposed. First, ribosome targeting could facilitate Gcn2p access to its substrate eIF2. Arguing against this idea is the observation that the kinase domain of Gcn2p-Hyper by itself efficiently phosphorylates eIF2 *in vivo* despite the absence of ribosome association (Qiu et al. 2002). Furthermore, as will be discussed further below, glucose deprivation activates Gcn2p eIF2 kinase activity and the Gcn2p-605 mutant that is blocked for ribosome association significantly phosphorylates eIF2 and induces *GCN4* translation (Yang et al. 2000). Therefore, ribosome association of Gcn2p is not obligate for eIF2 access *in vivo*. A second model explaining the necessity for ribosome targeting for Gcn2p activity involves its requirement for dimerization and trans-phosphorylation. Ribosome targeting could elevate localized concentrations of the eIF2 kinase, contributing to dimerization. Such a model has been proposed for the related eIF2 kinase PKR that participates in an anti-viral defense pathway in mammalian cells (Vattem et al. 2001). Opposing this model is the observation that the dimerization of Gcn2p through it carboxy terminus is quite stable independent of amino acid availability (J.N., unpublished observation). Therefore, a role for a dynamic equilibrium between Gcn2p monomers and dimers in the mechanism of eIF2 kinase regulation appears unlikely.

A third model for ribosome targeting of Gcn2p that emphasizes the interaction between this eIF2 kinase and the Gcn1p-Gcn20p complex revolves around the idea that levels of uncharged tRNA are best measured in the context of the ribosome itself. Gcn1p is suggested to be localized in proximity to the A site, and its role may reside in its ability to eject uncharged tRNA that enters the ribosome during elongation (Sattlegger and Hinnebusch 2000). Such evicted uncharged tRNA would be transferred by the Gcn1p-Gcn20p complex to the HisRS-related domain of Gcn2p, inducing the active conformation of this eIF2 kinase. While uncharged tRNAs have been shown to bind in a codon-dependent manner to the A site of eukaryotic ribosomes, the levels of uncharged tRNA required to facilitate such binding *in vivo* have not yet been resolved (Murchie and Leader 1978). The Gcn1p-Gcn20p may serve to enhance binding of uncharged tRNA to ribosomes. The amount of Gcn1p is much reduced compared to ribosomal levels in yeast, and therefore only a portion of total ribosomes are associated with Gcn1p. If Gcn1p is overexpressed in yeast, which would facilitate the proposed binding of uncharged tRNA to ribosomes, there is enhanced sensitivity to the aminoglycoside antibiotic paromomycin, a drug that reduces translation fidelity (Sattlegger and Hinnebusch 2000).

Strongly supporting the model that the Gcn1p-Gcn20p complex is critical for optimal activation of Gcn2p eIF2 kinase activity in response to amino acid starvation, is the observation that deletion of *GCN1* blocks eIF2 phosphorylation by Gcn2p and the resulting induction of translation of *GCN4* mRNA (Marton et al. 1993). As observed for *gcn2* mutants, including those removing the CNT domain of Gcn2p, deletion of either *GCN1* or *GCN20* render cells hypersensitive to growth inhibition in response to amino acid deprivation (Marton et al. 1993; Vazques de Aldana et al. 1995; Garcia-Barrio et al. 2000). This proposed regulatory linkage between Gcn2p and Gcn1p appears to be conserved throughout evolution, as orthologues for both proteins are found in a range of organisms, including fungi, *Caenorabditis elegans*, *Drosophila melanogaster*, *Arabidopsis*, and mammals. By contrast, orthologues for the transcription activator Gcn4p is restricted to certain fungi, although related basic-zipper (bZIP) transcriptional regulators may carry out an analogous function in *S. pombe* and higher eukaryotes.

7.4 Phosphorylation of eIF2 induces *GCN4* translational expression

Translational control of *GCN4* is the predominant mechanism regulating expression of this transcriptional activator in response to nutrient limitation. As described above, the event initiating induced *GCN4* translation is phosphorylation of eIF2 by Gcn2p. The ternary complex consisting of eIF2 coupled with GTP and initiator Met-tRNA$_i^{Met}$ participates in the ribosomal selection of the start codon. During this translation initiation process, GTP associated with eIF2 is hydrolyzed to GDP and eIF2 is released from the ribosome. Recycling of eIF2-GDP to the GTP-bound active form requires a guanine nucleotide exchange factor, eIF2B (Fig. 1). Gcn2p phosphorylation of the α subunit eIF2 at Ser51 converts this initiation factor from a substrate to an inhibitor of the eIF2B, thereby reducing the levels of eIF2-GTP available for translation initiation (Rowlands et al. 1988; Dever et al. 1995; Hinnebusch 1997). eIF2B consists of five polypeptide subunits designated eIF2B α-ε that are organized into catalytic and regulatory subcomplexes (Pravitt et al. 1998). Guanine nucleotide exchange is catalyzed by Gcd6p (ε) with the assistance of Gcd1p (γ). Phosphorylated eIF2 binds tightly to the regulatory subcomplex consisting of Gcn3p (α), Gcd7p (β) and Gcd2p (δ), preventing eIF2 association with the catalytic subcomplex and blocking GDP-GTP exchange (Rowlands et al. 1988; Pravitt et al. 1998; Krishnamoorthy et al. 2001).

Control of *GCN4* translation initiation is mediated by four upstream open reading frames (ORFs) located in the 5'- non-coding portion of the *GCN4* mRNA (Fig. 2). These upstream ORFs, numbered from 1 through 4, are each only two or three codons in length. Experiments involving analysis of different configurations of the *GCN4* mRNA leader fused to a reporter gene in yeast, and *in vitro* measurements of ribosome association at different locations along the leader of the *GCN4* mRNA support the following model (Hinnebusch 1997; Gaba et al. 2001). Translation of the *GCN4* mRNA begins in a cap-dependent fashion with the scanning ribosome initiating at the 5'-proximal ORF1. Upstream ORF1 functions as a positive-acting element in *GCN4* translational control by allowing ribosomes to reinitiate at downstream ORFs. The basis for the reinitiation capacity appears to reside in the termination context, whereby sequences 3'-to the ORF1 stop codon are proposed to facilitate the retention of the small ribosomal subunit with the *GCN4* mRNA (Grant and Hinnebusch 1994). Following translation of upstream ORF1, the 80S ribosomal subunit is proposed to decouple, with the 40S ribosomal subunit retaining its association with the ORF1 termination region of the leader of the *GCN4* mRNA. This small ribosomal subunit would resume scanning in a 5' to 3' direction along the leader of the *GCN4* mRNA. When eIF2-GTP is plentiful during the non-starved state, the small ribosomal subunit quickly requires the eIF2 ternary complex and, coupled with the 60S ribosome, reinitiates translation at upstream ORF2, ORF3, or ORF4. Following translation of one of these three upstream ORFs, the ribosome dissociates from the *GCN4* mRNA, thereby blocking expression of the *GCN4* coding region. When amino acids are limiting and eIF2-GTP levels are reduced, there is a delay in ribosome reinitiation following translation of upstream ORF1. The increased time required for reacquisition of eIF2-GTP coupled with Met-tRNA$_i^{Met}$ allows the 40S ribosomal subunit to scan through the negative-acting upstream ORFs 2, 3, and 4. While scanning the mRNA leader region from ORF4 to the initiation codon of the *GCN4* coding region, the ribosome acquires the eIF2 ternary complex, facilitating translational expression of *GCN4*.

During starvation for a single amino acid, such as that observed following the addition of 3-aminotriazole (3-AT)- an inhibitor of histidine biosynthesis to yeast cultures, there is enhanced eIF2 phosphorylation and Gcn4p synthesis accompanied by a reduction in both general translation and yeast growth. However, as judged by polysome profiles in sucrose gradient sedimentation experiments there is no significant accumulation of free ribosomal subunits, suggesting that there is not a block in translation initiation due to the levels of Gcn2p phosphorylation of eIF2 induced by the 3-AT inhibitor (Ramirez et al. 1991). This observation suggests that reduced general translation accompanying amino acid limitation is simply a function of the lowered levels of free amino acids, rather than lowered availability of eIF2-GTP required to sustain general translation initiation. Therefore, stimulation of *GCN4* translation can occur in response to a modest reduction in eIF2-GTP that does not impede general translation initiation. Perhaps reinitiation of translation that occurs in the *GCN4* mRNA leader is particularly sensitive to lowered levels of eIF2 ternary complex accompanying such amino acid limitations. Reduced translation initiation can occur when a yeast strain that is auxotrophic for amino acids is shifted from media containing complete amino acids to that deprived of all amino acids (Ashe et al. 2000). Under this severe starvation condition where the strain cannot synthesize its full complement of amino acids there appears to be enhanced activation of Gcn2p and hyperphosphorylation of eIF2 that would further reduce eIF2-GTP levels required to sustain general translation initiation. Constitutively active mutants of GCN2 or expression of high levels of mammalian eIF2 kinases PKR or PEK/Perk in yeast also lead to hyperphosphorylation of eIF2 and a general reduction in protein synthesis (Wek et al. 1990; Ramirez et al. 1992; Dever et al. 1993; Sood et al. 2000). In these examples of reduced general translation initiation by Gcn2p hyperphosphorylation of eIF2, there is still enhanced translation of *GCN4* mRNA, although there would be lowered levels of general translation including the synthesis of proteins encoded by genes transcriptionally induced by Gcn4p. Together, these studies indicate that a range of eIF2 phosphorylation levels can induce gene-specific translation, and this translational control can occur in the absence of a general protein synthesis defect.

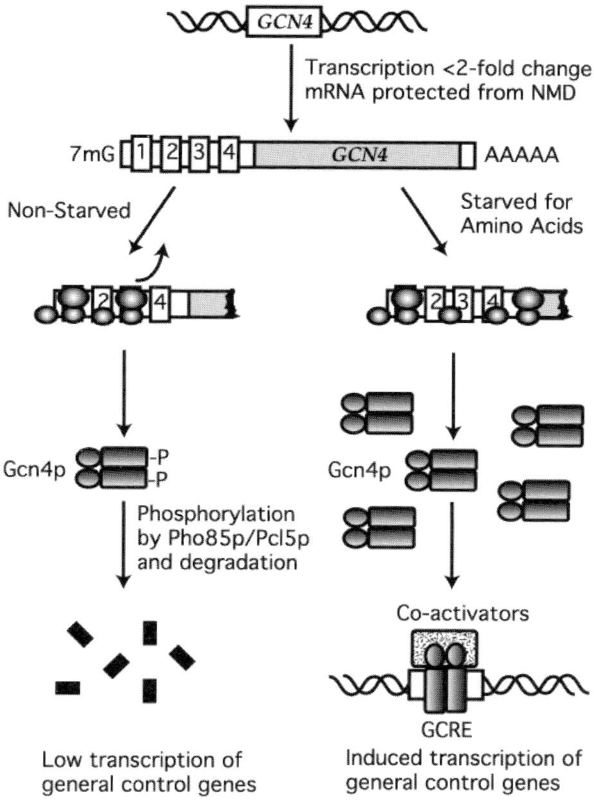

Fig. 2. Multiple regulatory mechanisms contribute to induced Gcn4p-directed transcription. In response to amino acid starvation there is increased synthesis of the "master regulator" Gcn4p. The mechanisms leading to increased expression of Gcn4p involves less than a two-fold increase in *GCN4* transcription in yeast cells starving for amino acids. *GCN4* mRNA contains four short ORFs (boxes 1-4) located upstream of the *GCN4* coding region (shaded box) that are important for translational control in response to amino acid limiting conditions. Included in this leader region of the *GCN4* mRNA is a stabilizer element that protects this transcript from degradation by the nonsense mediated decay (NMD) pathway. When yeast are not limiting for amino acids, the upstream ORFs reduce translation of the *GCN4* coding region. Gcn4p that is synthesized is phosphorylated by Pho85p/Pcl5p, ubiquitinated, and degradated via proteasomes. In response to amino acid limitations, phosphorylation of eIF2 by GCN2p leads to a reduction in eIF2-GTP levels that allows scanning ribosomes to proceed through the inhibitory ORFs 2-4, facilitating increased translation of the *GCN4* coding region. During nutrient-limiting conditions, Pcl5p levels are proposed to be reduced, blocking Pho85p-mediated phosphorylation and degradation of Gcn4p. Increased levels of Gcn4p lead to elevated binding to the general control response element (GCRE) located in the promoter of the Gcn4p-targeted genes. Gcn4p interacts with coactivator complexes that facilitate binding of the RNA polymerase II transcriptional apparatus, enhancing expression of genes subject to the general control.

Only after falling below a certain threshold level of eIF2-GTP is there a reduction in general translation.

Regulation of *GCN4* translation by eIF2 phosphorylation is also central to general control pathways in other fungi such as *Candida albicans* and *Neurospora crassa* (Paluh et al. 1988; Luo et al. 1995; Tripahti et al. 2002). However, in these other fungi examples the leader of the mRNAs encoding Gcn4p orthologues has only two upstream ORFs. The first upstream ORF functions similarly to the positive-acting ORF1 in yeast *GCN4* mRNA, while the second upstream ORF is the sole inhibitory element, preventing ribosomal reinitiation at the downstream coding region during the fed state. This two upstream ORF configuration has been constructed artificially in yeast *GCN4* mRNA by deleting ORFs 2 and 3 (Hinnebusch 1997). The *GCN4* mRNA leader that retains only upstream ORF1 and ORF4 mediates translational control in response to eIF2 phosphorylation, albeit at reduced levels compared to the wild type version of *GCN4* mRNA containing the full complement of inhibitory upstream ORFs. Translation of the related mammalian bZIP transcription factor Atf4 is also translationally regulated by a mechanism involving multiple upstream ORFs (Harding et al. 2000). Although it is not yet certain whether this regulatory process involves the leaky ribosome scanning as described above for *GCN4* in yeast.

7.5 Multiple regulatory mechanisms induce Gcn4p levels in response to starvation for amino acids

While translational control is a major mechanism inducing Gcn4p levels in response to nutrient depletion, regulation of the synthesis and stability of *GCN4* mRNA and protein turnover also contribute to the overall increase in Gcn4p concentrations. Elevated synthesis of *GCN4* mRNA in response to amino acid limitation in *S. cerevisiae* is modest, with less than a two-fold increase in *GCN4* transcription (Fig. 2). As discussed more fully below, other stress conditions can induce *GCN4* translation and glucose limitation or exposure to the drug rapamycin can induce as much as a three fold increase in *GCN4* transcription. By contrast increased synthesis of mRNA encoding Gcn4p orthologues in other fungi, such as *Candida albicans*, *Neurospora crassa*, and *Aspergillus nidulans*, is a significant contributor to overall expression of this transcriptional activator (Paluh et al. 1988; Hoffmann et al. 2001; Tripahti et al. 2002). For example the *GCN4* orthologue in *A. nidulans*, designated *cpcA*, has an eight-fold increase in its mRNA levels in response to eight hours of exposure to 10 mM 3-AT, compared to only a five-fold increase in the CpcA protein levels (Hoffmann et al. 2001). An important contributor to this increase in *cpcA* mRNA is autoregulation, whereby CpcA binds to its own gene promoter leading to a further amplification of expression.

The fact that the leader of the *GCN4* mRNA contains short ORFs that precede the *GCN4* coding region presents challenges to the stability of this mRNA. Transcripts that contain nonsense mutations within the protein coding region are degraded in yeast by the nonsense mediated decay (NMD) pathway, preventing the synthesis of truncated proteins (Wilusz et al. 2001). The NMD pathway degrades not only nonsense-containing mRNAs, but also frameshift and improperly-spliced transcripts, and mRNAs containing ORFs preceding the coding region. It is proposed that ribosomes pause at nonsense codons, promoting the assembly of a surveillance complex that upon translation termination scans towards the 3'-end of the transcript. An improper translation termination event is recognized if the surveillance complex detects a specific downstream sequence element (DSE) and associated proteins, leading to assembly of additional factors, including Hrp1p, that facilitate decapping of the transcript by Dcp1p. In the case of the *GCN4* transcript there is the presence of a stabilizer element (STE) 3' of the upstream ORF4 that associates with Pub1p and prevents signaling of the decapping pathway (Fig. 2) (Ruiz-Echevarria and Peltz 2000). This would maintain the stability of *GCN4* mRNA independent of the nutritional status of the cell. However, Pub1p binding with RNA does not appear to impede scanning ribosomes, and therefore does not prevent translation reinitiation at the *GCN4* coding region.

Gcn4p is predominantly in the nucleus where it is highly unstable with a half-life of less than 5 minutes (Pries et al. 2002). Upon amino acid starvation there is a stabilization of Gcn4p that, in combination with increased expression of *GCN4*, leads to elevated steady state levels of Gcn4p and enhanced transcriptional activation. Degradation of Gcn4p depends on its ubiquitination by the ubiquitin-conjugating enzyme Cdc34p in combination with the SCF^{Cdc4p} complex (Fig. 2) (Meimoun et al. 2000). Such ubiquitination directs Gcn4p to the proteasome where is it is degraded. Ubiquitination of Gcn4p is induced by the cyclin-dependent protein kinase Pho85p that, in conjunction with its regulatory subunit Pcl5p, targets Gcn4p for ubiquitination by specifically phosphorylating Gcn4p at residue Thr165 (Shemer et al. 2002). Central to the regulation of Pho85p phosphorylation of Gcn4p is the availability of Pcl5p. *PCL5* mRNA is induced in response to nutrient limitation by a mechanism involving transcriptional activation by Gcn4p. However, Pcl5p is thought to be labile, and it is suggested that translation of *PCL5* mRNA is low when there is reduced general translation at the onset of an amino acid starvation condition. Reduced Pcl5p levels would lower Pho85p phosphorylation of Gcn4p and insure the availability of this transcription factor at the onset of a starvation condition. With elevated levels of Gcn4p and increased expression of its target genes, amino acid levels would be replenished in yeast, contributing to increased synthesis of Pcl5p. Inherent in this model is delayed translation of *PCL5* mRNA relative to expression of Gcn4p and at least a portion of its target gene products. This timing of stressed-induced gene expression has not yet been well addressed experimentally. A second cyclin-dependent protein kinase Srb10p is also linked to Gcn4p turnover, and deletion of *SRB10* and *PHO85* together is required for maximum stability of Gcn4p (Chi et al. 2001). Srb10p may phosphorylate five distinct residues in Gcn4p, including Thr165. Similar to that described for Pho85p, such Srb10p phosphorylation is thought to mediate degradation of Gcn4p through Cdc34p and possibly the SCF^{Cdc4p} complex. However, regulation of Gcn4p levels by Srb10p appears to be controlled independent of the availability of amino acids.

7.6 Gcn4p mediates transcriptional activation by interfacing with the transcriptional machinery

Gcn4p is a member of the bZIP family of transcription factors. The bZIP region located at the extreme carboxy terminus of Gcn4p is important for Gcn4p dimerization and binding with DNA elements, termed general control response elements (GCREs), embedded in the promoter region of target genes (Fig. 2) (Hinnebusch 1992). While other members of the bZIP family such as mammalian Atf4 can heterodimerize with other bZIP proteins, Gcn4p is thought to function only as a homodimer (Hai and Hartman 2001). Such DNA binding can occur in the absence of nutrient limitation, contributing to the basal expression of Gcn4p regulated genes. This is best illustrated by the observation that while yeast cells deleted for *GCN4* are viable they can no longer grow without all amino acids supplemented in the medium. With the increased levels of Gcn4p during amino acid starvation, there is enhanced Gcn4p binding to GCREs and stimulation of transcription, frequently referred to in the literature as derepression. Activation of transcription involves the amino terminus of Gcn4p which contains seven clusters of hydrophobic residues interspersed among acidic residues (Jackson et al. 1996). These hydrophobic segments are essential for recruitment of multi-subunit protein complexes, collectively referred to as coactivators (Drydale et al. 1998; Swanson et al. 2003).

Genetic analysis of the viable mutants generated by the *Saccharomyces* Genome Deletion project indicates that at least seven different coactivator complexes can associate with Gcn4p and impact expression of genes subject to the general control pathway (Fig. 2) (Swanson et al. 2003). One of the best characterized examples of Gcn4p-coactivator interaction involves the SAGA complex, which contains the histone acetyltransferase subunit, Gcn5p (Berger and Sterner 2000). Acetylation of nucleosomal H3 and H2B by Gcn5p leads to remodeling of chromatin that exposes or masks binding sites for TATA-binding protein (TBP) and RNA polymerase II in core promoter regions. Along with Gcn5p, SAGA contains TBP-associated factors (TAFs) that directly contribute to recruitment of general transcription factors. Additional Gcn4p co-activator complexes are SWI/SNF and RSC that hydrolyze ATP to displace nucleosomes and alter the availability of protein binding sites in promoters (Natarajan et al. 1999; Swanson et al. 2003).

The precise contribution of these coactivators in Gcn4p-mediated induction of transcription is still not well understood. Clearly, portions of different coactivator complexes can contribute to activation by Gcn4p at individual target gene promoters. For example, mutations in multiple subunits of seven different coactivators, including SAGA, SWI/SNF, and RSC, lowered the induced levels of *HIS4* and *SNZ1* mRNA in response to amino acid limitation compared to transcription in wild type cells (Swanson et al. 2003). Surprisingly, among the SAGA subunits characterized, only Gcn5p was dispensable for increased expression of these Gcn4p target genes in response to amino acid starvation. By comparison, significant induction of *ARG1* expression required four coactivators, RSC, CCR4/NOT, SRB/MED, and PAF1 complex (THO/TREX), with SAGA and SWI/SNF being dispensable (Swanson et al. 2003). However, ChIP experiments measuring recruitment of Gcn4p-associated proteins to the chromatin of the *ARG1* promoter region indicated that SAGA and SWI/SNF1, in addition to SRB/MED, were strongly associated in response to amino acid starvation conditions. Less pronounced, albeit significant, Gcn4p-dependent immunoprecipitation in the *ARG1* promoter region was observed for coactivators SRB/MED and PAF1 complex. Together these results suggest that Gcn4p can recruit more coactivators to a given target promoter than is required for full expression in response to nutrient deprivation (Swanson et al. 2003). Given that Gcn4p activates hundreds of genes, Gcn4p may interact with many coactivators to overcome diverse regulatory arrangements in target promoters. It is unlikely that these large multi-subunit coactivator complexes reside at a given promoter simultaneously. Perhaps each coactivator binds transiently, contributing their specific functions at the promoter and dissociating prior to entry of a different coactivator complex. Furthermore, the subunit composition of coactivator complexes may vary between different promoter contexts, with certain subunits being dispensable for coactivator complex function or combining differentially to form diverse complex arrangements.

While GCREs are an important feature of Gcn4p-mediated activation of gene transcription, almost half of the genes with a fourfold or greater induction dependent on Gcn4p function had no recognizable binding element in their promoter region or sequences upstream of their translation start site (Natarajan et al. 2001; Hinnebusch and Natarajan 2002). It is certainly possible that these genes had GCREs in their coding region or 3'-untranslated regions that have functional significance for their transcriptional induction. An alternative explanation is that transcriptional control of these genes by Gcn4p is indirect. As described below, Gcn4p induces the expression of a large collection of transcriptional activators. Furthermore, Gcn4p could modulate transcription through protein-protein interactions that are independent of GCRE binding at a regulated gene. For example, Gcn4p could bind and inactivate transcriptional factors that mediate repression of genes void of GCREs.

While Gcn4p is predominantly viewed as an activator of transcription, DNA microarray analysis of Gcn4p-dependent gene expression indicates that Gcn4p can also contribute to repression of transcription (Natarajan et al. 2001). For example as described further below, there is a dependence on Gcn4p for repressed transcription of genes encoding ribosomal proteins or translation factors in response to amino acid starvation conditions. Most of these genes do not have recognizable GCREs in their promoter regions, indicating that Gcn4p probably contributes indirectly to their repressed expression. This is further supported by the observation that overexpression of Gcn4p in the absence of nutrient limitation does not lower transcription of these ribosomal protein genes. Collectively, expression of these genes are reduced during amino acid starvation conditions by mechanisms involving the transcriptional regulator Rap1p and signal pathways controlled by protein kinase A and Tor proteins (Moehle and Hinnebusch 1991; Neuman-Silberberg et al. 1995; Li et al. 1999; Natarajan et al. 1999; Schmelzle and Hall 2000; Crespo and Hall 2002). It is proposed that elevated levels of Gcn4p may enhance transcriptional repression in concert with these regulatory pathways by Gcn4p binding and sequestering transcription factors required for expression of ribosomal proteins and translation factors (Natarajan et al. 2001). Gcn4p is also reported to enhance expression of protein kinases and phosphatases, providing for a range of possible mechanisms that could modulate these signaling pathways. Finally, Gcn4p is linked to expression of a large number of other transcription factors that together could combine for direct and indirect Gcn4p control of diverse stress pathways.

7.7 Gcn4p coordinates expression of hundreds of genes in response to amino acid starvation

Gcn4p has been termed the "master regulator" of a five layered program of gene regulation designed to alleviate nutrient deprivation (Fig. 3) (Natarajan et al. 2001; Hinnebusch and Natarajan 2002). Certainly, the core layer of Gcn4p transcriptional control involves genes directly contributing to the synthesis of amino acids. As noted above, general control is a true cross-pathway stress response in that starvation for a single amino acid, such as histidine, induces the expression of genes directly involved in the synthesis of 19 of the 20 amino acids. The sole exception is cysteine, and in this example, Gcn4p directs the synthesis of cysteine pathway precursors serine and homocysteine. It has been confirmed for a large number of these amino acid biosynthetic genes that their encoded enzyme activities are induced as part of the general control program. However, in a recent DNA microarray study by Natarajan et al. (Natarajan et al. 2001) it was observed that of the 539 genes whose transcription requires Gcn4p for full induction in response to amino acid depletion only 73 are known to contribute to amino acid biosynthesis. Therefore, the influence of Gcn4p exceeds beyond core amino acid synthetic genes.

The second layer of gene regulation by Gcn4p involves intermediary metabolism related to amino acid biosynthesis and nutrition (Natarajan et al. 2001). For example, 16 genes function in the synthesis of vitamins that are important cofactors for enzymes in pathways related to amino acids. Expression of several genes encoding amino acid permeases are induced by Gcn4p, including the general amino acid permease Gap1, a basic amino acid permease Can1p, and a broad spectrum permease Agp1p. Of the 35 members of the mitochondrial carrier family involved in metabolite transport between this organelle and the cytoplasm, 10 are regulated transcriptionally by Gcn4p. Given that portions of the synthetic pathways for arginine, lysine and the branched chain amino acids are carried out in the mitochondria, it is rationalized that the availability of such transport systems is linked to the demands of the amino acid biosynthetic pathways. Another organelle associated with Gcn4p-directed gene expression is the peroxisome. The peroxisome has a primary role in the β-oxidation of fatty acids and detoxification which may be linked with a yeast cell strategy for coping with amino acid depletion. With regards to amino acid biosynthesis, lysine synthetic enzymes Lys1p and Lys4p are located in the peroxisome, and certain peroxisomal mutants in Pex8p

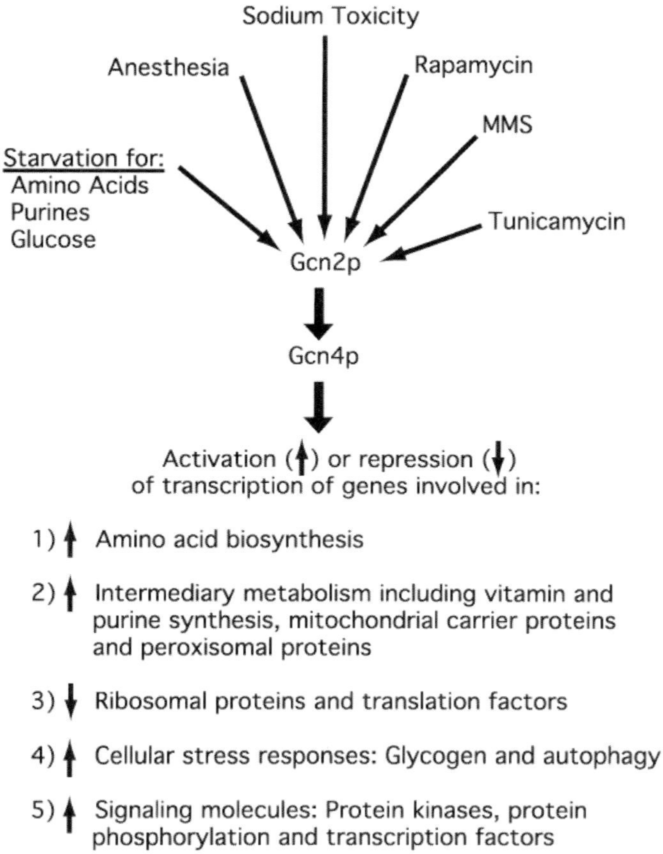

Fig. 3. Different stress conditions induce general control. A diverse array of cellular stresses increase eIF2 phosphorylation by Gcn2p, leading to a five layered program of gene regulation directed by Gcn4p. DNA microarray analysis of yeast starving for amino acids indicates that Gcn4p is required for the induction of 539 genes and for repression of ribosomal protein genes and those encoding related translation factors (Natarajan et al. 2001). Currently, we do not appreciate the contribution of Gcn4p in the gene expression profiles during these other stress conditions.

and Pex15p are impaired for synthesis of lysine (Geraght et al. 1999). Therefore, Gcn4p-directed expression of peroxisomal-related genes may contribute to enhanced lysine production.

It is of interest to note that Gcn4p also induces five purine biosynthetic genes in response to amino acid limitation, and as further discussed below purine deprivation is a potent inducer of eIF2 phosphorylation and *GCN4* translational control (Rolfes and Hinnebusch 1993; Natarajan et al. 2001). The physiological basis for this regulatory linkage may involve the metabolic overlap between the biosynthesis of certain amino acids and purines. For example, the purine ring of ATP is utilized early in the histidine biosynthetic pathway, and induced purine synthesis by Gcn4p may be important for supporting enhanced histidine production. The connection between histidine and purine synthesis is also highlighted by the regulation and function of *HIS7* encoding glutamine amidotransferase cyclase (Springer et al. 1996). This bifunctional enzyme catalyzes the fifth and sixth step in the histidine synthetic pathway and produces the by-product 5-aminoimidazole-4-carboxamide ribotide (AICAR) which is an important intermediate in the purine pathway. Regulation of *HIS7* involves Gcn4p binding to two GCREs, designated 1 and 2, in its promoter region that work synergistically to induce transcription. In response to adenine limitation, a second transcriptional activator Bas1p in complex with Bas2p is thought to bind GCRE-2 and activate *HIS7*. Both Gcn4p and the Bas1p/Bas2p complex are required for maximal expression of *HIS7* in response to combined starvation for amino acids and purines. *BAS1*, which functions to activate the transcription of multiple purine biosynthetic genes, is itself transcriptionally regulated by Gcn4p (Natarajan et al. 2001). Therefore, Gcn4p contributes to increased expression of *HIS7* both directly and through enhanced expression of other stress-related transcription factors.

The third layer of the Gcn4p-mediated program of gene regulation involves control of the translational machinery. As highlighted above, Gcn4p is required for reduced expression of a number of translation factors and 90 ribosomal protein genes (Natarajan et al. 2001). Such a reduction in the synthesis of the translational machinery would be appropriate with the reduced cellular growth rates associated with lowered amino acid availability. This regulatory strategy is analogous to the well described stringent response in *Escherichia coli* that represses synthesis of rRNA and subsequently ribosomal proteins in response to amino acid starvation (Cashel and Rudd 1987). It is interesting to note that in both *E. coli* and yeast, the signal for this repression is proposed to be placement of uncharged tRNA into the A sites of ribosomes. By contrast, Gcn4p enhances the expression of a number of different aminoacyl-tRNA synthetase genes, including *KRS1*, *ILS1*, and *MES1*, suggesting a mechanism to enhance the efficiency of aminoacylation of their corresponding cognate tRNAs during conditions of reduced amino acids (Natarajan et al. 2001). However, aminoacyl-tRNA synthetases are not uniformly induced by histidine limitation. In fact, genes encoding GlnRS, PheRS, and SerRS are repressed in cells exposed to 3-AT. Therefore, the role of Gcn4p and its impact on aminoacylation of different tRNA species has yet to be resolved.

A fourth layer of Gcn4p-mediated control involves broader themes in cellular stress responses. Starvation for various nutrients, including limiting nitrogen or carbohydrates, induces a process of autophagy that facilitates the bulk turnover of cytoplasmic material. Autophagosomes deliver cytoplasmic proteins and organelles to yeast vacuoles, which have many parallels to lysosomes in higher eukaryotes. In the vacuoles, proteins and organelles are degraded and reclaimed for later use (Reggiori and Klionsky 2002). Gcn4p induces the expression of 2 vacuolar proteases and 3 autophagy proteins, including the protein kinase Apg1p and its associated protein Apg13p that are instrumental for activating the autophagy process (Natarajan et al. 2001). The absolute requirement of *GCN4* for autophagy is still controversial. In response to histidine starvation, autophagy was reported to occur independent of Gcn4p function (Natarajan et al. 2001). By contrast, Talloczy et al. (2002) suggested that autophagy in response to nitrogen starvation is blocked in yeast strains devoid of *GCN4* or *GCN2*. The apparent conflict between the two studies may lie in the different yeast strains utilized or in the different nutrient stresses used to invoke autophagy. It has been observed that nitrogen limitation is a potent inducer of autophagy, while starvation for an amino acid such as tryptophan leads to accumulation of fewer autophagic bodies (Takeshig et al. 1992). A further complication is that while nitrogen starvation induces eIF2 phosphorylation by Gcn2p, there is no synthesis of Gcn4p (Grundmann et al. 2001). The mechanistic basis for the absence of Gcn4p expression during nitrogen deprivation is not understood, but it is reasoned that starvation for nitrogen would thwart amino acid biosynthesis.

General nutrient starvation also leads to accumulation of glycogen. This polymer of glucose begins to accumulate as general nutrients begin to be depleted, allowing yeast cells to accumulate carbohydrates to be utilized upon resumption of vegetative growth or during spore germination (Francois and Parrou 2001). Gcn4p contributes to the expression of several proteins involved in glycogen accumulation, including glycogen synthase, glycogenin, and the branching enzyme (Natarajan et al. 2001). Glycogen synthesis and turnover varies considerably during the time course of a nutrient-depletion study. Utilizing matched yeast cultures shifted from synthetic medium containing 2% to 0.05% glucose, a condition that increases eIF2 phosphorylation by Gcn2p and enhances *GCN4* translation, it was found that there was a 7-fold increase in glycogen levels after two hours of culture incubation (Yang et al. 2000). Accumulation of glycogen was similar between strains containing wild type or deleted *GCN2* function. However, following 22 hours of incubation in the glucose-deficient media, glycogen levels were more significantly reduced in the absence of Gcn2p activity, with the *gcn2* mutant cells having fourfold less glycogen than the wild type strain. Therefore, the general control has important roles in both amino acid and carbohydrate metabolism.

The fifth and final layer of Gcn4p-directed gene expression involves the induction of signaling proteins such as protein kinases, protein phosphatase catalytic and regulatory subunits, and transcription factors. Activation of these regulatory genes may allow for amplification of the general control response, and provide for a means of communication with other stress response pathways. In the case of protein kinases, it was noted above that Gcn4p induces expression of Apg1 protein kinase, which is important for eliciting autophagy (Natarajan et al. 2001). Autophagy is also positively regulated by Snf1 protein kinase, required for glucose derepression, and negatively impacted by nutrient sensing protein kinases Tor and Pho85p (Reggiori and Klionsky 2002). Therefore, multiple nutrient stress response pathways interconnect to control this catabolic trafficking process. Another protein kinase whose expression is induced by Gcn4p is Npr1p. Npr1p promotes the stabilization of Gap1p, a general amino acid permease, and the proteolysis of the tryptophan permease Tat2p (Schmelzle and Hall 2000). As described above Gcn4p activates *GAP1* expression, and the added *NPR1* expression may further contribute to the cellular uptake of amino acids. Npr1p is also inhibited by Tor-directed protein phosphorylation, further emphasizing cross pathway control of key nutrient regulatory steps. In the example of phosphatase-related genes, Gcn4p directs expression of Gip1p, a Type 1 protein phosphatase interacting protein required for synthesis of spore walls (Natarajan et al. 2001). Sporulation in yeast occurs in response to starvation for nitrogen in the absence of a fermentable carbon source, and the linkage between Gcn4p and spore wall formation may suggest that Gcn4p can contribute to a range of stress response options in yeast.

Gcn4p enhances expression of 26 different transcription factors involved in a broad range of stress response. The largest collection of transcription factors are involved in amino acid and purine biosynthesis, including Arg80p, Bas1p, Gln3p, Leu3p, Lys14p, Met4p, and Met28p (Natarajan et al. 2001). This suggests that the general control pathway can be fine tuned to accommodate transient cellular requirements for individual amino acids. As noted above Bas1p is critical for expression of purine biosynthetic genes and the histidine pathway gene *HIS7*. The example of Leu3p nicely illustrates this idea of superimposition of pathway specific regulation onto general control. Leu3p binds to promoter elements in a large number of genes involved in branched chain amino acid synthesis and represses transcription (Kohlhaw 2002). In the presence of the leucine biosynthetic precursor α-isopropylmalate (α-IPM), Leu3p becomes an activator of transcription. Levels of α-IPM are subject to the availability of leucine through feedback inhibition of α-IPM synthase, the *LEU4* product. Therefore, leucine levels would dictate whether Leu3p functions as an activator or repressor. Gcn4p also induces expression of genes encoding transcription factors involved in peroxisome proliferation (*PIP2*, and see discussion in the third layer of Gcn4p-direction gene expression), utilization of poor nitrogen sources (*GAT1* and *UGA3*), heat shock (*HSF1*), maltose catabolism (*MAL13*), and meiosis (*RIM101*) (Natarajan et al. 2001). Overall, this collection of transcription factors may work in concert with Gcn4p, insuring that the timing and content of stress gene expression is appropriately tailored to a mosaic of stress inputs.

7.8 The general control pathway and yeast physiological strategies

Yeast can synthesize each of its twenty amino acids *de novo*. However, in rich medium replete with amino acids, yeast import amino acids and reduce the levels of Gcn4p-directed gene expression. This insures that yeast do not synthesize biosynthetic enzymes unnecessarily, and facilitates a rapid doubling time. When there is an imbalance of amino acids in the medium, yeast will enhance biosynthetic genes to generate the required metabolites. In the laboratory, addition of 3-AT or sulfometuron methyl (SM), a chemical inhibitor of branched chain amino acid biosynthesis, induces high levels of Gcn4p and its target genes (Wek et al. 1995). It is curious that upon exposure to either inhibitor, yeast induces expression of not only genes required for the biosynthesis of the limiting amino acids but also those involved in synthesis of non-limiting amino acids. It is acknowledged that the above described DNA microarray analysis measured transcript levels, and many of these Gcn4p target genes have not yet been assessed for induced synthesis of their encoded proteins. Furthermore, given feedback inhibition for many of these biosynthetic pathways, expression of these many genes does not necessarily dictate that there is increased flux through each pathway. Nevertheless, it appears that yeast has coupled a central mechanism of induction of many biosynthetic pathways to a single signal- accumulation of uncharged tRNA.

We have proposed that an important goal of the general control is storage of nitrogen in the form of amino acids in response to severe nutrient deprivation (Yang et al. 2000). Upon glucose deprivation, yeasts increase their overall levels of free amino acids, with an elevation in the vacuolar amino acid pool and a concomitant depletion in the cytoplasmic amino acid levels. Concentrations of individual amino acids in the vacuole vary, with glutamate constituting nearly a third of the total pool, and arginine and alanine each constituting about 10% (Messenguy et al. 1980). Initially this increase in vacuolar amino acid levels is independent of *GCN2* activity (Yang et al. 2000). However, with longer periods of glucose limitation, the levels of vacuolar amino acids are much reduced in *GCN2*-deficient cells compared to wild type. As noted above, such glucose limitation increases eIF2 phosphorylation and *GCN4* translation, and these results suggest that induced general control contributes to the storage of amino acids when carbohydrates are limiting and there is reduced protein synthesis and cell growth. The storage of amino acids during nutrient limitation would provide yeast cells ready access to nitrogen when the carbohydrates become accessible again (Yang et al. 2000). Vacuolar storage of amino acids is also triggered in response to reduced assimilation of ammonia, indicating that this storage strategy is invoked in response to diverse nutrient stress conditions (Messenguy et al. 1980).

Accumulation of glycogen also occurs in yeast when glucose or amino acids begin to be depleted in the culture media and yeast cells have reduced their rate of growth (Francois and Parrou 2001). As highlighted above, the maintenance of accumulated glycogen levels is dependent on Gcn2p kinase activity, suggesting that general control has a broad role in storage of nutrients (Yang et al. 2000). It is noted that loss of *GCN4* is suggested to enhance glycogen accumulation as judged by iodine staining of cells grown on synthetic dextrose agar medium (Natarajan et al. 2001). This may suggest some differences between *GCN2* and *GCN4*-deficient strains. However, the timing in such iodine staining experiments is critical as glycogen synthesis and turnover change during growth phases. *GCN4*-deficient strains have a clear requirement for amino acid supplements for robust growth, and the absence of this transcription factor could enhance nutrient starvation signals that trigger early activation of glycogen synthase. This in turn would lead to earlier glycogen accumulation in *gcn*4 mutant strains compared to wild type cells. Similarly, vacuolar amino acid levels are enhanced in *gcn4* mutant cells grown to mid-logarithmic phase in synthetic medium in the absence of amino acid supplements as compared to wild type strains (Yang et al. 2000). With the addition of all twenty amino acids to the medium, there is a reduction in the vacuolar amino acid pool in the *gcn4* mutant, albeit the amino acid levels still remain significantly higher than that measured in non-starved wild type cells. Therefore, the loss of *GCN4* function is itself a stress that can trigger certain coping strategies. By comparison, *GCN2*-deficient cells, which retain some basal expression of Gcn4p, are not amino acid auxotrophs and can provide an important alternative tool to assess the contribution of the induced general control response to stress conditions.

Gcn4p orthologues are also required for long term strategies for dealing with nutritional stress in other fungi. For example in the dimorphic fungi *Candida albicans*, Gcn4p facilitates filamentous growth in response to amino acid starvation. Such filaments enable this fungi to forage for new resources when nutrients are depleted (Tripahti et al. 2002). In *Aspergillus nidulans*, the related Gcn4p transcription factor CpcAp functions in a cross pathway control, and serves to block formation of cleistothecia or fruit bodies when nutrients are limiting (Hoffmann et al. 2000). Formation of these complex reproductive structures consumes macromolecules and energy, and induced CpcAp expression would signal that this fungi lacks adequate nutrients to carry out this process.

7.9 Many different stress conditions activate Gcn2p eIF2 kinase activity

Phosphorylation of eIF2 by Gcn2p occurs in response to diverse nutrient limitations, including starvation for different amino acids, purines, or glucose (Fig. 3). Accumulation of uncharged tRNA is thought to be the activating signal for each of these starvation condition because yeast expressing gcn2-m2p, defective for binding to uncharged tRNA, are unable to induce eIF2 phosphorylation and mediate *GCN4* translation control. Purine starvation may elicit elevated uncharged tRNA levels by reducing ATP levels or the biosynthesis of certain amino acids, such as histidine, or by altering the processing of tRNA. In the case of glucose limitation, accompanying energy reductions or the lowering of amino acid levels in the cytoplasm that accompany vacuolar accumulation of amino acids could lead to increased uncharged tRNA. While 60S ribosomal association is thought to be obligate for activation of wild type Gcn2p during amino acid limitation, it appears to be largely dispensable in response to glucose limitation (Yang et al. 2000). Furthermore, the requirement for Gcn20p is not essential for induced *GCN4* expression during the deprivation for this carbohydrate. These results suggest that uncharged tRNA is an important signaling molecule that activates Gcn2p in response to many different nutritional limitations. However, there may be differences between the mechanisms by which uncharged tRNA is delivered to and recognized by Gcn2p during amino acid and carbohydrate deficiency.

In addition to nutrient limitation, many other stress conditions have been recently reported to activate Gcn2p eIF2 kinase activity. Exposure of yeast cells to high concentrations of sodium, volatile anesthetics, the immunosuppressant rapamycin, methyl methanesulfonate (MMS), or tunicamycin have been reported to induce eIF2 phosphorylation, and many of these conditions have also been shown to elevate *GCN4* translation (Fig. 3) (Goosens et al. 2001; Natarajan et al. 2001; Valenzuela et al. 2001; Palmer et al. 2002; Cherkasova and Hinnebusch 2003). Two fundamental questions come to mind concerning the induction of eIF2 phosphorylation by these diverse stress conditions. First, what are the mechanisms activating Gcn2p in response to these diverse stresses? And second what is the physiological rationale for inducing Gcn4p in response to this range of cellular stress conditions? Unfortunately, we do not yet have clear answers to these questions, although there are several clues that provide insight into likely explanations. These clues and insights are provided below.

The first question concerns the signals that activate Gcn2p in response to this diverse collection of stress conditions. In the examples of sodium toxicity, anesthetics and rapamycin there is a linkage to amino acid metabolism. Exposure to anesthetics and elevated concentrations of sodium are suggested to impair uptake of amino acids by yeast (Goosens et al. 2001; Palmer et al. 2002). Uptake of leucine or tryptophan are inhibited by anesthetics, and the enhanced levels of these amino acids to the medium can overcome the growth defect associated with anesthetics. Furthermore, yeast cells prototrophic for amino acid biosynthesis are resistant to anesthesia. Together these observations support the idea that the physiological action of anesthesia involves amino acid uptake in cells and nutrition availability. The example of activation of Gcn2p by high levels of sodium is complex. While elevated concentrations of either sodium or potassium reduce yeast uptake of phenylalanine and leucine, only sodium induces eIF2 phosphorylation and *GCN4* translational control (Goosens et al. 2001). Furthermore, general control in yeast strains prototrophic for amino acids is still activated by the addition of high sodium concentrations to the medium (K.S., unpublished observation). These results suggest that sodium stress is induced by a mechanism other than nutritional starvation due to reduced uptake of amino acids. The identity of one or more of these alternative sodium stress signals is currently not known. A curious note regarding sodium stress is that deletion of *GCN2*, or other general control genes such as *GCN1*, confer growth resistance to sodium (Goosens et al. 2001). Such sensitivity may indicate that sodium induces hyperphosphorylation of eIF2 and a block in translation. By deleting the eIF2 kinase, protein synthesis would not be impeded by sodium stress. By contrast, deletion of *GCN4* is reported to confer growth sensitivity to sodium (Pascual-Ahuir et al. 2001). Gcn4p activates transcription of *HAL1*, an important regulator of salt balance, through antagonism of Sko1p repressor. The apparent phenotypic difference whereby *GCN2* are growth resistant, and *gcn4* mutant cells are sensitive has also been reported for rapamycin treatment (Cherkasova and Hinnebusch 2003).

Tor has a central role in linking protein synthesis and cell growth and division to nutrient sufficiency, and rapamycin combines with the immunophilin-related protein Fpr1 to inhibit this protein kinase in yeast (Schmelzle and Hall 2000; Crespo and Hall 2002). Interestingly, rapamycin stimulates eIF2 phosphorylation by Gcn2p in non-starved cells by blocking Tor-mediated phosphorylation of Gcn2p at Ser577 (Cherkasova and Hinnebusch 2003). Such phosphorylation of Gcn2p is not thought to be directly by Tor, but rather through an unknown protein kinase that is downstream of Tor and Tap42p-regulated Type 2A and Type-2A-related protein phosphatases. Phosphorylation of Gcn2p at Ser577 reduces its binding to uncharged tRNA, while an alanine substitution at Ser577 contributes to induced Gcn2p eIF2 kinase activity independent of nutrient availability. Induction of

Gcn4p increases the expression of amino acid biosynthetic genes, and contributes in concert with Tor to activation of genes required for catabolism of poor nitrogen sources, as well as repression of genes encoding ribosomal proteins and translation factors. Thus, rapamycin mediates dephosphorylation and activation of Gcn2p, and this linkage between the Tor and Gcn2p provides a mechanism of crosstalk between different nutrient sensing pathways. It is noteworthy that the nutritional stresses characterized, amino acid or purine starvation, do not contribute to dephosphorylation of Gcn2p at Ser577 (Cherkasova and Hinnebusch 2003); therefore, the precise nutritional stress modulating Tor-directed control of general control remains to be determined. Uncharged tRNA may be a contributing signal to activation of Gcn2p by rapamycin, given that yeast containing *gcn2-m2* are blocked for induction of *GCN4* translation in response to rapamycin exposure. However, there is currently no evidence to support the idea that rapamycin treatment alters the efficiency of aminoacylation of tRNA.

The final examples of stress agents that activate Gcn2p that will be discussed are MMS and tunicamycin. MMS induces DNA damage, and DNA microarray analysis indicates that this alkylating agent increases expression of genes involved in synthesis and repair of DNA and detoxification, and represses those functioning in the synthesis of nucleotides, RNA and ribosomes (Jelinksy and Samson 1999). Interestingly, over 90 genes involved in amino acid biosynthesis are induced by twofold or more, suggesting a linkage between MMS and the general control pathway. Indeed, MMS enhances eIF2 phosphorylation by Gcn2p and *GCN4* translation (Natarajan et al. 2001). This induction mechanism is blocked in yeast containing the *gcn2-m2* mutation or defects in *GCN1* or *GCN20*, supporting the model that uncharged tRNA is a contributing signal to activation of Gcn2p in response to MMS. Furthermore, mutations in checkpoint proteins that are required to respond to DNA damage, e.g. Rap53p, do not reduce MMS induction of *GCN4* expression (Natarajan et al. 2001). This suggests that important signaling pathways required for repair of DNA damage are not involved in activation of Gcn2p.

Alkylation of proteins by MMS can lead to impaired activity of enzymes, such as aminoacyl-tRNA synthetases. However to our knowledge, there have not been any reports of increased uncharged tRNA levels in yeast treated with MMS. It is also noted that oxidized proteins are ubiquitinated and degraded in proteasomes, and impaired proteasome function in mammalian cells through drug treatment or by overexpression of proteins containing poly-glutamine sequences can lead to ER stress linked to the eIF2 kinase PEK/Perk (Nishitoh et al. 2002). One mechanism by which the ER secretory pathway manages misfolded protein in the ER lumen is to evict such proteins back to the cytoplasm for ubiquitin-mediated degradation. Proteasome dysfunction in the cytoplasm would block such protein degradation, contributing to a backup of misfolded protein in the ER lumen that elicits an ER stress response. Perhaps, MMS-damaged proteins in yeast also overload the proteasome, contributing to not only a cytoplasmic stress but also stress in the ER. In the case of the ER transmembrane protein PEK, the ER chaperone GRP78 binds to the lumenal portion of PEK and represses its cytoplasmic eIF2 kinase activity (Bertolotti et al. 2000; Ma et al. 2002). Misfolded protein in the lumen of the ER is proposed to titrate GRP78 from PEK, facilitating oligomerization and trans phosphorylation that induces eIF2 kinase activity. The cytoplasmic chaperone Hsp90p binds yeast Gcn2p and is proposed to have a critical role for Gcn2p maturation and regulation (Donze and Picard 1999). Following the model proposed for PEK and ER stress, perhaps Hsp90p inhibition of Gcn2p is relieved by accumulation of damaged proteins that selectively titrate the cytoplasmic chaperone from mature Gcn2p. Removal of Hsp90p from Gcn2p may contribute to an activated conformation and elevated eIF2 kinase activity. Altered proteasome function has also been proposed to alter efficient turnover of proteins that could modify Gcn4p activity or expression (Stitzel et al. 2001).

Tunicamycin blocks protein glycosylation in the ER and contributes to misfolded protein in this organelle (Harding et al. 2002; Kaufman et al. 2002). Unlike mammals, *S. cerevisae* has only a single eIF2 kinase Gcn2p. However, yeast does have Ire1p, a transmembrane ER protein kinase that shares sequence similarities with mammalian PEK in its ER lumenal regions (Sidrauski et al. 1998). Ire1p activates the expression of Hac1p, a transcriptional activator of genes important for ER protein folding and secretion. In mammals, PEK phosphorylation of eIF2 serves to reduce general translation, preventing further synthesis of secretory proteins that would further overload the ER. Additionally, PEK induces expression of the transcriptional activator Atf4 and its target genes in mammalian cells subjected to ER stress (Harding et al. 2000, 2003). Given that PEK is not present in yeast it was assumed that this portion of the ER stress pathway was absent from fungi. However with the observation that tunicamycin can activate Gcn2p eIF2 kinase activity, it is inviting to speculate that Gcn2p has assumed a translational regulation role during ER stress (Cherkasova and Hinnebusch 2003). Glucose deprivation in mammals also induces ER stress and PEK activity (Harding et al. 2002; Kaufman et al. 2002). Therefore, the observation discussed above that glucose deprivation can induce eIF2 phosphorylation in yeast may further support this ER stress linkage with Gcn2p. It is curious that a DNA microarray study measuring tunicamycin-induced gene expression in yeast did not identify Gcn4p-directed amino acid biosynthetic genes (Travers et al. 2000). This may indicate that during ER stress Gcn2p functions predominantly to regulate general translation and there is a decoupling of *GCN4* expression. Genetic variations between yeast strains could also impact the different contributions of the general control on ER stress responses.

Acknowledgments

The authors acknowledge research support from grants RO1GM49164 and R01GM643540 from the National Institutes of Health.

References

Anand M, Chakraburtty K, Marton MJ, Hinnebusch AG, Kinzy AG (2001) Functional interactions between yeast translation eukaryotic elongation factor (eEF) 1A and eEF3. J Biol Chem 278:6985-6991

Ashe MP, De Long SK, Sachs AB (2000) Glucose depletion rapidly inhibits translation initiation in yeast. Mol Biol Cell 11:833-848

Berger SL, Sterner DE (2000) Acetylation of histones and transcription-related factors. Microbiol Mol Biol Rev 64:435-459

Bertolotti A, Zhang Y, Hendershot L, Harding H, Ron D (2000) Dynamic interaction of BiP and the ER stress transducers in the unfolded protein response. Nat Cell Biol 6:326-332

Cashel M, Rudd KE (1987) The stringent response. In: Neidhardt FC, Ingraham JL, Magasanik B, Low KB, Umbarger HE (eds) *Escherichia coli and Salmonella typhimurium*: cellular and molecular biology. American Society for Microbiology, Washington, D.C, pp.1410-1438

Chakraburtty K (1999) Functional interaction of yeast elongation factor 3 with yeast ribosomes. Int J Biochem Cell Biol 31:163-173

Cherkasova VA, Hinnebusch AG (2003) Translational control by TOR and TAP42 through dephosphorylation of eIF2alpha kinase GCN2. Genes Dev 17:859-872

Chi Y, Huddlerston MJ, Zhang X, Young RA, Annan RS, Carr SA, Deshaies RJ (2001) Negative regulation of Gcn4 and Msn2 transcription factors by Srb10 cyclin-dependent kinase. Genes Dev 15:1078-1092

Connor JH, Weiser DC, Li S, Hallenbeck JM, Shenolikar S (2001) Growth arrest and DNA damage-inducible protein GADD34 assembles a novel signaling complex containing protein phosphatase 1 and inhibitor 1. Mol Cell Biol 21:6841-6850

Crespo JL, Hall MN (2002) Elucidating TOR signalling and rapamycin action: Lessons from *Saccharomyces cerevisiae*. Microbiol Mol Biol Rev 66:579-591

Dever TE, Chen JJ, Barber GN, Cigan AM, Feng L, Donahue TF, London IM, Katze MG, Hinnebusch AG (1993) Mammalian eukaryotic initiation factor 2 alpha kinases functionally substitute for GCN2 protein kinase in the *GCN4* translational control mechanism of yeast. Proc Natl Acad Sci USA 90:4616-4620

Dever TE, Feng L, Wek RC, Cigan AM, Donahue TF, Hinnebusch AG (1992) Phosphorylation of initiation factor 2 alpha by protein kinase GCN2 mediates gene-specific translational control of GCN4 in yeast. Cell 68:585-596

Dever TE, Yang W, Astrom S, Bystrom AS, Hinnebusch AG (1995) Modulation of $tRNA_i^{Met}$, eIF-2, and eIF-2B expression shows that *GCN4* translation is inversely is inversely coupled to the level of eIF-2·GTP·Met-$tRNA_i^{Met}$ ternary complexes. Mol Cell Biol 15:6351-6361

Dong J, Qiu H, Garcia-Barrio M, Anderson J, Hinnebusch AG (2000) Uncharged tRNA activates GCN2 by displacing the protein kinase moiety from a bipartite tRNA-binding domain. Mol Cell 6:269-279

Donze O, Picard D (1999) Hsp90 binds and regulates the ligand-inducible α subunit of eukaryotic translation initation factor kinase Gcn2. Mol Cell Biol 19:8422-8432

Drydale CM, Jackson BM, McVeigh R, Klebanow ER, Bai Y, Kokubo T, Swanson M, Nakatani Y, Weil PA, Hinnebusch AG (1998) The Gcn4p activation domain interacts specifically *in vitro* with RNA polymerase II holoenzyme TFIID, and the Adap-Gcn5p coactivator complex. Mol Cell Biol 18:1711-1724

Engelberg D, Klein C, Marinetto H, Struhl K, Karin M (1994) The UV response involving the RAS signalling pathway and AP-1 transcription factors is conserved between yeast and mammals. Cell 77: 381-390

Francois J, Parrou JL (2001) Reserve carbohydrate metabolism in the yeast *Saccharomyces cerevisiae*. FEMS Microbiol Rev 25:125-145

Gaba A, Wang Z, Krishnamoorthy T, Hinnebusch AG, Sachs MS (2001) Physical evidence for distinct mechanisms of translational control by open reading frames. EMBO J 20:6453-6463

Garcia-Barrio M, Dong J, Ulfano S, Hinnebusch AG (2000) Association of GCN1-GCN20 regulatory complex with the N-terminus of eIF2alpha kinase GCN2 is required for GCN2 activation. EMBO J 19:1887-1899

Geraght MT, Bassett D, Morrell JC, Gatto GJ, Bai J, Geisbrecht BV, Hieter P, Gould SJ (1999) Detecting patterns of protein distribution and gene expression in silico. Proc Natl Acad Sci USA 96:2937-2942

Goosens A, Dever TE, Pascual-Ahuir A, Serrano R (2001) The protein kinase Gcn2p mediates sodium toxicity in yeast. J Biol Chem 276:30753-30760

Grant CM, Hinnebusch AG (1994) Effect of sequence context at stop codons on efficiency of reinitiation in GCN4 translational control. Mol Cell Biol 14:606-618

Grundmann O, Mosch HU, Braus GH (2001) Repression of *GCN4* mRNA translation by nitrogen starvation in *S. cerevisiae*. J Biol Chem 276:25661-25671

Hai T, Hartman MG (2001) The molecular biology and nomenclature of the activating transcription factor/cAMP responsive element binding family of transcription factors: Activating transcription factor proteins and homeostasis. Gene 273:1-11

Harding HP, Zhang Y, Zeng H, Novoa I, Lu PD, Calfon M, Sadri N, Yun C, Popko B, Paules R, Stojdl DF, Bell JC, Hettmann T, Leiden JM, Ron D (2003) An integrated stress response regulates amino acid metabolism and resistance to oxidative stress. Mol Cell 11: 619-633

Harding HP, Calfon M, Urano F, Novoa I, Ron D (2002) Transcriptional and translational control in the mammalian unfolded protein response. Ann Rev Cell Dev Biol 18:575-599

Harding HP, Novoa I, Zhang Y, Zeng H, Wek R, Schapira M, Ron D (2000) Regulated translation initiation controls stress-induced gene expression in mammalian cells. Mol Cell 6:1099-1108

Hinnebusch AG (1992) General and pathway-specific regulatory mechanisms controlling the synthesis of amino acid biosynthetic enzymes in *Saccharomyces cerevisiae*. In: Jone EW, Pringle JR, Broach JR (eds) The molecular and cellular biology of the yeast *Saccharomyces*. Cold Spring Harbor Laboratory Press, Plainview, N.Y., pp 319-414

Hinnebusch AG (1997) Translational regulation of yeast GCN4. A window on factors that control initiator-tRNA binding to the ribosome. J Biol Chem 272:21661-21664

Hinnebusch AG, Natarajan K (2002) Gcn4p, a master regulator of gene expression, is controlled at multiple levels by diverse signals of starvation and stress. Eukaryot Cell 1:22-32

Hoffmann B, Valerius O, Andermann M, Braus GH (2001) Transcriptional autoregulation and inhibition of mRNA translation of amion acid regulator gene *cpcA* of filamentous fungus *Aspergillus nidulans*. Mol Biol Cell 12:2846-2857

Hoffmann B, Wanke C, LaPagia SK, Braus GH (2000) c-Jun and RACK1 homologues regulate a control point for sexual development in *Aspergillus nidulans*. Mol Microbiol 37:28-41

Jackson BM, Drysdale CM, Natarajan K, Hinnebusch AG (1996) Identification of seven hydrophobic clusters in GCN4 making redundant contributions to transcriptional activation. Mol Cell Biol 16:5557-5571

Jelinksy SA, Samson LD (1999) Global response of *Saccharomyces cerevisiae* to an alkylating agent. Proc Natl Acad Sci USA 96:1486-1491

Kaufman RJ, Scheuner D, Schroder M, Shen X, Lee K, Lin CY, Arnold SM (2002) The unfolded protein response in nutrient sensing and differentiation. Nat Rev Mol Cell Biol 3:411-421

Kohlhaw GB (2002) Leucine biosynthesis in fungi: Entering metabolism through the back door. Microbiol Mol Biol Rev 67:1-15

Krishnamoorthy T, Pavitt GD, Zhang F, Dever TE, Hinnebusch AG (2001) Tight binding of the phosphorylated α subunit of initiation factor 2 (eIF2α) to the regulatory subunits of guanine nucleotide exchange factor eIF2B is required for inhibition of translation initiation. Mol Cell Biol 21:5018-5030

Kubota H, Sakaki Y, Ito T (2000) GI domain-mediated association of the eukaryotic intiation factor 2alpha kinase GCN2 with its activator GCN1 is required for general amino acid control in budding yeast. J Biol Chem 275:20243-20246

Li B, Nierra CR, Warner JR (1999) Transcriptional elements involved in the repression of ribosomal protein synthesis. Mol Cell Biol 19:5393-5404

Luo Z, Freitag M, Sachs MS (1995) Translational regulation in response to changes in aminol acid availability in *Neurospora crassa*. Mol Cell Biol 15:5235-5345

Ma K, Vattem KM, Wek RC (2002) Dimerization and release of molecular chaperone inhibition facilitate activation of eukaryotic initiation factor-2 kinase in response to endoplasmic reticulum stress. J Biol Chem 277:18728-18735

Marton MJ, Crouch D, Hinnebusch AG (1993) GCN1, a translational activator of GCN4 in *Saccharomyces cerevisiae*, is required for phosphorylation of eukaryotic translation initiation factor 2 by protein kinase GCN2. Mol Cell Biol 13:3541-3556

Marton MJ, Vazquez de Aldana CR, Qui H, Chakraburtty K, Hinnebusch AG (1997) Evidence that GCN1 and GCN20, translational regulators of GCN4, function on elongating ribosomes in activation of eIF-2alpha kinase GCN2. Mol Cell Biol 17:4474-4489

Meimoun A, Holtzman T, Weissman Z, McBride HJ, Stillman DJ, Fink GR, Kornitzer D (2000) Degradation of the transcription factor Gcn4 requires the kinase Pho85 and the SCF^{CDC4} ubiquitin-ligase complex. Mol Biol Cell 11:915-927

Messenguy F, Colin D, ten Have JP (1980) Regulation of compartmentation of amino acid pools in *Sacharomyces cerevisiae* and its effects on metabolic control. Eur J Biochem 108:439-447

Moehle CM, Hinnebusch AG (1991) Association of RAP1 binding sites with stringent control of ribosomal protein gene transcription in *Saccharomyces cerevisiae*. Mol Cell Biol 11:2723-2735

Murchie MJ, Leader DP (1978) Codon-specific interaction of uncharged transfer-RNA with eukaryotic ribosomes. Biochim Biophys Acta 520:233-236

Natarajan K, Jackson BM, Zhou H, Winston F, Hinnebusch AG (1999) Transcriptional activation by Gcn4p involves independenet interactions with the SWI/SNF complex and the SRB/mediator. Mol Cell 4:657-664

Natarajan K, Meyer MR, Jackson BM, Slade D, Roberts C, Hinnebusch AG, Marton MJ (2001) Transciptional profiling shows that Gcn4p is a master regulator of gene expression during amino acid starvation in yeast. Mol Cell Biol 21:4347-4368

Neuman-Silberberg FS, Bhattacharya S, Broach JR (1995) Nutrient availability and the RAS/cyclic AMP pathway both induce expression of ribosomal protein genes in *Saccharomyces cerevisiae* but by different mechanisms. Mol Cell Biol 15:3187-3196

Nishitoh H, Matsuzawa A, Tobiume K, Saegusa K, Takeda K, Inoue K, Hori S, Kakizuka A, Ichijo H (2002) ASK1 is essential for endoplasmic reticulum stress-induced neuronal cell death triggered by expanded polyglutamine repeats. Genes Dev 16:1345-1355

Novoa I, Zeng H, Harding HP, Ron D (2001) Feedback inhibition of the unfolded protein response by GADD34-mediated dephosphorylation of eIF2alpha. J Cell Biol 153:1011-1022

Novoa I, Zhang Y, Zeng H, Jungreis R, Harding HP, Ron D (2003) Stress-induced gene expression requires programmed recovery from translational repression. EMBO J 22:1180-1187

Palmer LK, Wolfe D, Keeley JL, Keil RL (2002) Volatile anesthetics affect nutrient availability in yeast. Genetics 161:563-574

Paluh JL, Orback MJ, Legerton TL, Yanofsky C (1988) The cross-pathway control gene of *Neurospora crassa*, cpc-1, encodes a protein similar to GCN4 of yeast and the DNA-binding domain of the oncogene v-jun-encoded protein. Proc Natl Acad Sci USA 85:3728-2732

Pascual-Ahuir A, Serrano R, Proft M (2001) The Sko1p repressor and Gcn4p activator antagonistically modulate stress-mediated transcription in *Saccharomyces cerevisiae*. Mol Cell Biol 21:16-25

Pravitt GD, Ramiah KVA, Kimball SR, Hinnebusch AG (1998) eIF2 independently binds two distinct eIF2B subcomplexes that catalyze and regulate guanine-nucleotide exchange. Genes Dev 12:514-526

Pries R, Bomeke K, Irniger S, Grundmann O, Braus GH (2002) Amino acid-dependent Gcn4p stability regulation occurs exclusively in the yeast nucleuse. Eukaryot Cell 1:663-672

Qiu H, Dong J, Francklyn CS, Hinnebusch AG (2001) The tRNA-binding moiety in GCN2 contains a dimerization domain that interacts with the kinase domain and is required for tRNA binding and kinase activation. EMBO J 20:1425-1438

Qiu H, Garcia-Barrio MT, Hinnebusch AG (1998) Dimerization by translation initiation factor 2 kinase GCN2 is mediated by interactions of the C-terminal ribosome binding region and the protein kinase domain. Mol Cell Biol 18:2697-2711

Qiu H, Hu C, Dong J, Hinnebusch AG (2002) Mutations that bypass tRNA binding activate the intrinsically defective kinase domain in GCN2. Genes Dev 16:1271-1280

Qiu H, Hu J, Anderson J, Bjork GR, Sarkar S, Hopper AK, Hinnebusch AG (2000) Defects in tRNA processing and nuclear export induce *GCN4* translation independently of phosphorylation of the alpha subunit of eukaryotic intiation factor 2. Mol Cell Biol 20:2505-2516

Ramirez M, Wek RC, Hinnebusch AG (1991) Ribosome association of GCN2 protein kinase, a translational activator of the *GCN4* gene of *Saccharomyces cerevisiae*. Mol Cell Biol 11:3027-3036

Ramirez M, Wek RC, Vazquez de Aldana CR, Jackson BM, Freeman B, Hinnebusch AG (1992) Mutations activating the yeast eIF-2 alpha kinase GCN2: isolation of alleles altering the domain related to histidyl-tRNA synthetases. Mol Cell Biol 12:5801-5815

Reggiori F, Klionsky DJ (2002) Autophagy in the eukaryotic cell. Eukaryot Cell 1:11-21

Rolfes RJ, Hinnebusch AG (1993) Translation of the yeast transcriptional activator GCN4 is stimulated by purine limitation: implications for activation of the protein kinase GCN2. Mol Cell Biol 13:5099-5111

Romano PR, Garcia-Barrio MT, Zhang X, Wang Q, Taylor DR, Zhang F, Herring C, Mathews MB, Qin J, Hinnebusch AG (1998) Autophosphorylation in the activation loop is required for full kinase activity *in vivo* of human and yeast eukaryotic initiation factor 2α kinases PKR and GCN2. Mol Cell Biol 18:2282-2297

Rowlands AG, Panniers R, Henshaw EC (1988) The catalytic mechanism of guanine nucleotide exchange factor action and competitive inhibition by phosphorylated eukaryotic initation factor 2. J Biol Chem 263:5526-5533

Ruiz-Echevarria MJ, Peltz SW (2000) The RNA binding prtoein Pub1 modulates the stability of transcripts containing upstream open reading frames. Cell 101:741-751

Sattlegger E, Hinnebusch AG (2000) Separate domains in GCN1 for binding protein kinase GCN2 and ribosome are required for GCN2 activation in amio acid-starved cells. EMBO J 19:6622-6633

Schmelzle T, Hall MN (2000) TOR, a central controller of cell growth. Cell 103:253-262

Shemer R, Meimoun A, Holtzman T, Kornitzer D (2002) Regulation of the transcription factor Gcn4 by Pho85 cyclin Pcl5. Mol Cell Biol 22:5395-5404

Sidrauski C, Chapman R, Walter P (1998) The unfolded protein response: an intracellular signalling pathway with many surprising features. Trends Cell Biol 8:245-249

Sood R, Porter AC, Ma K, Quilliam LA, Wek RC (2000) Pancreatic eukaryotic initation factor -2α kinase (PEK) homologues in humans, *Drosophila melanogaster* and *Caenorhabditis elegans* that mediate translational control in response to ER stress. Biochem J 346:281-293

Springer C, Kunzler M, Balmelli T, Braus GH (1996) Amino acid and adenine cross-pathway regulation act through the same 5'-TGACTC-3' motif in the yeast HIS7 promoter. J Biol Chem 271:29637-29643

Stitzel ML, Durso R, Reese JC (2001) The proteasome regulates the UV-induced activation of the AP-1 like transcription factor Gcn4. Genes Dev 15:128-133

Swanson MJ, Qiu H, Sumibcay L, Krueger A, Kim SJ, Natarajan K, Yoon S, Hinnebusch AG (2003) A multiplicity of coactivators is required by Gcn4p at individual promoters *in vivo*. Mol Cell Biol 23:2800-2820

Takeshig K, Baba M, Tsuboi S, Noda T, Ohsumi Y (1992) Autophagy in yeast demonstrated with proteinase-deficient mutants and conditions for its induction. J Cell Biol 119:301-311

Talloczy Z, Jiang W, Virgin HW, Leib DA, Scheuner D, Kaufman RJ, Eskelinen E-L, Levine B (2002) Regulation of starvation- and virus-induced autophagy by the eIF2α kinase signaling pathway. Proc Natl Acad Sci USA 99:190-195

Travers KJ, Patil CK, Wodicka L, Lockhart DJ, Weissman JS, Walter P (2000) Functional and genomic analyses reveal an essential coordination between the unfolded protein response and ER-associated degradation. Cell 101:249-258

Tripahti G, Wiltshire C, Macaskill S, Tournu H, Budge S, Brown AJP (2002) Gcn4 coordinates morphogenetic and metabolic responses to amino acid starvation in *Candida albicans*. EMBO J 21:5448-5456

Valenzuela L, Aranda C, Gonzalez A (2001) TOR modulates *GCN4*-dependent expression of genes turned on by nitrogen limitation. J Bacteriol 183:2331-2334

Vattem K, Staschke KA, Wek RC (2001) Mechanism of activation of the double-stranded-RNA-dependent protein kinase, PKR: Role of dimerization and cellular localization in the stimulation of PKR phosphorylation of eukaryotic initiation factor -2α (eIF2α). Eur J Biochem 268:3674-3684

Vazques de Aldana CR, Marton MJ, Hinnebusch AG (1995) GCN20, a novel ATP binding cassette protein, and GCN1 reside in a complex that mediates activation of the eIF-2a kinase GCN2 in amino acid-starved cells. EMBO J 14:3184-3199

Vazquez de Aldana CR, Wek RC, Segundo PS, Truesdell AG, Hinnebusch AG (1994) Multicopy tRNA genes functionally suppress mutations in yeast eIF-2 alpha kinase GCN2: evidence for separate pathways coupling GCN4 expression to unchanged tRNA. Mol Cell Biol 14:7920-7932

Wek RC, Cannon JF, Dever TE, Hinnebusch AG (1992) Truncated protein phosphatase GLC7 restores translational activation of GCN4 expression in yeast mutants defective for the eIF-2 alpha kinase GCN2. Mol Cell Biol 12:5700-5710

Wek RC, Jackson BM, Hinnebusch AG (1989) Juxtaposition of domains homologous to protein kinases and histidyl-tRNA synthetases in GCN2 protein suggests a mechanism for coupling *GCN4* expression to amino acid availability. Proc Natl Acad Sci USA 86:4579-4583

Wek RC, Ramirez M, Jackson BM, Hinnebusch AG (1990) Identification of positive-acting domains in GCN2 protein kinase required for translational activation of *GCN4* expression. Mol Cell Biol 10:2820-2831

Wek SA, Zhu S, Wek RC (1995) The histidyl-tRNA synthetase-related sequence in eIF-2 alpha protein kinase GCN2 interacts with tRNA and is required for activation in response to starvation for different amino acids. Mol Cell Biol 15:4497-4506

Wilusz CJ, Wang W, Peltz SW (2001) Curbing the nonsense: the activation and regulation of mRNA surveillance. Genes Dev 15:2781-2785

Yang R, Wek SA, Wek RC (2000) Glucose limitation induces *GCN4* translation by activation of Gcn2 protein kinase. Mol Cell Biol 20:2706-2717

Zhu S, Sobolev AY, Wek RC (1996) Histidyl-tRNA synthetase-related sequences in GCN2 protein kinase regulate *in vitro* phosphorylation of eIF-2. J Biol Chem 271:24989-24994

Zhu S, Wek RC (1998) Ribosome binding domain of eukaryotic initiation factor-2 kinase GCN2 facilitates translation control. J Biol Chem 273:1808-1814

8 Tor-signaling and Tor-interacting proteins in yeast

Ted Powers, Ching-Yi Chen, Ivanka Dilova, Aaron Reinke, and Karen P. Wedaman

Abstract

The Tor (target of rapamycin) signaling pathway is an important mechanism used by eukaryotic cells to regulate their growth in response to nutrient-related environmental cues. Recent studies have revealed that the two Tor kinases in *S. cerevisiae*, Tor1p and Tor2p, regulate gene expression at several levels, including transcription, translation, intracellular protein trafficking, as well as protein stability. How the activity of each kinase is controlled remains to be elucidated, however, as does the nature of potential upstream regulatory signals. Here we review recent efforts to address these issues by focusing on two areas related to Tor signaling: (1) transcriptional control of genes required for the *de novo* biosynthesis of glutamate and glutamine and (2) characterization of interacting partners of Tor1p and Tor2p. These studies have converged in unanticipated ways to yield new insights into how these kinases may function both to receive as well as transmit nutritional information in yeast.

8.1 Introduction

Normal cell growth requires that cells adjust their metabolic activity according to nutrient availability and other environmental cues. Specialized signal transduction pathways exist which enable cells to perceive and integrate these cues in order to establish and/or maintain appropriate patterns of gene expression. Understanding how these pathways function is thus important for understanding both normal cellular behavior as well as the underlying basis of many human diseases, including cancer. One important signaling pathway used by all eukaryotic cells is the Tor (Target of rapamycin) kinase pathway. This pathway was discovered through the action of the antibiotic rapamycin, a potent inhibitor of T cell proliferation, which targets the large, evolutionarily conserved Tor kinase (Dennis et al. 1999; Schmelzle and Hall 2000; Raught et al. 2001; Rohde et al. 2001; Crespo and Hall 2003). Rapamycin inhibits the growth of a wide variety of cell types and organisms, including *S. cerevisiae*. Two highly homologous Tor kinases exist in this organism, encoded by the *TOR1* and *TOR2* genes, and both are inhibited by rapamycin. Treating yeast cells with rapamycin has several distinctive effects that mimic starvation, including inhibition of protein synthesis and ribosome biogenesis, cell cy-

Topics in Current Genetics, Vol. 7
J. Winderickx, P.M. Taylor (Eds) Nutrient-Induced Responses in Eukaryotic Cells

cle arrest at the G1/S boundary, onset of autophagy, and entry into G0. Superimposed on its role in nutritional control of cell growth, Tor2p has a separate, rapamycin-insensitive function in polarized cell growth and actin cytoskeletal dynamics.

Presently, we can define at least three important areas of Tor signaling in yeast that remain to be fully understood: First, we want to define the molecular components of each of the major branches downstream of Tor, as well as understand how these components are coordinately regulated according to extracellular nutritional conditions. Second, we wish to understand how these individual branches interact with other, Tor-independent, cell regulatory pathways. Third, we still know very little about how Tor kinase activity is regulated, in particular with respect to potential upstream regulatory signals. Here we review our efforts to address these issues by focusing on two specific areas related to Tor signaling: (1) transcriptional control of genes required for *de novo* biosynthesis of glutamate and glutamine and (2) characterization of interacting partners of Tor1p and Tor2p. As described below, these two different approaches have converged in an unanticipated way to yield new insights into how Tor may function to both receive upstream signals as well as control downstream target genes in yeast.

8.2 Scope of Tor signaling in yeast

In recent years, studies by many laboratories have been devoted to understanding how the Tor proteins control downstream events required for nutrient-regulated cell growth. Results of these studies have revealed that Tor1p and Tor2p regulate gene expression at several levels, including transcription, translation, intracellular protein trafficking, as well as protein stability (Schmelzle and Hall 2000; Raught et al. 2001; Rohde et al. 2001; Crespo and Hall 2003). Our initial studies demonstrated that Tor plays a direct role in regulating ribosome biogenesis in yeast (Powers and Walter 1999). In particular, we found that a functional Tor pathway is essential for continued transcription of r-protein genes, as well as for the synthesis and subsequent processing of 35S precursor ribosomal RNA. A number of independent studies have revealed that Tor activity regulates each of the three cellular RNA polymerases (Zaragoza et al. 1998; Cardenas et al. 1999; Hardwick et al. 1999). Moreover, recent studies have provided evidence that regulation of r-protein gene expression by Tor involves changes in histone acetylation and altered chromatin structure (Damelin et al. 2002; Rhode and Cardenas 2003).

Another major class of genes controlled by Tor is involved in carbon and nitrogen metabolism (Beck and Hall 1999; Cardenas et al. 1999; Hardwick et al. 1999; Komeili et al. 2000; Shamji et al. 2000). In particular, we have demonstrated that there is significant correlation between genes regulated by the amino acid glutamine, a preferred source of assimilable nitrogen, and genes controlled by Tor, suggesting Tor may provide an important link between nitrogen metabolism and cell growth in eukaryotic cells (Komeili et al. 2000). One prominent group of glutamine-repressed genes encodes permeases and enzymes involved in the uptake and catabolism, respectively, of alternative sources of nitrogen, for example urea. Studies by Hall and coworkers have demonstrated that many of these genes are regulated negatively by Tor, using a mechanism whereby the GATA-specific transcription factor Gln3p is sequestered in the cytoplasm (Beck and Hall 1999). A role for Tor in regulating Gln3p has been reported by two additional groups as well (Hardwick et al. 1999; Bertram et al. 2000).

A second prominent group of glutamine/Tor repressed genes identified by our microarray analyses are termed *RTG* target genes and are required for the *de novo* biosynthesis of glutamate and glutamine (Komeili et al. 2000). Butow and coworkers originally identified *RTG* target genes as they are controlled by a mitochondria-to-nucleus signaling pathway, or retrograde response pathway, that adjusts their transcription in response to the respiratory state of the cell (Liao et al. 1991; Liao and Butow 1993; Liu and Butow 1999). Many of these genes include enzymes involved in the TCA and glyoxylate cycles and are regulated by the heterodimeric bHLH/Zip transcription factors Rtg1p and Rtg3p (Liu and Butow 1999). A prototypical example of an *RTG* target gene is *CIT2*, which encodes a peroxisomal form of citrate synthase (Kim et al. 1986; Rosenkrantz et al. 1986; Liu and Butow 1999). As proposed originally by Butow and coworkers (Liu and Butow 1999), the likely importance of *RTG* target gene expression is to maintain adequate levels of α-ketoglutarate for the production of glutamate, which in turn is required for glutamine biosynthesis. The conversion of α-ketoglutarate to glutamate, and ultimately glutamine, represents the primary means by which carbon and nitrogen metabolism are linked in all cells (Magasanik 1992; Marzluf 1997; Magasanik and Kaiser 2002). The finding that Tor is intimately involved in this process underscores the importance of this pathway in regulating essential aspects of basic cellular metabolism. Moreover, these results are particularly relevant given that glutamine levels in humans play an important role in the progression of a number of diseases, including cancer, and are believed to be influenced, at least in part, by Tor (Iiboshi et al. 1999). Nevertheless, as described in detail below, the precise relationship between Tor and nitrogen metabolism is very complex and appears to involve multiple signaling pathways.

Tor also controls the expression and/or trafficking of amino acid transporters in yeast (Schmidt et al. 1998; Beck et al. 1999). In particular, Tor has been shown to be a positive regulator of the tryptophan permease, Tat2p, which is expressed when cells are grown under rich nutrient conditions. Thus, following nutrient starvation or upon inactivation of Tor by rapamycin, Tat2p is diverted from the late secretory pathway and targeted to the vacuole for degradation (Beck et al. 1999). Tor signaling also regulates the General Amino Acid Permease, Gap1p (Schmidt et al. 1998; Beck and Hall 1999). In contrast to Tat2p, Gap1p is repressed transcriptionally in the presence of a rich nitrogen source (e.g. glutamine) in a Tor-dependent manner (Beck and Hall 1999). Conversely, under poor nitrogen conditions, Gap1p is produced and transported to the plasma membrane (Roberg et al. 1997b). Interestingly, in the presence of glutamate, which in this instance may be considered an intermediate quality nitrogen source, Gap1p is synthesized and yet is targeted to the vacuole for degradation (Roberg et al. 1997b). The mechanism used for targeting both Tat2p and Gap1p to the vacuole involves their ubiquitination as well as utilizes many of the same components (Beck et al. 1999; Helliwell et al. 2001). Nevertheless, it has been controversial whether Tor signaling plays a direct role in Gap1p trafficking (Beck et al. 1999; Chen and Kaiser 2002; Magasanik and Kaiser 2002). As described in greater detail below, we and others have now discovered that Lst8p, a protein identified originally as a regulator of Gap1p trafficking (Roberg et al. 1997a), is a binding partner of both Tor1p and Tor2p, which raised the possibility that Tor may indeed play a role in Gap1p localization. Recent functional studies on the relationship between Tor signaling and Lst8p from the Kaiser lab have borne out this prediction entirely (Chen and Kiaser 2003). Control of intracellular trafficking of nutrient-regulated permeases may be a conserved feature of Tor signaling as rapamycin has been shown to affect regulated exocytosis of the insulin-responsive glucose transporter GLUT4 in mammalian cells (Bogan et al. 2001).

In addition to its role in nutritional control of cell growth, Tor2p has a separate and unique function not sensitive to rapamycin that is involved in polarized cell growth and actin cytoskeletal dynamics (Schmidt et al. 1996, 1997; Helliwell et al. 1998; Schmelzle and Hall 2000; Crespo and Hall 2003). This Tor2p unique function involves signaling to components in remodeling of actin at the site of bud emergence, including regulators of the Rho1p GTPase (Schmidt et al. 1997; Pruyne and Bretscher 2000). Recent results from Mike Hall's lab suggest that this may be a conserved activity in mammalian cells as well (Loewith et al. 2002). Interestingly, there is an emerging connection between components involved in the Tor2p-unique function and components involved in what has been termed the "cell integrity pathway" that monitors cell envelope/cell wall stability in response to osmotic and/or thermal stress (Heinisch et al. 1999; Schmelzle et al. 2002; Schmitz et al. 2002). Moreover, Tor1p as well as components downstream of rapamycin-sensitive Tor signaling have now also been implicated in the cell integrity pathway (Angeles de la Torre-Ruiz et al. 2002; Torres et al. 2002). Together these findings raise the important question concerning the extent of potential cross talk between Tor1p/Tor2p shared and Tor2p unique activities.

8.3 *RTG* target gene control: convergence of retrograde and Tor signaling

Given the essential role of *RTG* target gene expression in the anapleurotic supply of α-ketoglutarate for glutamate and glutamine biosynthesis, we have endeavored to understand Tor's role in this process. We have now established a regulatory link between Tor and the Rtg1p and Rtg3p transcription factors involving two proteins, Rtg2p and Mks1p, that appear to be dedicated primarily, if not exclusively, to this pathway (Komeili et al. 2000; Dilova et al. 2002). Remarkably, there is close correspondence between the mechanism we have detailed for Tor-dependent control of this pathway and that described by Butow and coworkers for retrograde control of *RTG* target gene expression (Sekito et al. 2000, 2002). Our present view is that these are likely to represent distinct yet intimately intertwined regulatory pathways that converge on the *RTG* target genes. Below we describe our present view of the role of Rtg2p and Mks1p and their relationship to both Tor as well as retrograde signaling.

8.3.1 Rtg2p and regulation of Rtg1p/Rtg3p nuclear import

In collaboration with Erin O'Shea's laboratory, we determined that a functional Tor pathway is required to retain the Rtg1p/Rtg3p complex in the cytoplasm when cells utilize glutamine as a sole nitrogen source (Komeili et al. 2000). Accordingly, when cells grown in media containing glutamine are treated with rapamycin, the Rtg1p/Rtg3p complex moves from the cytoplasm into the nucleus where it activates its respective target genes. We also found that rapamycin-induced nuclear entry of the complex requires Rtg2p, a cytoplasmic protein previously identified by Butow and coworkers as important for activation of *RTG* target gene expression (Liao and Butow 1993). Moreover, we determined that nuclear export of the complex requires the action of Msn5p, a member of the importin-β family of nuclear export factors (Komeili et al. 2000). Taken together, these results extend the finding of Beck and Hall (1999) that an important role of TOR is to couple nutritional signals to the subcellular localization of specific transcription factors.

The *RTG1-RTG3* genes were originally discovered by Butow and coworkers as required for growth when mitochondrial respiratory function is impaired, as occurs in ρ0 petite cells that have lost their mitochondrial DNA (Liao et al. 1991; Liao and Butow 1993). As alluded to previously, the *RTG* genes have also been linked to peroxisomal function, indicating that they regulate a number of distinct metabolic functions. Activation of *RTG* target gene expression in ρ0 petite cells requires the cytoplasmic to nuclear translocation of Rtg1p/Rtg3p in an Rtg2p dependent process, similar to what was described above for Tor (Sekito et al. 2000). Thus, these early studies suggested that both Tor and retrograde signaling converged on the same nuclear trafficking event to control the expression of these metabolic genes. The single exception to this overall similarity between these two responses concerned the phosphorylation state of Rtg3p. Thus, while retrograde-dependent induction of the pathway involves dephosphorylation of Rtg3p, we observed that rapamycin-induced activation of the pathway correlates with hyperphosphorylation of this protein (Komeili et al. 2000; Sekito et al. 2000). We have subsequently found that these differences can be explained by both the strains and media composition used in these published studies, an indication that additional levels of complexity exist in the regulation of this pathway (our unpublished results). These results are consistent with recent observations from Broach and coworkers that multiple levels of phosphorylation and dephosphorylation events are likely to accompany *RTG* target gene regulation (Duvel et al. 2003).

While nuclear localization of Rtg1p and Rtg3p is necessary for *RTG* gene activation, it appears by itself not to be sufficient. Evidence for this conclusion comes from an analysis of *msn5Δ* cells, where both Rtg1p and Rtg3p are localized constitutively within the nucleus (Komeili et al. 2000). Thus, if regulated access to the nucleus represented the primary means by which the activity of the Rtg1p/Rtg3p complex is controlled, then we expected to observe constitutive activation of their target genes in *msn5Δ* cells. However, no such increased expression of *RTG* target genes was observed compared to wild type cells. This observation is consistent with studies of other regulated transcription factors, namely, that constitutive nuclear localization does not necessarily result in gene activation and that other regulatory mechanisms are involved (Komeili and O'Shea 1999). Remarkably, however, we observed rapid induction of *RTG* target genes in *msn5Δ* cells following rapamycin treatment, demonstrating that despite its steady state nuclear localization, the Rtg1p/Rtg3p complex nevertheless remains responsive to changes in Tor signaling.

8.3.2 Mks1p: a negative regulator of Rtg1p/Rtg3p activity

An outstanding question has been the precise physiological role of Rtg2p. Results of several recent studies now indicate that it is likely to function by antagonizing the activity of another protein, Mks1p, which is itself a negative regulator of *RTG* target gene expression (Dilova et al. 2002; Sekito et al. 2002; Tate et al. 2002). We became interested in Mks1p following the report by Schreiber and coworkers suggesting that it acts downstream of Tor as a *positive* regulator of the *RTG* pathway (Shamji et al. 2000). This conclusion was based on results of a large scale microarray study wherein the *RTG* targets *CIT2* and *DLD3* failed to be induced in an *mks1Δ* strain following rapamycin treatment. However, in contrast to this conclusion, we subsequently found that deletion of the *MKS1* gene in each of three different strain backgrounds resulted in constitutive activation of the pathway. Similar results were also been reported by the Butow and Cooper laboratories, leading to the consensus picture that Mks1p is in fact a negative regulator of the *RTG* pathway (Sekito et al. 2002; Tate et al. 2002).

To explain the results of the Shamji et al. (2000), Cooper and coworkers suggested that these investigators had misinterpreted their own results, due to ambiguities inherently associated with interpreting difference ratios obtained from dual-color microarray experiments (Tate et al. 2002). However, our subsequent analyses of the specific *mks1Δ* strain isolate used by Shamji et al. (2000) revealed that, in addition to a deletion of the *MKS1* gene, this strain had likely acquired an additional mutation(s) that impairs activation of the *RTG* pathway. Thus, rather than being a cautionary tale for the interpretation of microarray data, this is instead an illustration of the need to vigorously confirm phenotypes of mutant strains obtained for large-scale studies.

In addition to *RTG* target gene regulation, *MKS1* has been identified in a number of genetic screens and has been linked to several diverse cellular processes, including ras signaling, lysine biosynthesis, and formation of the *URE3* yeast prion (Matsuura and Anraku 1993; Feller et al. 1997; Edskes et al. 1999; Edskes and Wickner 2000). It now appears that many of these results can be explained as secondary effects due to altered *RTG* target gene expression. For example, *MKS1* was found to be identical to *LYS80*, a negative regulator of lysine biosynthesis (Feller et al. 1997). Thus, when *mks1Δ* cells are grown in rich media, lysine biosynthetic genes are induced, due to elevated levels of α-ketoglutarate (Feller et al. 1997). Moreover, we have demonstrated the expression of lysine biosynthetic genes in cells requires a functional *RTG* pathway (Dilova et al. 2002). Similarly, the role of *MKS1* in production of *URE3*, the inactive prion form of the Ure2p regulatory protein, has been explained recently by altered *RTG* target gene expression (Sekito et al. 2002; Tate et al. 2002). Finally, microarray analysis demonstrates that a very limited scope of genes is affected when Mks1p is inactivated, consistent with the idea that this protein is dedicated primarily, if not exclusively, to the regulation of *RTG* target genes (Dilova et al. 2002).

8.3.3 Architecture of the *RTG* branch of Tor signaling

Based on results of epistasis analyses, we postulated a simple linear pathway in which Mks1p acts downstream of both Tor and Rtg2p as well as upstream of Rtg1p/Rtg3p (Dilova et al. 2002) (Fig. 1A). Consistent with this model was the observation that the Rtg1p/Rtg3p complex enters the nucleus when the *MKS1* gene is deleted and *RTG* target genes become constitutively activated (Dilova et al. 2002; Sekito et al. 2002). This model also accounted for the observation that Rtg2p becomes dispensable for growth in *mks1Δ* cells in the absence of exogenously supplied glutamate or glutamine. Indeed, the ability to bypass the glutamate auxotrophy of an *rtg2Δ* strain formed the basis of a genetic selection that enabled Butow and coworkers to identify *mks1* mutants (Sekito et al. 2002). We note that this same genetic screen led to the identification of *LST8* (Liu et al. 2001), which we will discuss again in a later section. Finally, this scheme can explain why in *rtg2Δ* cells, *RTG* targets cannot be induced by rapamycin; in the absence of Rtg2p, Mks1p becomes a constitutive inhibitor of the pathway.

A number of additional observations were inconsistent with the formal linear pathway described above. First, our analyses of the phosphorylation states of both Rtg3p and Mks1p reveal that these proteins are responsive to Tor signaling in the absence of their immediate upstream regulators. Thus, we observe rapamycin-induced dephosphorylation of Mks1p that is independent of the presence of Rtg2p (Dilova et al. 2002). Similarly, we observe rapamycin-induced hyperphosphorylation of Rtg3p in the absence of both Mks1p as well as Rtg2p (Komeili et al. 2000) (our unpublished results). Based on these results we concluded that additional functional interactions were likely to exist between Tor and individual components involved in *RTG* target gene expression (Dilova et al. 2002).

Recent results from the Butow lab reveal important insights into the molecular mechanism by which Rtg2p and Mks1p collaborate to regulate Rtg1p/Rtg3p activity (Liu et al. 2003). These investigators have found that dynamic physical interactions between Mks1p and Rtg2p, as well as between Mks1p and both Bmh1p and Bmph2p, the latter being the yeast homologues of mammalian 14-3-3 proteins, play key roles in this system (Fig. 1B). In particular, Mks1p forms mutually exclusive interactions with either Rtg2p or Bmh1p/Bmh2p in a manner that is regulated

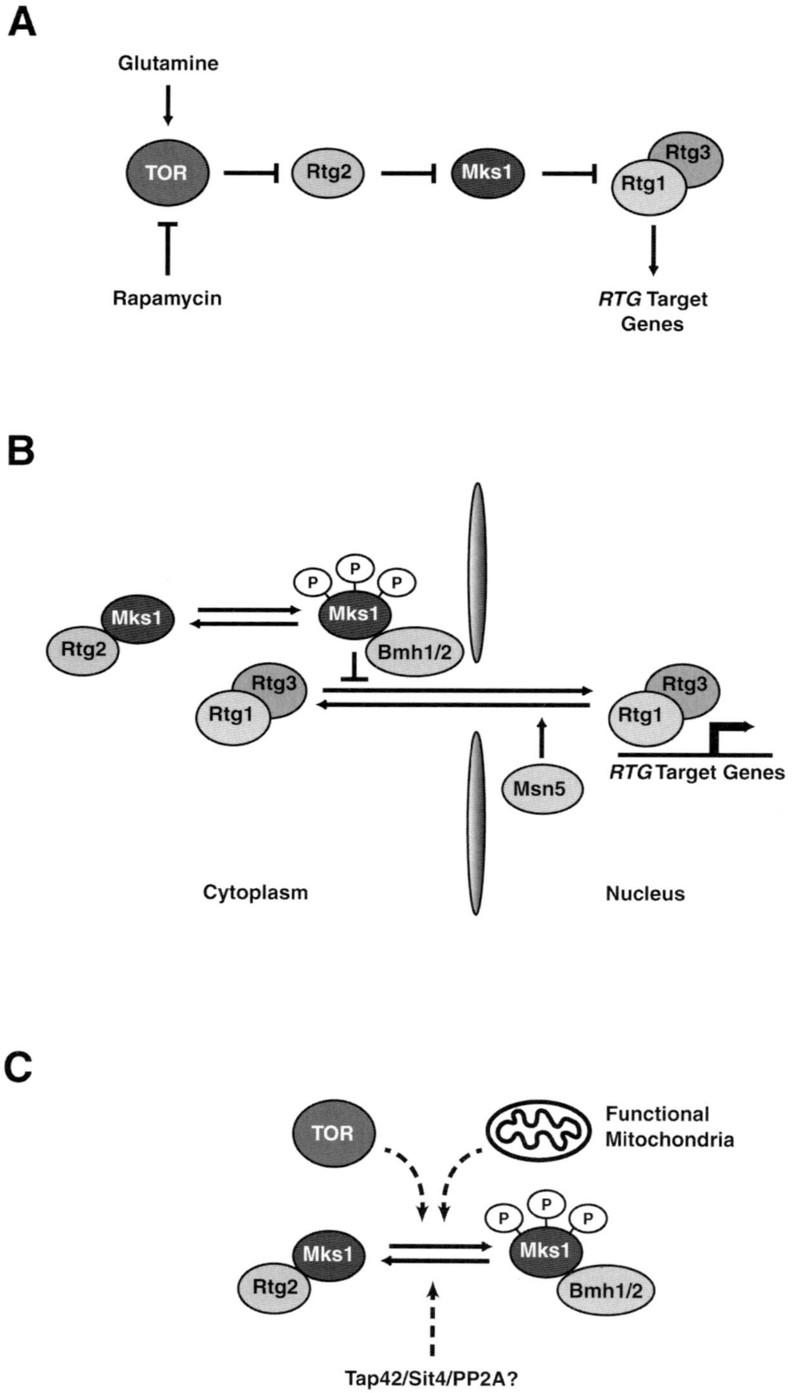
A
Glutamine
TOR
Rtg2
Mks1
Rtg3
Rtg1
Rapamycin
RTG Target Genes
B
P
P
P
Mks1
Rtg2
Mks1
Bmh1/2
Rtg3
Rtg1
Rtg3
Rtg1
Msn5
RTG Target Genes
Cytoplasm
Nucleus
C
TOR
Functional Mitochondria
P
P
P
Mks1
Rtg2
Mks1
Bmh1/2
Tap42/Sit4/PP2A?

Fig. 1. An evolving view of the *RTG* branch of the Tor kinase pathway. **(A)** Initial model for the role of Tor in rapamycin-dependent control of *RTG* target gene expression based on functional analyses of the indicated components (Dilova et al. 2000). **(B)** Interactions between Mks1p and Rtg2p and between Mks1p and Bmh1p/Bmh2p, the yeast 14-3-3 proteins, are regulated by differential phosphorylation of Mks1p (Liu et al. 2003). According to this scheme, dephosporylated Mks1p interacts strongly with Rtg2p, making Mks1p unavailable for interactions with Bmh1p/Bmp2p. (C) Tor signaling and wild type mitochondria positively influence Mks1p phosphorylation and thereby inhibit *RTG* target gene expression. A speculative role for the Type 2A phosphatases and their associated regulatory protein Tap42 is also indicated (after Düvel et al. 2003). In (A)-(C), arrows indicate positive regulation whereas bars indicate inhibitory interactions.

by Mks1p phosphorylation. Thus, when Mks1p is hyperphosphorylated, it interacts preferentially with Bmh1p or Bmh2p to form a complex that inhibits nuclear localization of Rtg1p/Rtg3p. By contrast, dephosphorylation of Mks1p results in its association with Rtg2p, where it is prevented from forming an inhibitory complex with the 14-3-3 proteins. According to this scheme, starvation for glutamate/glutamine, rapamycin treatment, or mitocondrial dysfunction all result in dephosphorylation of Mks1p and promote its association with Rtg2p. This model emphasizes that Mks1p phosphorylation provides a focal point for controlling *RTG* target gene activation (Fig. 1B). Moreover, these results help resolve one of the inconsistencies with the linear pathway described in Figure 1A. Thus, glutamate/glutamine and Tor need not signal through Rtg2p but can modulate this pathway by influencing Mks1p directly. A role for Bmh1p and Bmh2p in the regulation of Rtg3p had been demonstrated previously by van Heusden and coworkers (van Heusden and Steensma 2001).

At present, the identities of the kinase and phosphatase that act upon Mks1p or Rtg3p remain unknown. Interestingly, Mks1p becomes hyperphosphorylated in *rtg2*Δ cells, suggesting Rtg2p may play a role in Mks1p dephosphorylation (Dilova et al. 2002; Sekito et al. 2002). Consistent with this possibility, Rtg2p contains an Hsp70-like ATP binding domain and displays homology to certain bacterial phosphatases (Liao and Butow 1993; Koonin 1994). Moreover, it has been determined that a mutation in the ATP binding domain of Rtg2p is important for the inhibitory activity of this protein (Liu et al. 2003). Whether Rtg2p is a bona fide phosphatase that acts upon Mks1p remains to be determined, however. Even if it turns out that Rtg2p plays a direct role in this process, it is clear that one or more additional phosphatases must exist that act upon Mks1p. Evidence for this conclusion comes from the fact that glutamine starvation as well as rapamycin treatment both result in significant dephosphorylation of Mks1p within *rtg2*Δ cells (Dilova et al. 2002). These phosphatases could include the type 2A related protein phosphatases Pph21p, Pph22p, and Sit4p, along with their regulator partner Tap42p, which have recently been linked to rapamycin dependent induction of *RTG* target genes (Duvel et al. 2003).

A number of recent results add complexity to our current view of the role of Tor in *RTG* target gene regulation. First, in collaboration with Mike Hall's lab, we have presented evidence that glutamine can act as a signal to inhibit *RTG* target gene expression in a manner that correlates with Tor activity but that is independent of glutamate (Crespo et al. 2002). By contrast, Cooper and coworkers have demonstrated that Tor activity can be uncoupled from *RTG* target gene regulation under conditions where cells use proline or glutamate as a sole nitrogen source (Tate et al. 2002; Tate and Cooper 2003). Thus, when cells are grown on these nitrogen sources, the *RTG* target genes *CIT2* and *DLD3* are repressed in a rapamycin-insensitive manner, raising the possibility that a Tor-independent mechanism of repression operates under these conditions. We too have observed that *RTG* target gene expression switches to a rapamycin-insensitive mode when either proline or glutamate is used as a sole nitrogen source (our unpublished results). Indeed, we have found that the metabolic signals governing *RTG* target gene regulation are remarkably complex, with glutamate, glutamine, as well as ammonia each serving as distinct signals that collaborate to regulate *RTG* target gene expression (our unpublished results). Involvement of Tor appears to be acutely sensitive to the intracellular balance of glutamate and glutamine, where the availability of ammonia plays a key role in determining whether these genes are repressed *via* a rapamycin-sensitive mechanism. Together these results suggest that both Tor dependent as well as Tor independent mechanisms that respond to overlapping yet distinct nutritional cues will turn out to play important roles during *RTG* target gene regulation. It is tempting to speculate that Tor independent control of *RTG* target genes represents a retrograde-specific pathway that responds to glutamate. In any event, all nutrient-based signals appear to ultimately converge on Mks1p, where we find that a direct correlation exists between the degree of Mks1p dephosphorylation and the extent of *RTG* target gene activation (our unpublished results).

8.4 Tor signaling and the role of distinct membrane associated Tor1p- and Tor2p-containing protein complexes

The above discussion illustrates many of the challenges associated with taking a "bottom up" approach to understand Tor signaling, where one begins with a downstream target, in this case the Rtg1p/Rtg3p transcription factor complex, and attempts to work "upstream", building molecular links to the Tor kinases. This has proven difficult in part because these proteins are not involved in isolated linear pathways but are instead integrated with other cellular activities, for example retrograde signaling in the case of *RTG* target gene expression. With these issues in mind, we and others have realized that it is important to complement these efforts with a "top down" approach, whereby proteins that interact directly with Tor1p and Tor2p are identified, with the aim of finding regulators and/or potential targets of Tor activity. Here we review the overall architecture of the Tor proteins and then discuss a number of recently described interactions with a novel set of Tor-associated proteins in yeast.

8.4.1 Molecular architecture of the Tor proteins

The Tor proteins are members of an expanding family of PI-kinase related kinases (PIKKs) that include a number of proteins involved in DNA damage and cell cycle checkpoint regulation, for example ATM, ATR, Rad3, and DNA-PK (Thomas and Hall 1997; Bosotti et al. 2000; Schmelzle and Hall 2000; Jacinto and Hall 2003). Members of this family are extremely large, ranging from ~2,000 to greater than 4,000 amino acids in length, and each contain a conserved C-terminal kinase domain that is structurally related to PI-3 and/or PI-4 kinase (Bosotti et al. 2000). To date no lipid kinase activity has been demonstrated for members of this family and it is generally believed they are likely to be strict protein kinases. Each subfamily is proposed to possess a number of additional domains that impart substrate specificity, aid in establishing proper intracellular localization, as well as govern interactions with other proteins (Bosotti et al. 2000).

For the Tor subfamily, where each member averages about 2400 amino acids in length, there are at least three additional structural and/or functional regions, including a large N-terminal domain consisting of multiple HEAT sequence repeats, named after the proteins in which they were first identified: the Huntington's disease protein, EF3, the A subunit of PP2A, and Tor (Andrade and Bork 1995; Schmelzle and Hall 2000). Structural studies of several HEAT repeat-containing proteins have demonstrated these repeats form supra-helical structures that act as protein docking surfaces for interacting ligands (Chook and Blobel 1999; Cingolani et al. 1999). Thus, it was immediately realized that these repeats in Tor were likely to be important for protein-protein interactions crucial for its function (Schmelzle and Hall 2000). Moreover, there have been reports that these repeats are important for interactions with other proteins, as well as membranes, although the precise nature of these interactions or their potential role in Tor signaling are only beginning to be characterized (Sabatini et al. 1999; Bertram et al. 2000; Kunz et al. 2000; Wu et al. 2002).

Adjacent to the kinase domain is the rapamycin binding domain (termed FRB). This region interacts with rapamycin in combination with the ubiquitous and highly conserved prolyl isomerase FKBP and a single amino acid change of a conserved serine within the FRB is sufficient to confer resistance to rapamycin (Helliwell et al. 1994; Cardenas and Heitman 1995; Chen et al. 1995). It has been proposed that the FKBP/rapamycin complex inhibits access of an effector protein to Tor that is required for downstream events leading to continued cell growth and cell cycle progression (Schmelzle and Hall 2000; Crespo and Hall 2003). An unanswered question is whether interactions between FKBP and Tor ever take place in the absence of rapamycin. Structural studies indicate that the FRB domain makes more extensive contacts with rapamycin than with FKBP, suggesting interactions between the two proteins are likely to be of low affinity if they occur at all in the absence of drug (Choi et al. 1996). In addition, other than rapamycin resistance, no significant phenotype has been reported to accompany mutations in *FPR1*, the gene encoding FKBP in yeast. Nevertheless, it remains possible that FKBP is involved in Tor function and that a redundant activity exists in yeast.

A final domain that has been characterized lies between the HEAT and FRB regions. This domain was first termed the Toxic Effector Domain and was identified by the fact that it induces a G1 arrest when over-expressed in yeast, a result very reminiscent of rapamycin treatment (Alarcon et al. 1999). One suggested explanation for this arrest is that over-expression of this domain titrates an essential substrate and/or regulator of Tor. Independently, this region was identified by sequence analysis as a conserved motif that exists within all of the PIK family members and has been termed the FAT domain, derived from the names of the three major groups that possess this domain [Frap (aka Tor), Atm, and Trapp] (Bosotti et al. 2000). In addition, a second conserved motif among these family members at their extreme C-terminus has been named the FATC domain (Bosotti et al. 2000).

Interestingly, a recent study entailing extensive sequence analyses of the PIKK family, including Atm, Atr, and Tor, has led to the provocative conclusion that, with the exception of the kinase domain, the entire rest of these proteins are composed essentially entirely of HEAT repeats (Perry and Kleckner 2003). This conclusion would more than double the number of previously identified HEAT repeats in the Tor proteins and, furthermore, would extend this motif throughout both the Toxic/FAT as well as FRB domains. While a conclusive examination of this hypothesis will most likely require determining the physical structures of these proteins, for example by x-ray crystallography, it immediately implies a conserved mechanism of action for members of the PIKK family beyond what has been appreciated to date. Moreover, this model suggests that clues into the nature of protein-protein interactions between Tor and its partners may be provided by studies of these other family members.

8.4.2 Evidence for protein-protein interactions with Tor: clues from higher eukaryotes

As described above, both sequence analysis as well as functional studies provided circumstantial evidence for interactions between different regions of Tor with one or more possible ligands. Independently, two approaches have been used successfully in both yeast and mammalian systems to identify proteins that interact physically with Tor, namely two-hybrid and direct biochemical methods. The first report of an interacting partner of Tor was for mammalian Tor (mTor), where a two-hybrid study led to the identification of Gephyrin, a protein previously identified as required for normal clustering of glutamate receptors in neuronal cells (Sabatini et al. 1999). More recently, *in vitro* approaches have led to the identification of Raptor as well as Tsc1 and Tsc2, the latter two being tumor suppressor proteins involved in the autosomal dominant disorder Tuberous sclerosis, as additional interacting partners of mTor (Gao et al. 2002; Hara et al. 2002; Inoki et al. 2002; Kim et al. 2002). Whereas Gephyrin and Raptor are positive regulators of mTor activity, both Tsc1 and Tsc2 appear to antagonize the ability of mTor to signal to downstream targets, including p70 ribosomal protein S6 kinase and the eIF4E binding protein, 4E-BP1, both of which are regulators of translational initiation. Together these reports help lay a foundation for understanding how Tor activity will turn out to be regulated by interacting partners. With the exception of Raptor, however, these other described proteins do not have apparent homologues in yeast, indicating that species specific interactions are likely to play an important role in Tor signaling.

8.4.3 Tor protein complexes in yeast: composition and function

Biochemical approaches have now been used successfully to identify a number of Tor-interacting proteins in yeast. In particular, Mike Hall's group used ion-exchange chromatography, followed by immuno-affinity selection, to characterize two distinct Tor-containing protein complexes, which they have termed Tor Complex One (TORC1) and TORC2 (Loewith et al. 2002). TORC1 is proposed to contain Tor1p or Tor2p along with Kog1p (the yeast homologue of Raptor) and Lst8p. TORC2 contains Tor2p (but not Tor1p) as well as Lst8p and three additional proteins, Avo1p-Avo3p. Except for Avo2p (and Tor1p), these proteins are each encoded by an essential gene in yeast. Homologues of Kog1p and Lst8p exist throughout eukaryotes, suggesting that TORC1 represents a conserved aspect of Tor signaling. Consistent with this conclusion, the mammalian homologue of Lst8p (termed mLst8p or GβL) has been shown recently to be a positive regulator of mTor activity and to work in conjunction with Raptor (Kim et al. 2003). A determination of predicted protein-interacting motifs possessed by these interacting partners reveals enormous potential for multiple and complex protein-protein interactions within these complexes, including multiple HEAT, WD, and ankyrin repeat motifs (Loewith et al. 2002; Jacinto and Hall 2003).

A number of observations are consistent with the proposal that TORC1 is responsible for carrying out rapamycin-sensitive Tor1p/Tor2p shared functions whereas TORC2 is responsible for Tor2p unique activities (Loewith et al. 2002) (Fig. 2). First, biochemical experiments demonstrate that Tor1p, Tor2p, Kog1p, and Lst8p are specifically co-precipitated by an affinity-tagged version of Fpr1p in the presence of rapamycin (Loewith et al. 2002). By contrast, no co-precipitation of Avo1p-Avo3p is observed, consistent with their involvement in a rapamycin-insensitive function. Second, gene shut-off experiments (the functional equivalent of disrupting essential genes) demonstrate that depletion of Kog1p, a component exclusive to TORC1, causes a number of effects that mimic rapamycin treatment, including characteristic changes in gene expression, an overall decrease in protein synthesis, and a G1 cell cycle arrest (Loewith et al. 2002). By contrast, depletion of Avo1p, a TORC2 specific component, instead results in actin depolarization and growth arrest at various stages of the cell cycle, two hallmarks of defects in the Tor2p-unique activity. Furthermore, as with a number of *tor2* alleles, these *avo1* phenotypes can be suppressed by overexpression of components in the signaling pathway linking Tor2p to actin dynamics, including RHO2 and PKC1 (Loewith et al. 2002).

In independent work, in collaboration with John Yates we have used affinity purification and mass spectrometry to identify Tor2p-interacting proteins and isolated what most likely corresponds to TORC2 (Wedaman et al. 2003). More recently we have identified two additional proteins, Tco89p and Bit61p, and have determined that they are components of TORC1 and TORC2 respectively (our unpublished results). In general, our results are very consistent with interactions described by Hall and coworkers, with the exception that we have not been able to detect significant interactions between Tor2p and Kog1p (Wedaman et al. 2003). In this context, it is notable that only relatively weak co-immunoprecipitation between Tor2p and Kog1p was observed in the Loewith et al. (2002) study and, moreover, only in cells that were deleted for *TOR1*. Together these findings suggest that the majority of Tor2p is normally associated with Lst8p and Avo1p-Avo3p in the form of TORC2. Given that Tor2p can apparently completely substitute for Tor1p in wild type cells, a corollary to this conclusion is that sub-stoichiometric association between Tor2p proteins and their partners may be sufficient to carry out essential TORC1-dependent activities. An alternative explanation is that significant differences exist in the relative stability of interactions observed in these types of studies between Kog1p and Tor1p *versus* Kog1p and Tor2p. Here it is relevant that interactions between mTor and Raptor were reported to be relatively unstable and influenced by the nutritional state of the cell (Kim et al. 2002). These uncertainties nevertheless underscore the present need to precisely determine the precise stoichiometries and stabilities of different Tor protein complexes.

8.4.4 Lst8p as a component of both TORC1 and TORC2

Several lines of evidence support a model wherein Lst8p acts as a component of both TORC1 and TORC2 (Fig. 2). First, co-immunoprecipitation studies demonstrate that Lst8p interacts with both Tor1p and Tor2p (Loewith et al. 2002; Chen and Kiaser 2003; Wedaman et al. 2003). Additional immunoprecipitation experiments show that Lst8p also interacts with Kog1p as well as with Avo1p-Avo3p, specific components of TORC1 and TORC2, respectively (Loewith et al. 2002). Functional inactivation of *LST8*, either by gene shut off experiments or through the generation of temperature-sensitive *lst8* alleles, results in defects related to both the rapamycin sensitive Tor1p/Tor2p shared as well as Tor2p unique functions (Loewith et al. 2002; Chen and Kiaser 2003). Finally, at the ultrastructural level as well as in cell fractionation studies, a significant portion of Lst8p colocalizes with the Tor proteins (Chen and Kiaser 2003; Wedaman et al. 2003).

The finding that Lst8p is a component of TORC1 is important as it brings together a number of previous observations linking Tor signaling and amino acid metabolism. As mentioned previously, *LST8* was identified originally by Kaiser and coworkers in a genetic screen for components of the exocytic pathway and has been implicated in the regulated intracellular sorting of Gap1p (Roberg et al. 1997a). Independently, Butow and coworkers identified *LST8* as a negative regulator of *RTG* target gene expression by acting through the retrograde signaling pathway (Liu et al. 2001). The finding that Lst8p interacts with the Tor proteins further emphasizes the mechanistic similarities between Tor signaling and retrograde control of *RTG* target gene expression. An outstanding question, however,

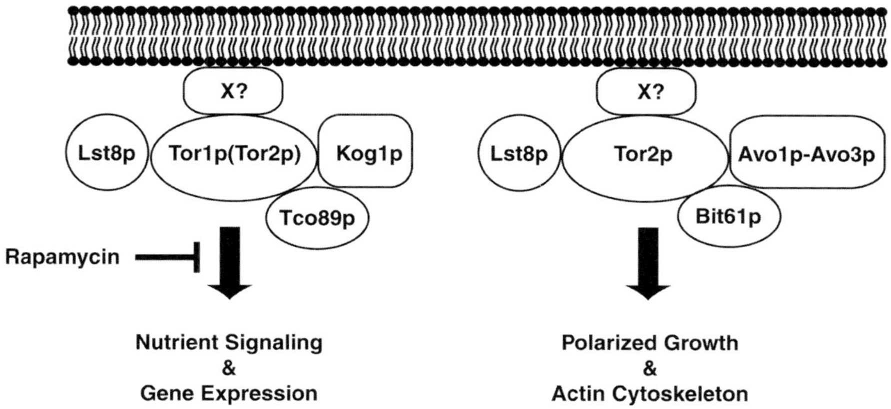

Fig. 2. Model for involvement of Tor interacting proteins and membrane localization of Tor signaling events. Indicated are proteins demonstrated to interact with Tor1p and/or Tor2p (Loewith et al. 2002; Wedaman et al. 2003; Chen and Kaiser 2003; our unpublished results). This model is adapted from Loewith et al. (2002) who has demonstrated the complex on the left (TORC1) interacts with and is inhibited by rapamycin whereas the complex on the right (TORC2) is not (Note that Bit61p and Tco89p are not reported by Loewith et al.) Also indicated is the proposed existence of additional factor(s) (X) that act to link the Tor protein complexes to internal membranous sites. These factors could include additional cytosolic proteins and/or integral membrane proteins.

has been whether Lst8p's role in Gap1p trafficking *versus RTG* target gene regulation result from the same or rather distinct activities of this protein. This question has now been largely answered by Kaiser and coworkers, who have demonstrated that effects on Gap1p sorting are indirect consequences of Lst8p-dependent regulation of amino acid levels via the *RTG* pathway (Chen and Kiaser 2003).

At present, it is unclear what precise mechanistic role Lst8p may play in the function of either TORC1 or TORC2. By analogy to *in vitro* analysis of mLst8p, it is likely that this protein interacts with the Tor proteins adjacent to their kinase domains and, moreover, regulates their kinase activity (Kim et al. 2003). This protein consists almost entirely of seven WD repeats that are predicted to form a circular seven-bladed propeller structure (Roberg et al. 1997a; Liu et al. 2001). By analogy to other WD-repeat containing proteins, these propellers form surfaces for interactions with other proteins and, thus, it has been suggested that Lst8p may act as a molecular link between the Tor proteins and other ligands (Loewith et al. 2002; Chen and Kiaser 2003; Jacinto and Hall 2003). These ligands could include already identified components of TORC1 and TORC2 or, alternatively, other proteins that have yet to be identified. Furthermore, Lst8p is unique in that, apart from Tor2p, it is the only protein known at present to be a component of both TORC1 and TORC2. Accordingly, it represents a likely candidate that could promote "crosstalk" between the Tor1p/Tor2p shared and Tor2p unique activities that would facilitate coordinate control of cell growth and cell polarity in yeast.

8.4.5 Membrane localization of Tor protein complexes

It has become increasingly clear that the intracellular location of signaling components is often coupled to their function as well as regulation. Accordingly, there is great interest in determining the precise intracellular location of Tor1p and Tor2p as well as their interacting partners. An early immunofluorescence study suggested Tor2p was associated with the vacuole (Cardenas and Heitman 1995). By contrast, immunofluorescence detection of plasmid-expressed, epitope-tagged versions of *TOR1* and *TOR2* indicated the majority of these proteins were localized at the plasma membrane (Kunz et al. 2000). This conclusion was consistent with a concomitant cell fractionation analysis of endogenous Tor1p and Tor2p, which showed that the majority of these proteins associated with distinct membrane populations, one of which coincided with plasma membrane markers (Kunz et al. 2000). In addition, it was noted in this study that a second population of Tor-associated membranes existed, although the identity of this compartment was not determined.

In collaboration with Michael McCaffery, we have recently used immunoelectron microscopy (IEM) to examine the location of the Tor proteins as well as one of their common interacting partners, Lst8p, in ultrathin cyrosections (Wedaman et al. 2003). The results demonstrate that these proteins are localized in distinct punctate clusters in regions that are adjacent to, yet apparently distinct from, the plasma membrane, as well as within the cell interior. In many instances, these clusters are associated with membranes that, at the ultrastructural level, appear most similar to characteristic tracks and/or tubules that have been attributed to membranes of the endocytic pathway (Rieder et al. 1996; Mulholland et al. 1999; Wang et al. 2001). These results are similar to studies of mammalian cells that indicate mTor is also associated with internal membranes, suggesting this too is a conserved aspect of Tor signaling. Moreover, recent cell fractionation and floatation analyses support an association of Tor1p and Tor2p with membranes associated with the exocytic and/or endocytic pathways (Chen and Kiaser 2003).

A potential connection between Tor and endosomes is attractive given that this compartment is the site of a growing number of signaling events related to the control of cell growth and morphogenesis in yeast as well as in higher eukaryotes (Harsay and Schekman 2002; Seto et al. 2002). A connection to the endosomal system is also provided by a three-way convergence of events related to endocytosis, actin dynamics, and sphingolipid biosynthesis (Friant et al. 2001; Anderson and Jacobson 2002; Dickson and Lester 2002). Specifically, a number of recent studies indicate that sphingolipid-mediated signaling is linked to proper endocytosis as well as actin cytoskeletal organization (Friant et al. 2001; Anderson and Jacobson 2002; Dickson and Lester 2002). In this context, it is significant that Avo3p, a component of TORC2, was originally identified in a genetic screen for components involved in sphingolipid biosynthesis (Beeler et al. 1988) (note: in this study *AVO3* is called *TSC11*). Interestingly, *TOR2* was also identified in this study, emphasizing a potential link, albeit indirect, between Tor signaling and endocytosis (Beeler et al. 1988). Most significantly, a specific link between Tor and endocytosis has now been established recently by Hicke and coworkers, who have

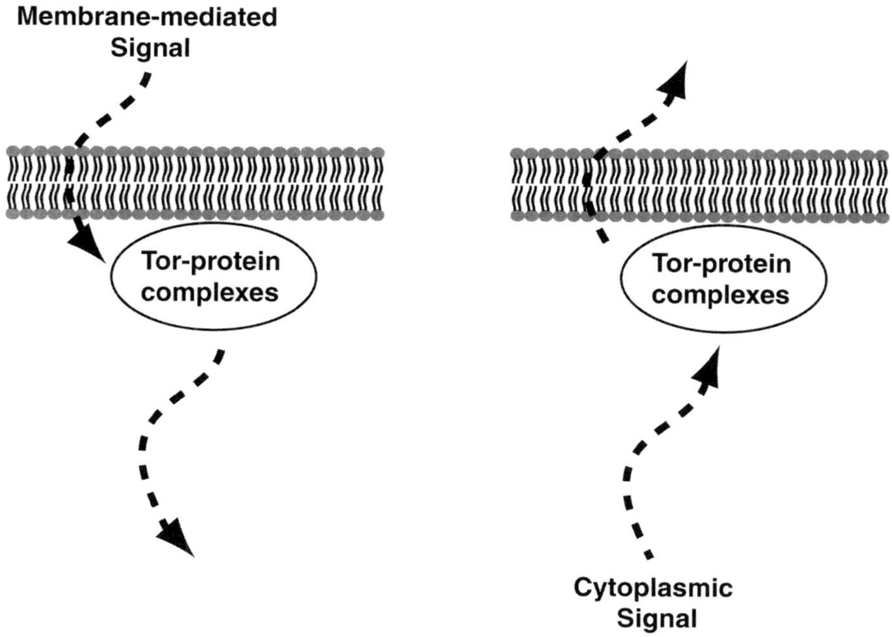

Fig. 3. Potential relationships between Tor signaling and the membrane-association of Tor-containing protein complexes. See text for details.

found that a specific mutation in Tor2p, G2128R, results in defects in the internalization of alpha factor receptor (deHart et al. 2003). Further studies indicate that this internalization defect is attributable to defects in a Tor2p-unique activity (e.g. TORC2), in particular the loss of Tor2p-mediated activation of the cell integrity pathway dependent on the Rho1p GTPase (deHart et al. 2003).

Despite this advance in our understanding of the potential relationship between TORC2 and events related to endocytosis, it remains unclear what role membranes may play with respect to TORC1 and rapamycin-sensitive Tor1p/Tor2p shared functions. One possibility is that TORC1 receives signals (i.e. nutrient status) via its association with membranes and then transmits this information to downstream cytoplasmic targets (Fig. 3). Indeed, this possibility has been suggested with respect to the relationship between amino acid metabolism and Tor1p/Tor2p/Lst8p mediated regulation of *RTG* target gene expression (Chen and Kiaser 2003). Alternatively, it is possible that TORC1 responds to a cytoplasmic signal that is then conveyed to its target membrane (Fig. 3). Further insight into these possibilities will require both an understanding of the molecular composition the Tor-membrane junction as well as well as its regulation.

8.5 Conclusions

The study of Tor signaling in yeast appears to be entering a critical phase. On the one hand, the scope of cellular activities affected by this pathway seems to ever increase as different investigators discover that their biological process of interest is perturbed by addition of rapamycin to yeast cells. Furthermore, it is becoming increasingly clear that Tor converges with many other signaling pathways involved in nutrient-based control of cell growth, cell polarity, and maintenance of cellular integrity and, in a few cases, shared molecular links have been identified (one example is Mks1p and Lst8p for Tor and retrograde regulation). On the other hand, we still understand very little about how Tor kinase activity is regulated in yeast. The discovery of interacting partners of Tor1p and Tor2p should allow the field to take the next step and determine what components and/or events lie immediately upstream as well as downstream of these protein assemblies. Here is it immensely important that many of the Tor-interacting proteins that have been identified in yeast are conserved in higher eukaryotes and associate with mTor, which will enable "cross feeding" between experimental systems. However, it is now key that bona fide substrates for Tor1p and/or Tor2p kinase activity be identified in yeast, which will enable the development of accurate *in vitro* systems in which to explore the regulation of Tor kinase activity. We believe that such systems will be crucial in order to gain insight into the molecular mechanisms that control this pathway. In addition, we believe that further studies into the nature of Tor-membrane interactions will provide valuable insights into the physiological role of Tor signaling. Given the rapid pace of this field and the powerful genetic and molecular tools available for yeast, it can be predicted that these issues will be tackled in the very near future.

Acknowledgement

Work in the Powers' lab is supported by a Basil O'Connor Starter Research Award from the March of Dimes and by National Science Foundation Grant MCB-1031221. We thank Ron Butow and Linda Hicke for communication of their results prior to publication.

References

Alarcon CM, Heitman J, Cardenas ME (1999) Protein kinase activity and identification of a toxic effector domain of the target of rapamycin TOR proteins in yeast. Mol Biol Cell 10:2531-2546

Anderson RG, Jacobson K (2002) A role for lipid shells in targeting proteins to caveolae, rafts, and other lipid domains. Science 296:1821-1825

Andrade MA, Bork P (1995) Heat repeats in the Huntington's disease protein. Nat Genet 11:115-116

Angeles de la Torre-Ruiz M, Torres J, Arino J, Herrero E (2002) Sit4 is required for proper modulation of the biological functions mediated by Pkc1 and the cell integrity pathway in *Saccharomyces cerevisiae.* J Biol Chem 277:33468-33476

Beck T, Hall MN (1999) The TOR signalling pathway controls nuclear localization of nutrient-regulated transcription factors. Nature 402:689-692

Beck T, Schmidt A, Hall MN (1999) Starvation induces vacuolar targeting and degradation of the tryptophan permease in yeast. J Cell Biol 146:1227-1238

Beeler T, Bacikova D, Gable K, Hopkins L, Johnson C, Slife H, Dunn T (1988) The *Saccharomyces cerevisiae TSC10*/YBR265w gene encoding 3-ketosphinganine reductase is identified in a screen for temperature-sensitive suppressors of $Ca2^{+}$-sensitive *csg2Δ* Mutant. J Biol Chem 273:30688-30694

Bertram PG, Choi JH, Carvalho J, Ai W, Zeng C, Chan T-F, Zheng XFS (2000) Tripartite regulation of Gln3p by TOR, Ure2p, and phosphatases. J Biol Chem 275:35727-35733

Bogan JS, McKee AE, Lodish HF (2001) Insulin-responsive compartments containing GLUT4 in 3T3-L1 and CHO cells: Regulation by amino acid concentrations. Mol Cell Biol 21:4785-4806

Bosotti R, Isacchi A, Sonnhammer EL (2000) FAT: a novel domain in PIK-related kinases. Trends Biochem Sci 25:225-227

Cardenas M, Heitman J (1995) FKBP12-rapamycin target TOR2 is a vacuolar protein with an associated phosphatidylinositol-4 kinase activity. EMBO J 14:5892-5907

Cardenas ME, Cutler NS, Lorenz MC, Di Como CJ, Heitman J (1999) The TOR signaling cascade regulates gene expression in response to nutrients. Genes Dev 13:3271-3279

Chen EJ, Kaiser CA (2002) Amino acids regulate the intracellular trafficking of the general amino acid permease of *Saccharomyces cerevisiae.* Proc Natl Acad Sci USA 99:14837-14842

Chen EJ, Kiaser CA (2003) LST8 negatively regulates amino acid biosynthesis as a component of the TOR pathway. J Cell Biol 161:1-15

Chen J, Zheng XF, Brown EJ, Schreiber SL (1995) Identification of an 11-kDa FKBP12-rapamycin-binding domain within the 289-kDa FKBP12-rapamycin-associated protein and characterization of a critical serine residue. Proc Natl Acad Sci USA 92:4947-4951

Choi J, Chen J, Schreiber SL, Clardy J (1996) Structure of the FKBP12-rapamycin complex interacting with the binding domain of human FRAP. Science 273:239-242

Chook YM, Blobel G (1999) Structure of the nuclear transport complex karyopherin-beta2-Ran-GppNHp. Nature 399:230-237

Cingolani G, Petosa C, Weis K, Muller CW (1999) Structure of importin-beta bound to the IBB domain of importin-alpha. Nature 399:221-229

Crespo JL, Hall MN (2003) Elucidating TOR signaling and rapamycin action: lessons from *Saccharomyces cerevisiae.* Microbiol Mol Biol Rev 66:579-591

Crespo JL, Powers T, Fowler B, Hall MN (2002) The TOR-controlled transcription activators GLN3, RTG1 and RTG3 are regulated in response to intracellular levels of glutamine. Proc Natl Acad Sci USA 99:6784-6789

Damelin M, Simon I, Moy TI, Wilson B, Komili S, Tempst P, Roth FP, Young RA, Cairns BR, Silver PA (2002) The genome-wide localization of Rsc9, a component of the RSC chromatin-remodeling complex, changes in response to stress. Mol Cell 9:563-573

deHart AKA, Schnell JD, Allen DA, Tsai J-Y, Hicke L (2003) Receptor internalization in yeast requires the Tor2-Rho1 signaling pathway. Mol Biol Cell 14:4676-4684

Dennis PB, Fumagalli S, Thomas G (1999) Target of rapamycin (TOR): balancing the opposing forces of protein synthesis and degradation. Curr Opin Genet Dev 9:49-54

Dickson RC, Lester RL (2002) Sphingolipid functions in *Saccharomyces cerevisiae*. Biochim Biophys Acta 1583:13-25

Dilova I, Chen C-Y, Powers T (2002) Mks1 in concert with TOR signaling negatively regulates RTG target gene expression in *S. cerevisiae*. Curr Biol 12:389-395

Duvel K, Santhanam A, Garrett S, Schneper L, Broach JR (2003) Multiple roles of Tap42 in mediating rapamycin-induced transcriptional changes in yeast. Mol Cell 11:1467-1478

Edskes HK, Hanover JA, Wickner RB (1999) Mks1p is a regulator of nitrogen catabolism upstream of Ure2p in *Saccharomyces cerevisiae*. Genetics 153:585-594

Edskes HK, Wickner RB (2000) A protein required for prion generation: [URE3] induction requires the Ras-regulated Mks1 protein. Proc Natl Acad Sci USA 97:6625-6629

Feller A, Ramos F, Pierard A, Dubois E (1997) Lys80p of *Saccharomyces cerevisiae*, previously proposed to as a specific repressor of Lys genes, is a pleiotropic regulatory factor identical to Mks1p. Yeast 13:1337-1346

Friant S, Lombardi R, Schmelzle T, Hall MN, Riezman H (2001) Sphingoid base signaling via Pkh kinases is required for endocytosis in yeast. EMBO J 20:6783-6792

Gao X, Zhang Y, Arrazola P, Hino O, Kobayashi T, Yeung RS, Ru B, Pan D (2002) Tsc tumour suppressor proteins antagonize amino-acid-TOR signalling. Nature Cell Biol 4:699-704

Hara K, Maruki Y, Long X, Yoshino K, Oshiro N, Hidayat S, Tokunaga C, Avruch J, Yonezawa K (2002) Raptor, a binding partner of Target of Rapamycin (TOR), mediates TOR action. Cell 110:177-189

Hardwick JS, Kuruvilla FG, Tong JK, Shamji AF, Schreiber SL (1999) Rapamycin-modulated transcription defines the subset of nutrient-sensitive signaling pathways directly controlled by the Tor proteins. Proc Natl Acad Sci USA 96:14866-14870

Harsay E, Schekman R (2002) A subset of yeast vacuolar protein sorting mutants is blocked in one branch of the exocytic pathway. J Cell Biol 156:271-285

Heinisch JJ, Lorberg A, Schmitz H-P, Jacoby JJ (1999) The protein kinase C-mediated MAP kinase pathway involved in the maintenance of cellular integrity. Mol Micro 32:671-680

Helliwell SB, Howald I, Barbet N, Hall MN (1998) TOR2 is part of two related signaling pathways coordinating cell growth in *Saccharomyces cerevisiae*. Genetics 148:99-112

Helliwell SB, Losko S, Kaiser CA (2001) Components of a ubiquitin ligase complex specify polyubiquitination and intracellular trafficking of the general amino acid permease. J Cell Biol 153:649-662

Helliwell SB, Wagner P, Kunz J, Deuter-Reinhard M, Henriquez R, Hall MN (1994) TOR1 and TOR2 are structurally and functionally similar but not identical phosphatidylinositol kinase homologues in yeast. Mol Biol Cell 5:105-118

Iiboshi Y, Papst PJ, Hunger SP, Terada N (1999) L-asparaginase inhibits the rapamycin-targeted signaling pathway. Biochem Biophys Res Comm 260:534-539

Inoki K, Li Y, Zhu T, Wu J, Guan K-L (2002) TSC2 is phosphorylated and inhibited by Akt and supresses mTOR signalling. Nature Cell Biol 4:699-704

Jacinto E, Hall MN (2003) Tor signalling in bugs, brain, and brawn. Nat Rev Mol Cell Biol 4:117-126

Kim D-H, Sarbassov DD, Ali SM, King JE, Latek RR, Erdjument-Bromage H, Tempst P, Sabatini DM (2002) mTOR interacts with Raptor to form a nutrient-sensitive complex that signals to the cell growth machinery. Cell 110:163-175

Kim D-H, Sarbassov DD, Ali SM, Latek RR, Guntur KVP, Erdjument-Bromage H, Tempst P, Sabatini DM (2003) GβL, a positive regulator of the rapamycin-sensitive pathway required for the nutrient-sensitive interaction between Raptor and mTor. Mol Cell 11:895-904

Kim K-S, Rosenkrantz MS, Guarente L (1986) *Saccharomyces cerevisiae* contains two functional citrate synthase genes. Mol Cell Biol 6:1936-1942

Komeili A, O'Shea EK (1999) Roles of phosphorylation sites in regulating activity of the transcription factor Pho4. Science 284:977-980

Komeili A, Wedaman KP, O'Shea EO, Powers T (2000) Mechanism of metabolic control: Target of rapamycin signaling links nitrogen quality to the activity of the Rtg1 and Rtg3 transcription factors. J Cell Biol 151:863-878

Koonin EV (1994) Yeast protein controlling inter-organelle communication is related to bacterial phosphatases containing the Hsp70-type ATP-binding domain. Trends Biochem Sci 19:156-157

Kunz J, Schneider U, Howald I, Schmidt A, Hall MN (2000) HEAT repeats mediate plasma membrane localization of Tor2p in yeast. J Biol Chem 275:37011-37020

Liao X, Butow RA (1993) *RTG1* and *RTG2*: two yeast genes required for a novel path of communication from mitochondria to the nucleus. Cell 72:61-71

Liao X, Small WC, Srere PA, Butow RA (1991) Intramitochondrial functions regulate nonmitochondrial citrate synthase (*CIT2*) expression in *Saccharomyces cerevisiae*. Mol Cell Biol 11:38-46

Liu Z, Butow RA (1999) A transcriptional switch in the expression of yeast tricarboxylic acid cycle genes in response to a reduction or loss of respiratory function. Mol Cell Biol 19:6720-6728

Liu Z, Sekito T, Epstein CB, Butow RA (2001) RTG-dependent mitochondria to nucleus signaling is negatively regulated by the seven WD-repeat protein Lst8p. EMBO J 20:7209-7219

Liu Z, Sekito T, S'pirek M, Thornton J, Butow RA (2003) Retrograde signaling is regulated by the dynamic interaction between Rtg2p and Mks1p. Mol Cell 12:410-411

Loewith R, Jacinto E, Wullschleger S, Lorberg A, Crespo JL, Bonenfant D, Oppliger W, Jenoe P, Hall MN (2002) Two TOR complexes, only one of which is rapamycin sensitive, have distinct roles in cell growth control. Mol Cell 10:457-468

Magasanik B (1992) Regulation of nitrogen utilization. In: Jones EW, Pringle JR, Broach JR (eds) The molecular and cellular biology of the yeast *Saccharomyces*: gene expression. Cold Spring Harbor Laboratory Press, Plainview, NY, pp 283-317

Magasanik B, Kaiser CA (2002) Nitrogen regulation in *Saccharomyces cerevisiae*. Gene 290:1-18

Marzluf GA (1997) Genetic regulation of nitrogen metabolism in the fungi. Microbiol and Mol Biol Rev 61:17-32

Matsuura A, Anraku Y (1993) Characterization of the MKS1 gene, a new negative regulator of the ras-cyclic AMP pathway in *Saccharomyces cerevisiae*. Mol Gen Genetics 238:6-16

Mulholland J, Konopka J, Singer-Kruger B, Zerial M, Botstein D (1999) Visualization of receptor-mediated endocytosis in yeast. Mol Biol Cell 10:799-817

Perry J, Kleckner N (2003) The ATRs, ATMs, and TORs are giant HEAT repeat proteins. Cell 112:151-155

Powers T, Walter P (1999) Regulation of ribosome biogenesis by the rapamycin-sensitive TOR-signaling pathway in *Saccharomyces cerevisiae*. Mol Biol Cell 10:987-1000

Pruyne D, Bretscher A (2000) Polarization of cell growth in yeast. J Cell Sci 113:571-585

Raught B, Gingras AC, Sonenberg N (2001) The target of rapamycin (TOR) proteins. Proc Natl Acad Sci USA 98:7037-7044

Rieder SE, Banta LM, Kohrer K, McCaffery JM, Emr SD (1996) Multilamellar endosome-like compartment accumulates in the yeast vps28 vacuolar protein sorting mutant. Mol Biol Cell 7:985-999

Roberg KJ, Bickel S, Rowley N, Kaiser CA (1997a) Control of amino acid permease sorting in the late secretory pathway of *Saccharomyces cerevisiae* by *SEC13*, *LST4*, *LST7*, and *LST8*. Genetics 147:1569-1584

Roberg KJ, Rowley N, Kaiser CA (1997b) Physiological regulation of membrane protein sorting late in the secretory pathway of *Saccharomyces cerevisiae*. J Cell Biol 137:1469-1482

Rohde JR, Heitman J, Cardenas ME (2001) The TOR kinases link nutrient sensing to cell growth. J Biol Chem 276:9583-9586

Rohde JR, Cardenas ME (2003) The tor pathway regulates gene expression by linking nutrient sensing to histone acetylation. Mol Cell Biol 23:629-635

Rosenkrantz MS, Alam T, Kim K, Clark BJ, Srere PA (1986) Mitochondrial and nonmitochondrial citrate synthase in *Saccharomyces cerevisiae* are encoded by distinct homologous genes. Mol Cell Biol 6:4509-4515

Sabatini DM, Barrow RK, Blackshaw S, Burnett PE, Lai MM, Field ME, Bahr BA, Kirsch J, Betz H, Snyder SH (1999) Interaction of RAFT1 with gephryin required for rapamycin-sensitive signaling. Science 284:1161-1164

Schmelzle T, Hall MN (2000) TOR, a central controller of cell growth. Cell 103:253-262

Schmelzle T, Helliwell SB, Hall MN (2002) Yeast protein kinases and the RHO1 exchange factor TUS1 are novel compoents of the cell integrity pathway in yeast. Mol Cell Biol 22:1329-1339

Schmidt A, Beck T, Koller A, Kunz J, Hall MN (1998) Thr TOR nutrient signalling pathway phosphorylates NPR1 and inhibits turnover of the tryptophan permease. EMBO J 17:6924-6931

Schmidt A, Bickle M, Beck T, Hall MN (1997) The yeast phosphatidylinositol kinase homolog TOR2 activates RHO1 and RHO2 via the exchange factor ROM2. Cell 88:531-542

Schmidt A, Kunz J, Hall MN (1996) TOR2 is required for organization of the actin cytoskeleton. Proc Natl Acad Sci USA 93:13780-13785

Schmitz H-P, Huppert S, Lorberg A, Heinisch JJ (2002) Rho5p downregulates the yeast cell integrity pathway. J Cell Sci 115:3139-3148

Sekito T, Liu Z, Thornton J, Butow RA (2002) RTG-dependent mitochondria-to-nucleus signaling is regulated by Mks1 and is linked to formation of yeast prion [URE3]. Mol Biol Cell 13:795-804

Sekito T, Thorton J, Butow R (2000) Mitochondria-to-nuclear signaling is regulated by the subcellular localization of the transcription factors Rtg1p and Rtg3p. Mol Biol Cell 11:2103-2115

Seto ES, Bellen HJ, Lloyd TE (2002) When cell biology meets development: endocytic regulation of signaling pathways. Genes Dev 16:1314-1336

Shamji AF, Kuruvilla FG, Schreiber SL (2000) Partitioning the transcriptional program induced by rapmycin among the effectors of the Tor proteins. Curr Biol 10:1574-1581

Tate JJ, Cooper TG (2003) Tor1,2 regulation of retrograde gene expression in *S. cerevisiae* derives indirectly as a consequence of alterations in ammonia metabolism. J Biol Chem (in press).

Tate JJ, Cox KH, Rai R, Cooper TG (2002) Mks1p is required for negative regulation of retrograde gene expressioin in *Saccharomyces cerevisiae* but does not affect nitrogen catabolite repression-sensitive gene expression. J Biol Chem 277:20477-20482

Thomas G, Hall MN (1997) TOR signalling and control of cell growth. Curr Opin Cell Biol 9:782-787

Torres J, Di Como CJ, Herrero E, Angeles de la Torre-Ruiz M (2002) Regulation of the cell integrity pathway by rapamycin-sensitive TOR function in budding yeast. J Biol Chem 277:43495-43502

van Heusden GPH, Steensma HY (2001) 14-3-3 Proteins are essential for regulation of RTG3-dependent transcription in *Saccharomyces cerevisiae*. Yeast 18:1479-1491

Wang G, McCaffery JM, Wendland B, Dupre S, Haguenauer-Tsapis R, Huibregtse JM (2001) Localization of the Rsp5p ubiquitin-protein ligase at multiple sites within the endocytic pathway. Mol Cell Biol 21:3564-3575

Wedaman KP, Reinke A, Anderson S, Yates JI, McCaffery JM, Powers T (2003) Tor kinases are in distinct membrane-associated protein complexes in *Saccharomyces cerevisae*. Mol Biol Cell 14:1204-1220

Wu S, Mikhailov A, Kallo-Hosein H, Hara K, Yonezawa K, Avruch J (2002) Characterization of ubiquilin 1, an mTOR-interacting protein. Biochim Biophys Acta 1542:41-56

Zaragoza D, Ghavidel A, Heitman J, Schultz MC (1998) Rapamycin induces the G_0 program of transcriptional repression in yeast by interfering with the TOR signaling pathway. Mol Cell Biol 18:4463-4470

9 Integrated regulation of the nitrogen-carbon interface

Terrance G. Cooper

9.1 Abstract

Significant progress has been made in understanding the integrated regulation of nitrogen and carbon metabolism in *Saccharomyces cerevisiae*. A major contribution to that understanding has been identification of Tor1/2 influence on the intracellular localization of Gln3/Gat1 and Rtg1/3, the transcription factors that mediate nitrogen catabolite repression-sensitive and retrograde gene expression, respectively. This chapter discusses the three areas of investigation, which together form one of the carbon-nitrogen regulatory interfaces. **(i)** Nitrogen catabolite repression or NCR, the means through which cells selectively utilize good nitrogen sources in preference to those that support less good growth. **(ii)** Retrograde gene expression that provides the α-ketoglutarate carbon backbone of glutamate when the tricarboxylic acid cycle is inoperative. **(iii)** Torl/2, a master cellular regulator, reported to be the critical link connecting environmental nutritional conditions with the transcriptional responses they trigger.

9.2 Introduction

The first eukaryotic genome sequenced was that of *Saccharomyces cerevisiae* completed in 1996. Several years and the effort of hundreds of researchers worldwide accomplished the task in an impressive technical and collaborative *tour-de-force.* Today a single high-throughput center can complete the job in about a week or so as recently reported for the S. *paradoxus, S. bayanus*, and *S. mikatae* genomes. These powerful modern technologies have moved investigation of cell-wide regulatory networks from conjecture to solid reality. However, these enormous leaps in capability incur equally large challenges of interpretation and only by surmounting them will we be able to elucidate the molecular mechanisms through which cell-wide regulatory networks operate and communicate. Not only must increasing volumes of data be managed and creatively sorted, but primary and secondary consequences of experimental perturbation must also be clearly distinguished. Living cells are quintessential examples of just-in-time manufacturing, producing needed metabolites in precisely the right amounts at just the time they are needed. The breadth, sensitivity and flexibility of small molecule detection systems and multi-tiered responses that cells make to them stretch the imagination.

Topics in Current Genetics, Vol. 7
J. Winderickx, P.M. Taylor (Eds) Nutrient-Induced Responses in Eukaryotic Cells

The purpose of this chapter is to explore several regulatory interfaces between carbon and nitrogen metabolism. We demonstrate how meeting the demands of one metabolic pathway influences others around it and the difficulties these secondary influences generate when trying to distinguish the direct and indirect consequences of experimental manipulation. The specific examples will be drawn from: **(i)** nitrogen catabolite repression or NCR, the means through which cells selectively utilize good nitrogen sources in preference to those that support less good growth. **(ii)** Retrograde gene expression that provides the α-ketoglutarate carbon backbone of glutamate when the tricarboxylic acid cycle is inoperative, and **(iii)** Torl/2, a master cellular regulator, reported to be the critical link connecting environmental nutritional conditions with the transcriptional responses they trigger.

9.3 Nitrogen catabolite repression

Yeast cells in nature are continuously faced with feast or famine, i.e., good and poor nitrogen sources. *Saccharomyces cerevisiae,* like most microorganisms, transports, accumulates, and utilizes good nitrogen sources in preference to poor ones. This selectivity is achieved through nitrogen catabolite repression (NCR), which is the preeminent control exerted over nitrogen catabolic gene expression. When nitrogen is in excess transcription of genes encoding proteins needed to transport and degrade poor nitrogen sources is minimal. As good nitrogen sources become limiting, or only poor ones are available, NCR is relieved and transcription of these genes increases.

9.4 GATA-family transcription factors regulate NCR-sensitive transcription

9.4.1 Transcriptional activators, Gln3 and Gat1

NCR-sensitive transcription is mediated by UAS_{NTR} elements, with the sequence GATAA at their core, and GATA-family transcriptional activators, Gln3 and Gat1, each possessing a highly conserved GATAA-binding zinc-finger motif (C-X2-C-N-C-X2-C). Residues at the C-terminus of this GATA zinc-finger and just beyond contact four nucleotides on each strand of the DNA, seven in the major and one in the minor groove (Omichinski et al. 1993). Given the high degree of homology in the DNA binding regions of Gln3 and Gat1, it is not surprising that both proteins participate in transcriptional activation of most NCR-sensitive genes, albeit to differing degrees (Coffman et al. 1996). The evolutionary advantage and differential regulation that derives from the operation of these two very similar proteins is just beginning to be investigated (Shamji et al. 2000; Beck and Hall 1999).

9.4.2 Transcriptional repressors, Dal80 and Deh1

In addition to transcriptional activators, Gln3 and Gatl/Nill, a pair of transcriptional repressors, Dal80 and Dehl/Gzf3/Nil2, also regulate NCR-sensitive gene expression (Hoffman-Bang 1997). Dal80 and Dehl are also GATA-family member proteins whose DNA binding sites are highly homologous to those of Gln3 and Gat1. In fact, Gln3 and Dal80 bind some of the same promoter GATA sequences (Cunningham et al. 1994) and regulate many, but not all, genes in parallel (Daugherty et al. 1993). *DAL5,* for example, is scarcely regulated by Dal80, while *DAL7* or *GAT1* are highly regulated by it (Rai et al. 1999). Yet transcription of all three genes is highly Gln3/Gat1-dependent. This differential regulation derives from the fact that Gln3 binds to a single GATA sequence while Dal80, which dimerizes through a leucine zipper motif at its C-terminus, requires two GATA sequences correctly spaced and oriented to one another (Cunningham and Cooper 1993; Svetlov and Cooper 1998); Deh1, also forming a dimer *in vivo,* likely exhibits similar DNA binding requirements, though this has not yet been demonstrated.

Dal80 represses transcription by competing with Gln3/Gatl for binding to their target GATA sequences (Coffman et al. 1997; Cunningham et al. 2000). Such a competitive control mechanism can be successful only if amounts of activator and repressor proteins are tightly coordinated. This is achieved in S. *cerevisiae* through transcriptional regulation of *GAT1, DAL80,* and *DEH1* (Coffman and Cooper 1996; Coffman et al. 1997; Rowan et al. 1997). All three gene promoters contain multiple GATA sequences and their expression is Gln3-dependent as well as regulated by each of the remaining GATA-factors (Fig. 1) (Coffman and Cooper 1996; Coffman et al. 1997; Rowan et al. 1997). Such reciprocal regulation appears complex, but is in fact a simple feedback regulated circuit in which expression of each GATA-factor gene, except *GLN3,* is autogenously and cross-regulated by all of the GATA-factors (Coffman et al. 1997).

9.5 Physiological significance of competitive GATA-activator/repressor regulation

S. cerevisiae responds in two ways to a transition from an environment containing good nitrogen sources in excess supply to one with limiting good nitrogen or only poor nitrogen sources. In one case, overall nitrogen supply is the regulating determinant, while in the second it is overall nitrogen supply limitation additionally coupled with the presence of a specific alternative, poor nitrogen source. In both cases, competitive Dal80/Gln3/Gatl binding to DNA moderates the cell's response to the environmental change, preventing wasteful over reaction. In the first case, the onset of nitrogen limitation permits Gln3 to become functional (the mechanism will be discussed later), resulting in increased *GAT1* expression (Fig. 1). Gatl production is further enhanced and accelerated by its autogenous regulation. Once available and functional, Gln3 and Gatl together mediate expression of the NCR-sensitive genes required to scavenge and utilize poor nitrogen sources.

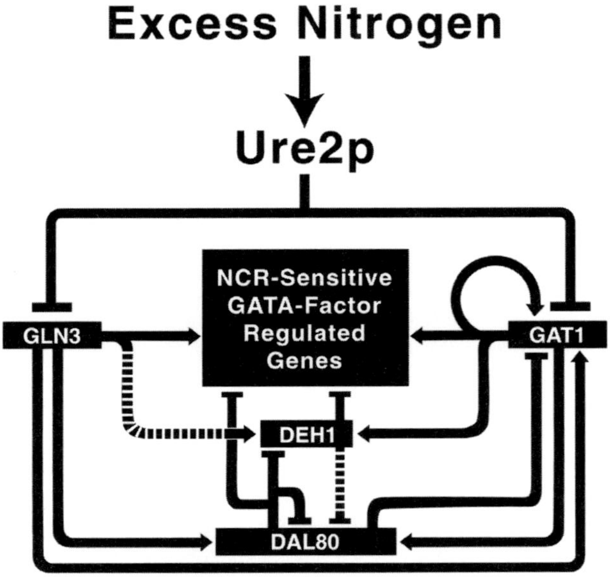

Fig. 1. Regulatory circuit of NCR-sensitive, autogenous, and reciprocal regulation of GATA-transcription factor gene expression. Also shown is GATA-factor regulation of NCR-sensitive gene expression *per se*. Arrows and bars indicate positive and negative regulation, respectively. Dashed lines indicate weak regulation. Redrawn from Coffman et al. (1997).

However, at times of nitrogen limitation, it would be a disadvantageous for a cell to produce a large array of proteins at high level. This situation is avoided by the fact that the requirements for expression of these NCR-sensitive genes, i.e., functional Gln3 and Gat1, are also those needed for *DAL8O* expression. As Dal80 production increases, GATA-factor-mediated gene expression is reduced by its competition with Gln3 and Gatl for DNA binding. Included among the genes down-regulated in this way is *GAT1* itself, which further diminishes NCR-sensitive expression. Thus autogenous and reciprocal activation/repression of GATA-factor gene expression brings the regulon to steady state at a much lower level than would occur in its absence (Cunningham et al. 2000).

In the second case, Dal80 similarly functions for inducible NCR-sensitive genes, such as *DAL4, DAL1, DAL7, DUR1,2* and *DUR3* (Daugherty et al. 1993). These genes encode the permeases and limiting catabolic enzymes needed to degrade of allantoin and urea, two poor nitrogen sources (Cooper 1982, 1996). Here, as above, it is advantageous for the cell to produce only small amounts, of these proteins in its search for alternative nitrogen sources. The promoters of these inducible genes are structured such that Dal80 out competes Gln3/Gatl for DNA binding in the absence of inducer and the genes are expressed only at minimal levels (Chisholm and Cooper 1982).These basal levels are further regulated by NCR as described in case one above. If allantoin or urea is present in the environment, however, it is transported into the cell and degraded by these basal levels of Dal and Dur proteins. This degradation results in production of the allantoin pathway inducer, which is the last intermediate in the degradative pathway (Cooper 1996). The transcription factor required for inducer-dependent transcription, Dal82, binds to an upstream induction sequence situated near the GATA elements. A direct or indirect interaction between Dal82 and Gln3 and/or Gat1 then occurs and enhances Gln3 binding to the promoter, which shifts the competitive Dal80:Gln3 balance in favor of Gln3 an high level *DAL* and *DUR* gene expression ensues (van Vuuren et al. 1991). This is but one of multiple examples of this form of transcriptional regulation, the goal of which appears to be the ability to produce small amounts of proteins required to scan the environment for any poor nitrogen source available and then increase the cell's transport and degradative capacity to utilize it if a sufficient and stable supply exists.

9.6 Genomic analysis of NCR-sensitive, GATA-factor-mediated transcription

The broadest and most uniformly regulated set of genes whose transcription is mediated by the GATA-factors are associated with uptake and catabolism of nitrogenous compounds. In addition, genomic analyses have greatly expanded the variety of genes that require GATA-factors for expression (Cox et al. 1999; Cardenas et al. 1999; Hardwick et al. 1999; Bertram et al. 2000; Shamji et al. 2000). Investigation of direct physiological relationships, if any, of these newly identified genes is just beginning. As demonstrated by analysis of retrograde gene expression described later in this chapter, rigorously establishing these relationships will be challenging, in no small part, due to: **(i)** the high frequency of indirect relationships as changes in one cellular process influence a chain of others, and **(ii)** the marked influences on gene expression of small changes in growth conditions and/or genetic composition that occur from one laboratory to another.

9.7 Mechanism of nitrogen catabolite repression

9.7.1 Ure2-dependent regulation of Gln3 is responsible for NCR-sensitive transcription

Great progress has been made in identifying some of the molecular events associated with NCR. The first discovered and central participant in NCR is Ure2, whose genetic locus was identified by mutations that permit uptake of a pyrimidine intermediate (**URE**idosuccinate) with ammonia as nitrogen source (Drillien and Lacroute 1972; Drillien et al. 1973); this does not occur in ammonia-grown wild type cells. "Ammonia repression" (ammonia repression was replaced by the term NCR when the phenomenon was found to occur with other nitrogen sources as well) is abrogated in *ure2* mutants and the phenotype was found to be highly pleiotropic extending to many nitrogen catabolic enzymes (Hoffman-Bang 1999; ter Schure et al. 2000; Cooper 1996; Turoscy and Cooper 1987).

Studies measuring synthetic capacity to produce allophanate hydrolase, product of the NCR-sensitive gene *DUR1,2,* and the effects of the RNA synthesis inhibitor, lomofungin, reported in the 1970s suggested that NCR occurred at the level of transcription (Lawther and Cooper 1973; 1975; Lawther et al. 1975). It was not until a decade later, when cloned genes became available, that NCR was definitively shown to correlate with decreased *DUR1,2* mRNA levels (Cooper et al. 1983). As already noted above, the second major participant in NCR is Gln3, first identified as a mutant locus that converted a *gln1* (encoding glutamine synthetase) bradytroph into a full glutamine auxotroph (Mitchell and Magasanik 1984). Subsequently, production of NAD-glutamate dehydrogenase activity (Courchesne and Magasanik 1988), and *GLN1, DAL5, DAL7, DUR1,2* and *CAR1* mRNAs (Benjamin et al. 1989; Cooper et al. 1990) were found to require functional Gln3. A rigorous correlation between Gln3 and NCR-sensitive transcription occurred with analysis of a synthetized DNA fragment, containing three copies of an 11 bp sequence from the *DAL5* promoter (CGATAAGAGTC), that was necessary and sufficient for Gln3-dependent, NCR-sensitive transcription (Bysani and Cooper 1991; Cooper et al. 1990, 1989; Rai et al. 1989).

The relative positions of Gln3 and Ure2 in the regulatory pathway controlling NCR-sensitive transcription derives from the observation that *gln3* mutations are epistatic to those at *ure2* (Fig. 1) (Courchesne and Magasanik 1988). Ure2 possesses significant homology to members of the theta-family glutathione S-transferases, leading to the suggestion that Ure2 might regulate Gln3 by glutathionating it (Coshigano and Magasanik 1991). Although such glutathionation could not be demonstrated, Gln3 does form a stable complex with Ure2 (Blinder et al. 1996; Beck and Hall 1999; Bertram et al. 2000). The homology with glutathione S-transferases, however, is not fortuitous. Recently, *ure2* mutants have been demonstrated to possess the characteristics found in many cell types with glutathione S-transferase gene defects (Rai et al. 2003). The glutationation substrates of *S. cerevisiae* Ure2 appear to be metal ions such as nickel and cadmium, which likely accounts for this trait not being discovered earlier. Although not discussed in this chapter, Ure2 is also a prion precursor. The major regions of Ure2 required for prion formation and NCR are separable, prion formation being primarily localized to the N-terminal *65* amino acids, and Gln3 binding the remaining C-terminal portion of Ure2 (Wickner et al. 1999; Kulkarni et al. 2001). Although the preponderance of published data argue that nitrogen metabolism does not play an important role in prion formation, there is one report concluding that it is negatively regulated by glutamate (Sekito et al. 2002).

9.7.2 NCR is achieved by regulated intracellular localization of Gln3

Insight into the mechanism of NCR derived from physiological investigation of *CAN1* (arginine permease) expression (Coffman et al. *1995;* Cox et al. 2000), and

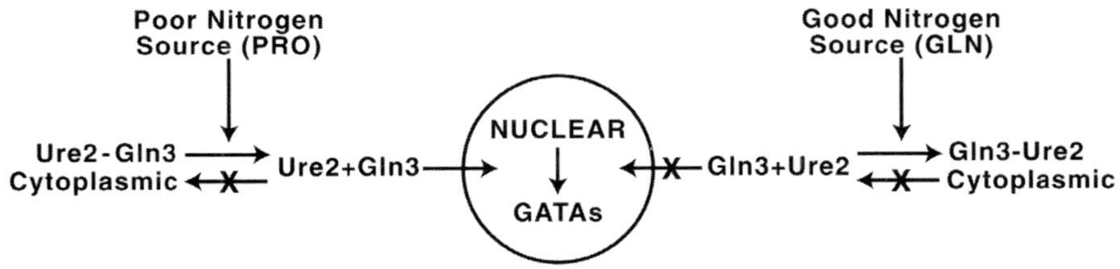

Fig. 2. Intracellular localization of Gln3 and Ure2 in response to growth with poor (Pro) and good (Gln) nitrogen sources, conditions that support of high and low NCR-sensitive gene expression, respectively.

from experiments with the growth inhibitor rapamycin (Cardenas et al. 1999; Hardwick et al. 1999; Bertram et al. 2000; Beck and Hall 1999). The physiological experiments showed that GATA sequences in the *CAN1* promoter serve as surrogate TATA elements in two circumstances: **(i)** growth with a nitrogen source eliciting high level NCR (e.g. glutamine), or **(ii)** in a *gln3Δgat1Δ* double mutant (Coffman et al. 1995; Cox et al. 2000). These data argued that GATA sequences are unoccupied by GATA-family transcription factors during NCR, which could occur for one or both of two reasons: **(i)** nuclear Gln3 and Gat1 are modified and hence unable to bind their target sequences. **(ii)** Gln3 and Gat1 are not in the nucleus under conditions of NCR. Intracellular localization of *GFP-GLN3* and *GFP-GAT1* supported the latter explanation (Cunningham et al. 2000; Cox et al. 2000). When NCR-sensitive transcription is high, both proteins are nuclear, and when low or when Ure2 is over-produced, both are excluded from the nucleus (Fig. 2) (Cunningham et al. 2000).

9.8 Rapamycin-induced NCR-sensitive gene expression

Although the above physiological experiments correlate GATA-factor localization and ability to mediate NCR-sensitive transcription *in vivo,* they don't provide insight into possible biochemical mechanisms associated with the correlation. Such mechanistic information derived from studies of the immunosuppressant and antineoplastic drug, rapamycin. Rapamycin, isolated from *Streptomyces hygroscopicus* found on Easter Island, is a lipophilic macrolide that binds to a small (12 kDa) peptidyl-prolyl isomerase or peptidylprolyl rotamase, FKBP12/Rbpl (Heitman et al. 1991). This complex binds to and potently inhibits Tori and Tor2 (Heitman et al. 1991), which share some homology with phosphatidylinositol-3 lipid kinases and possess ser/thr protein kinase activity (Cardenas and Heitman 1995; Jiang and Broach 1999).

The relation between Tor proteins and NCR was independently reported by four groups, the seminal observation being that expression of NCR-sensitive genes greatly increases following addition of rapamycin to cultures growing in rich media (Cardenas et al. 1999; Hardwick et al. 1999; Bertram et al. 2000; Beck and Hall 1999). Rapamycin treatment also decreases expression of a large number of genes, including those required for protein synthesis (Cardenas et al. 1999; Hardwick et al. 1999).

Rapamycin-induced NCR-sensitive transcription correlates with intracellular localization of Gln3, which is cytoplasmic when cells are grown in rich medium and accumulates in the nucleus following rapamycin treatment. (Bertram et al. 2000; Beck and Hall 1999). Further, Gln3 electrophoretic migration also correlates with its intracellular localization. Following rapamycin-treatment, Gln3 migrates more rapidly and in a manner similar to that observed when the extract is treated with phosphatase (Bertram et al. 2000; Beck and Hall 1999). In some, but not all, laboratories, Ure2 migration behaves similarly, i.e., its migration increases upon treating cells with rapamycin, arguing it too is dephosphorylated (Cardenas et al. 1999; Hardwick et al. 1999; Bertram et al. 2000; Beck and Hall 1999).

Taken together, these observations led to a model in which excess nutrient availability positively regulates Torl/2, which in turn promotes Gln3 (and Ure2) complex formation, phosphorylation, and Gln3 exclusion from the nucleus (Fig. 3) (Cardenas et al. 1999; Hardwick et al. 1999; Bertram et al. 2000; Beck and Hall 1999). When Torl/2 are inactivated by rapamycin, Gln3 phosphorylation is lost, or does not occur, which in turn results in its ability to access the nucleus and activate GATA factor-mediated transcription (Fig. 3) (Cardenas et al. 1999; Hardwick et al. 1999; Bertram et al. 2000; Beck and Hall 1999).

9.9 Gln3 structure and intracellular distribution

9.9.1 Gln3 functional domains

Mapping the functional regions of Gln3 has been undertaken to investigate the detailed molecular events associated with Gln3 function and regulation. So far seven regions have been identified. C-terminal amino acids 1-365 represent the smallest peptide that can interact with Ure2 and be excluded from the nucleus (Kulkarni et al. 2001), but a smaller region (amino acids 102-150) will give a weakly positive two-hybrid interaction with Ure2 (Carvaiho and Zheng 2003). A short, 13 amino acid region predicted to form an α-helix (positions 126-138) is responsible for transcriptional activation (Svetlov and Cooper 1997). The GATA binding site is situated between amino acids 306-330, and a C306S substitution destroys the ability of Gln3 to complement a *gln3Δ* (Svetlov and Cooper 1997). A nuclear export sequence occurs between amino acids 336-345 and a Torl binding region was localized to amino acids 600-667 (Kulkami et al. 2001; Carvalho and Zheng 2003). The precise nature of the nuclear localization sequence (NLS) remains to be settled. The existence of multiple regions (positions 343-360, 388-394 and 571-577) participating in nuclear localization best accounts for available data (Kulkarni et al. 2001; Carvalho and Zheng 2003). This conclusion is predicated upon the facts that: **(i)** data support the participation of all three regions in the nuclear

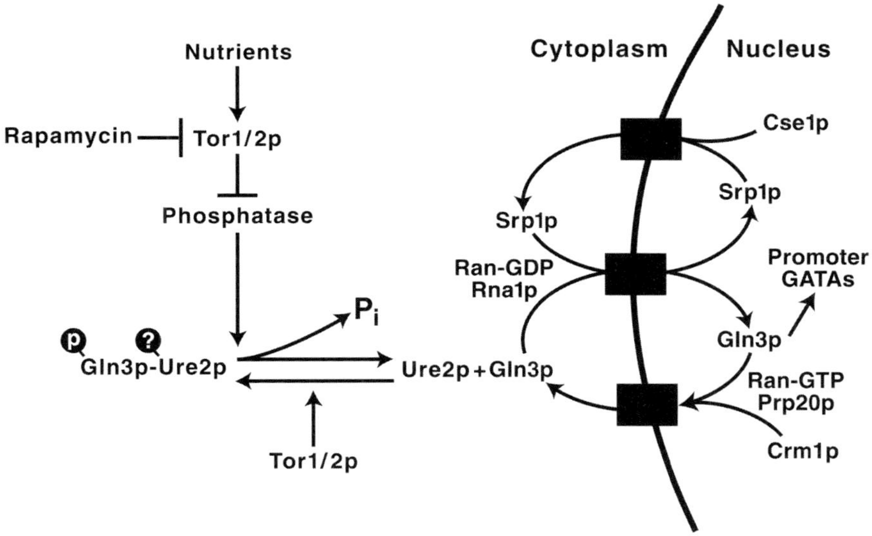

Fig. 3. Left Panel. Model summarizing the regulatory pathway whereby rapamycin and Tor regulate GATA-factor mediated transcription. The model incorporates elements proposed by both Beck and Hall (1999) and Bertram et al. (2000).Right Panel. Molecular interactions associated with Gln3 movement into and out of the nucleus. Redrawn from Carvalho et al. (2001).

localization process, **(ii)** none of the three regions can be shown to be both necessary and sufficient for regulated partitioning of Gln3 between nucleus and cytoplasm, and **(iii)** deletion of either of the first two regions results in rapamycin resistance, the phenotype of a *gln3Δ*. N- and C-termini of the Ure2 region required to exclude Gln3 from the nucleus are situated between Ure2 positions 101-151 and 330-346, respectively (Kulkarni et al. 2001). This coincides with the region reported earlier to be required for NCR-sensitivity of ureidosuccinate uptake (Wickner et al. 1999).

9.9.2 Nuclear transport of Gln3

Srpl, Csel, and Rnal are required for nuclear localization of Gln3, and loss of Crml leads to constitutive nuclear localization of Gln3 (Carvalho et at. 2001; Gorlich and Kutay 1999). Based on known functions of the above four proteins, Gln3 is proposed to interact with Srp1, and the complex transported into the nucleus. Gln3 can bind to promoter GATA-sequences and activate transcription, or complex with Crml and exit the nucleus (Fig. 3) (Carvalho et al. 2001). Srpl recycles back to the cytoplasm in association with Csel. Ran, in its GTP or GDP form, determines the directionality of nuclear import/export. Nuclear Ppr20 maintains Ran in its GTP form, while cytoplasmic Rnal maintains Ran in its GDP form (Corbett and Silver 1997; Mattaj and Englmeier 1998). The Rnal requirement for nuclear import of Gln3 explains very early results showing that expression of multiple NCR-sensitive genes is lost in an *rna1* mutant (Carvallho et al. 2001; Bossinger and Cooper 1976).

9.9.3 Gln3 is not uniformly distributed in the cytoplasm

Events associated with Gln3 in the cytoplasm have focused mainly on phosphorylation and complex formation with Ure2. However, close inspection of fluorescently labeled Gln3 reveals that it is not uniformly distributed throughout the cytoplasm. Gln3 appears as dots around the cell periphery, ringing the vacuole, or as tubes and networks of tubes (Fig. 4, top). In cells where some fluorescently labeled Gln3 still remains in the nucleus, i.e., following a shift from a poor to a rich nitrogen source, fluorescent projections can be seen to emanate from the nucleus (Fig. 4A, F) (Cox et al. 2002). When the images in Figure 4 (top) are rendered in 3-dimensions, the overall distribution of Gln3 (Green) appears similar to what would be expected if it (green) was contained within or attached to an organized cytoplasmic organelle. The nucleus (blue image, stained with DAPI) appears to come in close contact with several of the protrusions from this system (Fig. 4, bottom) (Cox et al.2002).

9.9.4 An intact actin cytoskeleton is required for nuclear accumulation of Gln3

Non-uniform cytoplasmic distribution of Gln3 discussed above could result from a variety of interactions including association with a cytoplasmic vesicular system or components of the cytoskeleton. Latrunculin, a drug that inhibits actin polymerization, was used to determine whether the actin cytoskeleton participates in intracellular Gln3 localization and movement (Cox et al. 2004). Latrunculin-treatment prevents nuclear accumulation of Gln3 and NCR-sensitive transcription in cells shifted from ammonia to pro medium. In contrast, latrunculin does not prevent cytoplasmic Gln3 accumulation when cells are shifted from proline to glutamine medium. These data suggest the actin cytoskeleton is required for nuclear entry of Gln3 in response to limiting nitrogen (Cox et al. 2004).

9.10 Tor1/2 participation in the regulation of Gln3 localization

9.10.1 Gln3 phosphorylation

Several protein kinases, Torl, Tor2, Tap42 among them, are reported to affect GATA-factor-activated transcription. Three observations are offered as evidence to support the position that Torl/2 itself phosphorylates Gln3 (Bertram et al.

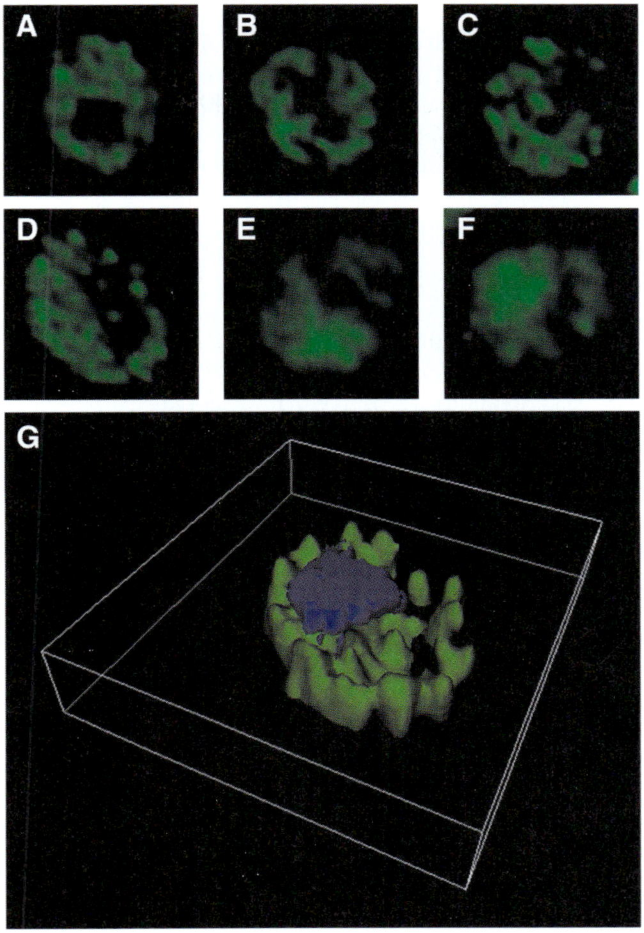

Fig. 4. Panels A-D. Non-uniform, cytoplasmic distribution of Gln3 in cells growing in minimal glutamine medium. Panels E and F. Intracellular Gln3 distribution one minute after cells were transferred from proline to glutamine medium. Here, much of the Gln3 is still situated within the nucleus Panel G. Three dimensional rendering of Gln3 localization in the cytoplasm of glutamine grown cells. Blue area is DAPI-positive material indicating the position of the nucleus. Taken from Cox et al. (2002).

2000): **(i)** rapamycin inhibition of Torl/2p decreases Gln3 (and in some cases Ure2) phosphorylation, brings about nuclear accumulation of Gln3p, and increases GATA-factor-mediated transcription (Hardwick et al. 1999; Bertram et al. 2000; Beck and Hall 1999). **(ii)** Mutations inactivating Tor kinase activity generate the same outcomes as rapamycin treatment (Bertram et al. 2000). **(iii)** Tor1, or a HEAT repeat motif-containing fragment of Tor1, interacts with Gln3, Gat1, and Ure2 in a 2-hybrid assay; Tor2 is reported to bind similarly (Bertram et al. 2000 "data not shown"). Additionally, Torl/2 associate with the plasma membrane and potentially with a vesicular fraction in the cell via the HEAT repeats (Kunz et al. 2000). This observation generates several questions: **(i)** Do Tor1/2 HEAT repeats interact both with cell membrane proteins, and Gln3? If so, simultaneously? Sequentially? **(ii)** Do Gln3 and Gat1 co-localize with Tor1 and Tor2 at the cell membrane? The answer to this last question is largely no (Cox and Cooper 2004). **(iii)** Do any of the membrane proteins, share areas of homology with Gln3 and Gat1? Therefore, present data are consistent with Tor1/2 phosphorylating Gln3, but as yet are far from compelling.

9.10.2 Gln3 dephosphorylation

Torl/2 have also been hypothesized to regulate Gln3 intracellular localization, and hence ability to function, through dephosphorylation. Two type 2A phosphatases have been suggested to dephosphorylate Gln3 in response to rapamycin-treatment, Sit4 and Pph3 (Beck and Hall 1999; Bertram et al. 2000). According to one model, Sit4 is proposed to be the primary phosphatase in Gln3 dephosphorylation. Torl/2 are proposed to phosphorylate Tap42, a protein kinase, when excess nutrients are available, which then complexes with Sit4, thereby inactivating it (Fig. 5, left side) (Beck and Hall 1999). With limiting nutrients or rapamycin addition, Torl/2 cease to function, Tap42 can no longer complex with Sit4, releasing it to dephosphorylate Gln3-P. Although, the model appears to accommodate existing data rather well, several unresolved issues remain: **(i)** Northern blots document that increased expression of *GAP1* and *MEP2* and repressed expression of ribosomal protein gene *RPS26* in response to rapamycin-treatment still occurs in a *tap42-11* mutant, which has lost its ability to respond to rapamycin inhibition of Torl/2 (Cardenas et al. 1999 "data not shown"). This would not be a result expected from the proposed model. However, observations similar to those seen in the Northern blots were not seen in genome-wide analyses (Cardenas et al. 1999 "data not shown"). **(ii)** Tap42 complexes with Pph21/22 to form an active phosphatase responsible for promoting protein synthesis (Fig. 5, right side) (Sutton et al. 1991), whereas with Sit4, complex formation with Tap42 is hypothesized to inactivate phosphatase activity thereby preventing it from dephosphorylating Gln3 (Fig. 5, left side) (Beck and Hall 1999). The latter hypothesis is not supported by the conclusion from genetic experiments that Tap42 positively regulates both Pph21/22 and Sit4 function (Di Como and Arndt 1996). **(iii)** Less than 5% of Sit4 is estimated to stably associate with Tap42 when cells are grown in rich medium. This, in the above model, is the condition under which Sit4 would be expected to be completely complexed and inactive. Otherwise, the free Sit4 would dephosphorylate Gln3-P, resulting in GATA-mediated gene expression in rich medium, which doesn't occur experimentally. Elsewhere rapamycin-induced *GAP1* expression is reported to be partially dependent on Pph3p, and in the *tap42-11* mutant to depend upon the rapamycin concentration used in the assay (Bertram et al. 2000 "data not shown"). Supporting the Sit4 model are genomic data demonstrating that Tap42 is required for rapamycin-induced nitrogen catabolic gene expression

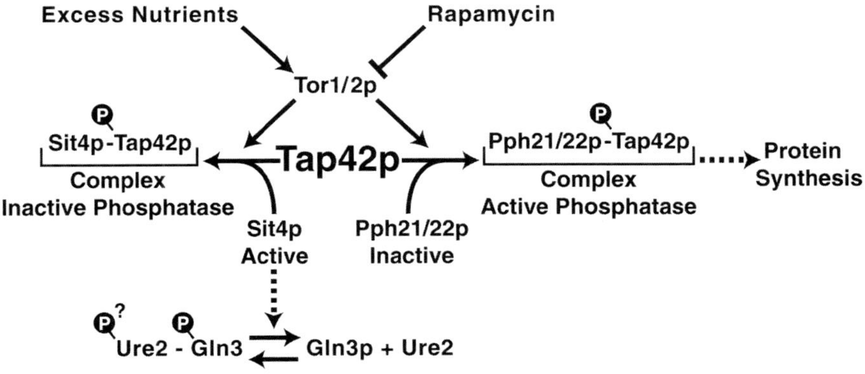

Fig. 5. Summary of Tap42 phosphorylation by Torl/2 and the molecules with which phosphorylated Tap42 interacts. Left side of the diagram describes the model of Beck and Hall (1999) and the right side is the model of Jiang and Broach (1999).

(Shamji et al. 2000). In sum, candidates for the phosphatase(s) regulating intracellular Gln3 localization have been identified, but much remains to be done before a convincing, comprehensive picture emerges.

9.11 Ure2 and Mksl participation in Tor1/2-mediated regulation

9.11.1 Ure2 participation in the Tor1/2 regulatory pathway

As already mentioned, Ure2 is the central negative regulator of NCR-sensitive, Gln3/Gatl-mediated transcription. Here, the focus will be mechanism oriented. Ure2 forms a complex with Gln3/Gat1, and directly or indirectly prevents nuclear entry of the transcriptional activators (Coshigano and Magasanik 199l; Beck and Hall 1999; Bertram et al. 2000; Shamji et al. 2000). It is unlikely, however, that Ure2 prevents nuclear entry by tethering Gln3/Gatl to a cytoplasmic structure, because Ure2 itself does not appear to be tethered. If a heterologous nuclear localization sequence is fused to its N-terminus along with LexA, LexA-Ure2 enters the nucleus, binds to a LexA binding site, Gln3 binds to DNA-bound Ure2, and the Gln3-Ure2 complex activates transcription (Kulkarni et al. 2001). Therefore, the Gln3 activation region remains functional in the nuclear Ure2-Gln3 complex.

The mechanism by which Ure2 prevents nuclear localization of Gln3/Gat1 remains controversial. Unresolved is identification of the direct causative event that prevents nuclear accumulation of Gln3. One model posits phosphorylation to be the direct determinant. Here, Ure2 facilitates Gln3 phosphorylation or stabilizes the phosphorylated form of Gln3 and it is this phosphorylated form that cannot be transported into the nucleus (Bertram et al. 2000). Alternatively, Gln3-Ure2 complex formation is the primary determinant. Here, phosphorylation facilitates Ure2-

Gln3 complex formation, and it is this complex that cannot be transported into the nucleus (Beck and Hall 1999).

There is evidence to support both models. Four observations support phosphorylation as the direct determinant: **(i)** Gst-Ure2 binds to both hyperphosphorylated and phosphatase-treated Gln3, or Gln3 derived from rapamycin-treated cells (Bertram et al. 2000). **(ii)** Tor1 interacts with Gln3 in the absence of Ure2, while conversely, Tor1 interacts with Ure2 only when Gln3 is present (Bertram et al. 2000). **(iii)** Free Gln3 is more resistant to *in vitro* phosphatase-treatment than Ure2-bound Gln3 (Bertram et al. 2000). **(iv)** Expression of multiple genes remains NCR-sensitive in a *ure2Δ* (Coffman et al. 1995, 1994, 1996). Consistent with complex formation being the direct determinant are: **(i)** *ure2* mutants possess a very strong phenotype for many genes, i.e., their expression becomes much less sensitive to NCR (Coffman et al. 1995, 1994, 1996). **(ii)** Gln3 localizes to the nucleus in a *ure2* mutant provided with nitrogen-rich medium in the absence of rapamycin and hence the signal for Gln3 phosphorylation. **(iii)** The amount of Gln3 isolated as a GST-Ure2-Gln3 complex decreases with time after treating cells with rapamycin (Beck and Hall 1999).

In neither case are present data compelling or do they eliminate the competing hypothesis, perhaps suggesting a third possibility, i.e., elements of both mechanisms may be physiologically relevant. Three observations are consistent with this possibility: **(i)** the expression of multiple genes remains NCR-sensitive in both *ure2Δgln3Δ* and *ure2Δgat1Δ* strains (Coffman et al. 1994, 1996). Therefore, Ure2 isn't absolutely required for NCR. It was this early observation with the *gln3Δure2Δ* that suggested components beyond Gln3 and Ure2 were involved in GATA-mediated, NCR-sensitive gene expression (Coffman et al. 1995, 1994, 1996). **(ii)** Ure2 effectively excludes a GFP-$Gln3_{1\text{-}487}$ truncation protein from the nucleus. This protein fragment lacks the region reported to bind Tor1 (residues 510-720) and hence the ability to be phosphorylated (Kulkarni et al. 2001). Therefore, more than phosphorylation is needed for nuclear exclusion. **(iii)** Overexpression of *URE2* results in nuclear exclusion of Gln3/Gat1 and loss of NCR-sensitive gene expression in the absence of a cellular signal (putatively phosphorylation) indicating nitrogen is in excess, i.e., it occurs with proline as nitrogen source (Cunningham et al. 2000; Cox et al. 2000; Kulkarni et al. 2001). Therefore, either Gln3 can be phosphorylated without the cellular signal to Tor1/2 and Gln3, or it is not required for nuclear exclusion of Gln3 in the presence of excess Ure2. It is conceivable that both mechanisms contribute to nuclear exclusion of Gln3 or the actual mechanism remains to be identified.

9.11.2 Mks1 and its negative regulation of Ure2

The most recently reported regulator of GATA-mediated gene expression is Mks1. *mks1* (**M**ulticopy compensator of A **K**inase **S**uppression) mutations were first isolated in an attempt to identify genes whose products function downstream of the RAS-cAMP pathway and protein kinase A (Matsuura and Anraku 1993). The *MKS1* locus was also identified as LYS80, the mutation of which caused overexpression of the *LYS* genes (Feller et al. 1997; Ramos and Wiame *1985;* Edskes et al. 1999). Physiological studies demonstrated that increased lysine production in *lys80* mutants derives from increased TCA cycle enzyme production, which increases intracellular α-ketoglutarate, and thereby lysine biosynthesis. From these observations, Mks1 was concluded to negatively regulate TCA cycle gene expression (Feller et al. 1997), an interpretation subsequently confirmed by others (Dilova et al. 2002; Sekito et al. 2002; Tate et al. 2002).

Participation of Mks1 in the regulation of Gln3/Gat1-mediated transcription derived from the observation that *MKS1* is a high-copy suppressor of the NCR-sensitivity of urcidosuccinate (USA) uptake mediated by the *DAL5* permease (Edskes et al. 1999; Edskes and Wickner 2000; Turoscy and Cooper 1987). Mksl is distal to Ure2 in the NCR regulon because *ure2* mutations are epistatic to *mks1* mutations (Edskes et al. 1999). This and other experiments led to the following regulatory pathway: $^{+}NH_4$ ⊣ Mksl ⊣ Ure2 ⊣ Gln3/Gatl → *DAL5* (Edskes et al. 1999; Edskes and Wickner 2000), i.e., Mksl is a negative regulator of Ure2 and hence positive regulator of Gln3/Gatl. This interpretation was confirmed by Schreiber's genome-wide analyses, which also led to the conclusion that Mksl is a positive regulator of rapamycin-induced, Gln3- and Rtgl/3-dependent gene expression (Shamji et al. 2000). In addition, these authors concluded Tap42 negatively regulates Mksl, Tap42 being positively regulated by Torl/2, i.e., Torl,2 → Tap42 ⊣ Mksl (Shamji et al. 2000).

As the Torl/2 pathway was studied further, data began to accumulate that raised questions about the role of Mksl in Torl/2 signal transduction (Cooper 2002). Among the most vexing observations were: **(i)** Mks1 is reported to be a negative regulator of Ure2 function, yet Ure2 dephosphorylation occurs normally in an *mks1* mutant (Shamji et al. 2000). **(ii)** Only a small peptide from the middle of the Mksl (M245 to T347) appears to be required for Mksl regulation of Ure2 (Edskes et al. 1999). **(iii)** Tap42 is phosphorylated by Torl/2 (Jiang and Broach 1999), and a *TAP42* mutant allele, *tap42-11,* which is unable to respond to rapamycin, was isolated. Processes downstream of Tap42 lose their rapamycin-responsiveness in *tap42-11.* For example, rapamycin-treatment results in dephosphorylation of Gln3 in wild type but not in a *tap42-11* mutant (Beck and Hall 1999). However, rapamycin-induced Ure2 dephosphorylation occurs normally in the *tap42-11* mutant (Shamji et al. 2000). It was incongruities such as these that prompted reinvestigation of the relationship between Mksl and Ure2 (Cooper 2002). Those studies demonstrated that Mksl is not a direct participant in the control of NCR-sensitive gene expression and has no direct influence on Ure2 function (Tate et al. 2002). However, understanding the observed indirect relationships whereby Mksl can influence NCR-sensitive gene expression must be deferred until control of retrograde gene expression has been described.

9.12 Retrograde gene expression and its control

A primary interface of carbon and nitrogen metabolism in *Saccharomyces cerevisiae* occurs at the early reactions of the tricarboxylic acid (TCA) cycle, and a group of enzymes encoded by the retrograde genes, *CIT2, ACO1, IDH1/2,* and *DLD3* (Fig. 6). These genes are highly expressed in ammonia-grown cells and minimal when glutamate is the nitrogen source (Liao et al. 1991; Liao and Butow 1993; Chelstowska and Butow *1995;* Velot et al. 1996; Zhegchang and Butow 1999). Although mitochondrial damage was associated with its discovery, this regulon has an important function in wild type cells. α-ketoglutarate, required for glutamate synthesis when ammonia is the nitrogen source, derives from the TCA cycle in cells that are growing respiratively (Fig. 6). During fermentation, however, the TCA cycle is inactive and cells must then rely on the retrograde enzymes to synthesize citrate and convert it to α-ketoglutarate when utilizing ammonia or compounds degraded to it. It is this supply and demand for α-ketoglutarate that links carbon and nitrogen regulation and complicated studies investigating the mechanisms through which these major metabolic pathways are controlled.

9.12.1 Small molecule to which retrograde grade gene expression responds

Central component(s) any metabolic regulon are the small molecule(s) signals to which it responds. Glutamate was the first candidate for this function and was reported to be a negative regulator of retrograde gene expression (Liao et al. 1991; Liao and Butow 1993; Chelstowska and Butow 1995; Velot et al. 1996; Zhegchang and Butow 1999). Next, it was the quality of the nitrogen source provided that was concluded to be the important determinant, with retrograde gene expression being minimal with preferred nitrogen sources (glutamate and glutamine) and highest with poor ones (urea) (Komeili et al. 2000). Consistent with this suggestion, only glutamate and glutamine of 17 amino acids tested could negatively-regulate *CIT2-lacZ* expression (Liu et al. 2001). Further, repression by these amino acids occurred only in wild type *SSY1* cells prompting the conclusion that Ssyl was the sensor responsible for transmitting a signal indicating the presence of these amino acids (Liu et al. 2001). These conclusions were questioned, however, by the finding that *CIT2* expression is undetectable with the poor nitrogen source, proline, but is clearly present with ammonia, a good nitrogen source (Tate et al. 2002). *CIT2* expression does not correlate with the quality of the nitrogen source, i.e., its ability to elicit NCR, but with the product to which it was degraded, i.e., ammonia or glutamate (Tate et al. 2002). The next proposal came from the demonstration that L-methionine sulfoximine (MSX), a glutamine synthetase inhibitor, causes glutamine depletion, nuclear localization of Rtg3, and high-level retrograde gene expression (Crespo et al. 2002). Here, glutamine was concluded to be the molecule to which both retrograde and NCR-sensitive expression responds (Crespo et al. 2002). Although these inhibitor experiments appear convincing, the

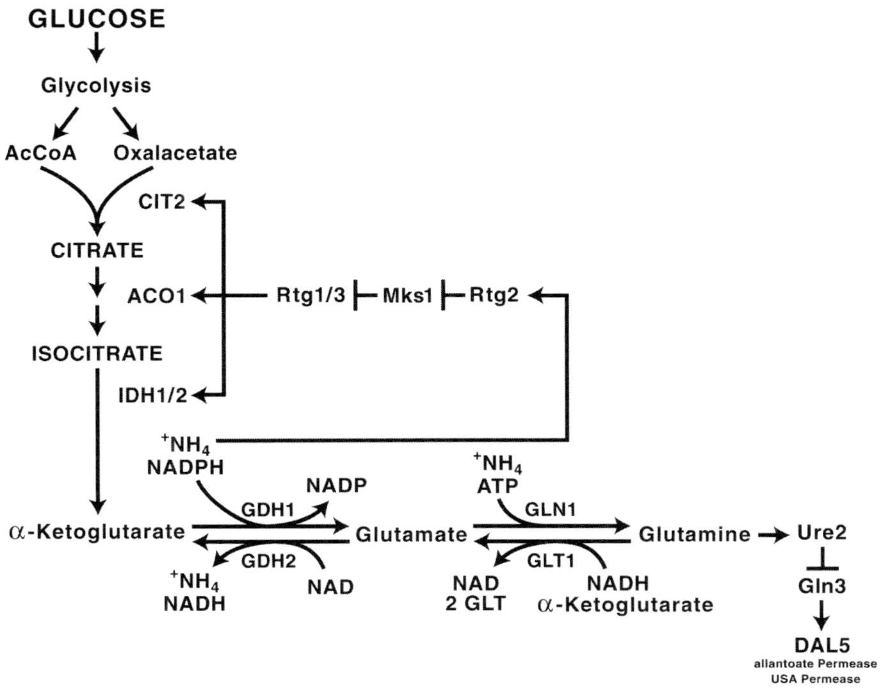

Fig. 6. Retrograde pathway for synthesis of a-ketoglutarate and its use in the assimilation of ammonia into glutamate and glutamine. Genes encoding the enzymes of this pathway are indicated adjacent to the arrows. Proteins (Ure2, Gln3, Rtgl-3, Mksl) associated with transcription control of the pathway are also indicated. Arrows and bars connecting these regulatory proteins indicate positive and negative control, respectively.

identity of the signal molecule is not settled. MSX is not glutamine synthetase-specific, the premise of these experiments. MSX is also a known substrate or inhibitor of γ–glutamylcysteine synthetase, L-amino acid oxidase, glutamine transaminase, and γ–cystathionase (Richman et al. 1973; Cooper et al. 1976; Griffith and Meister 1978). In fact, MSX is most accurately considered a glutamate analogue which likely inhibits, to varying degrees, most enzymatic reactions for which glutamate is a substrate. That more than the intended effects of MSX likely occurred in the above experiments is indicated by the observation that MSX-treatment decreased glutamine levels by 13 μmole/g of protein compared with a three-fold greater 42 μmole/g of protein increase in glutamate.

The most recent experiments investigating the small molecule to which retrograde transcription responds were predicated on a prediction of the MSX experiments, i.e., if glutamine regulates both retrograde and NCR-sensitive transcription, then reporter gene expression for the two regulons should respond in parallel when various compounds are provided as sole nitrogen source, taking into account that the two regulons may not be equally sensitive to glutamine. Unfortunately, the predicted correlation could not be demonstrated when NCR-sensitive and retrograde gene expression levels were compared in cells provided with various nitrogen sources (Tate and Cooper 2003). For example, NCR-sensitive transcription is low with serine or ammonia and high with proline as sole nitrogen source, whereas retrograde gene expression is just the reverse. With urea, both NCR-sensitive and retrograde gene expression are high. Therefore, NCR-sensitive and retrograde gene expression do not correlate across the range of nitrogen sources tested. An extensive correlation was found, however, between *CIT2* expression and intracellular ammonia levels both of which are lowest with glutamate and proline as nitrogen source and increase incrementally up to 14-fold depending upon the intracellular ammonia level generated by the various nitrogen sources provided (Tate and Cooper 2003). Further, addition of urea (degraded to ammonia) or the non-metabolized ammonia analogue, methylamine, to cells growing in proline medium, where *CIT2* expression is undetectable, results in appearance of *CIT2* expression without altering the intensity of NCR or proline uptake and utilization (Tate and Cooper 2003). α-ketoglutarate levels show an equally strong but inverse correlation with *CIT2* expression and ammonia, i.e., they are lowest with ammonia or urea and up to 29-fold higher with glutamate as sole nitrogen source. In contrast, with these large changes in ammonia and α-ketoglutarate levels, glutamate concentrations change no more than four-fold in the same experiments (Tate and Cooper 2003). Since both ammonia and α-ketoglutarate exhibit changes that are commensurate in magnitude with changes in *CIT2* expression, either might be considered a candidate to be the nutritional signal for retrograde transcription. Comparison of *CIT2* expression in wild type and *ure2Δ* strains, however, favors ammonia. *CIT2* expression dramatically increases in a *ure2Δ* relative to wild type, but only when ammonia is the nitrogen source, i.e., this increase does not occur with proline, glutamate, or glutamine. Consistent with this phenotype, *MEP* gene (encoding the ammonia permeases) expression increases in a *ure2Δ,* as does ammonia transport and intracellular ammonia levels, which increase seven-fold. α-ketoglutarate levels, on the other hand, are nearly identical in wild type vs. *ure2Δ* and glutamate levels change less than two-fold. The above observations argue that ammonia is a more likely candidate than glutamate or glutamine for the regulatory molecule to which retrograde gene expression responds (Fig. 6) (Tate and Cooper 2003). According to this model, the tricarboxylic acid (TCA) cycle isn't needed for energy metabolism in cells growing fermentatively and is shut down. However, a need for the TCA cycle intermediate, α-ketoglutarate, remains even during fermentation if ammonia or compounds degraded to it are the only available source of nitrogen. In this situation, the cell must produce sufficient α-ketoglutarate to support ammonia assimilation, which is achieved by ammonia-stimulated retrograde transcription.

9.12.2 Ammonia controls GDH2 expression beyond its role in NCR

On the other hand, when glutamate or compounds degraded to glutamate is provided as sole nitrogen source, cells must produce ammonia needed for the biosynthesis of glutamine and other nitrogenous compounds. Here, the objective is to produce sufficient ammonia to meet the cell's biosynthetic needs, but no more, since at high concentration, ammonia is toxic and could result in an ammonia-glutamate futile cycle. The critical point of regulation is NAD-glutamate dehydrogenase (encoded by *GDH2),* which catalyzes the main route of ammonia production when glutamate is provided as nitrogen source (Fig. 6). It has been previously accepted that NCR-dependent regulation of *GDH2* expression prevents excess ammonia production in this situation (22). However, a second more sensitive mode of regulation has been detected, which acts even before there is sufficient nitrogen flux to generate demonstrable NCR (Tate and Cooper 2003). This mode of regulation depends upon ammonia itself to signal down-regulation of *GDH2* expression. Supporting this conclusion is the observation that *GDH2* expression is totally inhibited when urea, a compound that doesn't elicit strong NCR, is provided as nitrogen source. Therefore, intracellular ammonia plays pivotal dual roles, regulating the interface of nitrogen and carbon metabolism at the level of both ammonia assimilation and production (Tate and Cooper 2003).

9.13 The Retrograde transcription regulatory elements, Rtg1/3 and Mks1

9.13.1 Rtg1 and Rtg3, a transcriptional activator

Retrograde gene expression is mediated by transcription factors Rtgl/3, and two factors whose biochemical functions are unknown, Rtg2 and Mksl (Liao et al. 1991; Liao and Butow 1993; Chelstowska and Butow 1995; Velot et al. 1996; Zhegchang and Butow 1999). The level of retrograde transcription correlates with intracellular localization of Rtgl/3; Rtgl/3p is nuclear when transcription is high and cytoplasmic when it is low (Sekito et al. 2000; Komeili et al. 2000). In addition, nuclear localization of Rtgl/3 and retrograde transcription are also induced by rapamycin (Komeilli et al. 2000), leading to the conclusion that Torl/2 is responsible for transmitting the nutritional signal to the more proximal retrograde regulatory elements (Komeilli et al. 2000).

Correlating with intracellular localization, Rtg3, but not Rtgl, phosphorylation levels change in response to rapamycin-treatment, but the nature of this phosphorylation remains somewhat controversial (Komeili et al. 2000). In one report, Rtg3 is phosphorylated both in rapamycin-treated and -untreated cells, but it is the hyper-phosphorylated species that correlates with nuclear localization of Rtgl/3 (Komeili et al. 2000). In contrast, Rtg3 phosphorylation decreases in ρ^o strains, lacking functional mitochondria and expressing the retrograde genes, compared with ρ^o cells in which expression is minimal (Sekito et al. 2000). Therefore, while retrograde gene expression is high in both rapamycm-treated cells and ammonia grown ρ^o mutants, Rtg3 phosphorylation levels elicited in these two cases appear to be different. These phosphorylation differences have been attributed to differing strains and media, but a detailed explanation of them has not yet appeared (Wedaman et al. 2003).

9.13.2 MKS1, a negative regulator

Mksl is a retrograde regulator that has been independently discovered to influence other metabolic regulons, one of which was a *lys80* mutant that overproduces the lysine genes. From the perspective of retrograde gene expression, the important characteristic of *lys80 (mksl)* mutants is their increased accumulation of ct-ketoglutarate and overproduction of TCA cycle enzymes (Feller et al. 1997). This led to the suggestion that Mksl was a negative regulator of TCA cycle enzyme production (Feller et al. 1997). However, it was the genomic analysis of Shamji et al. that first identified a relationship between Mksl and the retrograde genes. These investigators reported Mksl to be a positive regulator of retrograde and NCR-sensitive transcription (Shamji et al. 2000), the latter confirming an interpretation made earlier in studies of the prion precursor, Ure2, the negative regulator of NCR-sensitive transcription (Edskes et al. 1999). Three simultaneous and independent studies subsequently demonstrated Mks1 to be a strong negative rather than positive regulator of retrograde transcription, thus explaining overproduction of the early TCA cycle enzymes and accumulation of α-ketoglutarate in *lys80* mutants (Tate et al. 2002; Sekito et al. 2002; Dilova et al. 2002).

Recently, Mksl has been shown to be a phosphoprotein (Sekito et al 2002; Dilova et al. 2002). Under conditions where retrograde transcription is high, for example following rapamycin-treatment, Mksl migrates more rapidly than when transcription is low. Mksl phosphorylation levels do not however, seem to differ in wild type cells growing in SCD medium with and without added glutamine. The most striking changes in Mks1 phosphorylation levels occur in an *rtg2Δ*. Here, Mks1 migrates more slowly and as multiple distinct species (Sekito et al 2002; Dilova et al. 2002). Rapamycin addition, markedly increases the rate of Mks1 migration in the *rtg2Δ* supporting the suggestion that phosphorylation levels have decreased here as well (Sekito et al 2002; Dilova et al. 2002). Finally, deletion of *MKS1* results in a decreased phosphorylation of Rtg3 (Sekito et al. 2002).

9.13.3 Rtg2, a positive regulator

Rtg2, a positive regulator of overall retrograde gene expression, is the least well studied of the retrograde regulatory proteins. Its position in the pathway derives from the demonstration that *mks1* mutations are epistatic to *rtg2* mutations (Pierce et al. 2001; Sekito et al. 2002; Tate and Cooper 2003), and the fact that rapamycin-induced *CIT2* expression does not occur in *rtg2* mutants (Dilova et al. 2002). The latter observation prompted the conclusion that Rtg2 acts between Torl/2 and

Mksl in the order Torl/2 —| Rtg2 —| Mks1 —| Rtgl/3 (Pierce et al. 2001; Dilova et al. 2002; Sekito et al. 2002). Consistent with Mks1 directly following Rtg2 in the regulatory pathway is the demonstration that Rtg2 and Mks1 can form a complex (Sekito et al. 2002). However, the position and nature of the regulatory contribution of Rtg2 to retrograde gene control has recently been questioned by experiments with *lst8* mutants (Chen and Kaiser 2003). Lst8, a WD-repeat protein, was first related to the retrograde regulon when *lst8* mutants were shown to express *CIT2* even with glutamate as nitrogen source (Lin et al. 2001). This observation led to the proposition that Lst8 was a negative regulator of Rtgl/3-mediated gene expression situated between Rtg2 and Rtgl/3 in the order Rtg2 —| Lst8 —| Rtgl/3 (Liu et al. 2001); this model was proposed before Mksl was known to be a negative regulator of Rtgl/3-mediated transcription. The pertinent question is whether Lst8 enters the regulatory pathway above or below Mks1. Some alleles of *lst8* restore *CIT2* expression to an *rtg2Δ*, leading to the suggestion that Lst8 enters the retrograde control pathway below Mks1, acting more directly as a negative regulator of Rtgl/3 itself (Chen and Kaiser 2003). Lst8 is similarly hypothesized to be a negative regulator of Gln3 (Chen and Kaiser 2003). Lst8 has been found in endosomal/Golgi compartments of the cytoplasm and to be in complexes with Torl/Tor2 and several other proteins (Wedaman et al. 2003; Loewith et al. 2002; Chen and Kaiser 2003).

9.14 Tor1/2 control of retrograde gene expression

9.14.1 Tor1/2 regulation is an indirect consequence of its effects on nitrogen metabolism

Torl/2 regulation of the retrograde regulon is predicated on the observation that retrograde transcription markedly increases when cells are treated with rapamycin (Komeili et al. 2000). Moreover, rapamycin-treatment results in nuclear localization of Rtgl/3, and changes in the phosphorylation levels of Rtg3 and Mksl (Dilova et al. 2002; Sekito et al. 2002). In fact, data on phosphorylation profiles and nuclear localization of the retograde transcription activator, Rtg3, closely parallel those reported for the NCR-sensitive activator, Gln3.

On the other hand, several observations now seriously question whether Torl/2 is a direct or indirect participant in the regulation of retrograde transcription: **(i)** in wild type Σ1278b-derived cells, retrograde gene expression responds to rapamycin-treatment in a nitrogen-dependent manner, i.e., rapamycin induces high level *CIT2* expression with ammonia, but not proline or glutamate, as nitrogen source (Tate and Cooper 2003). In all models of Torl/2 control, Torl/2 is the recipient of the nutritional signal and therefore situated below it in the regulatory pathway. Rapamycin, being a direct inhibitor of Torl/2, leads to the expectation that rapamycin should induce retrograde transcription irrespective of the nitrogen source provided. **(ii)** Although rapamycin is unable to induce retrograde gene expression in wild type glutamate-grown cells, induction does occur in a glutamate-grown *gln3ΔgatlΔ* mutant. **(iii)** Deletion of *URE2* strongly increases retrograde gene expression with ammonia, but not proline, glutamate, or glutamine as nitrogen source. Again the response of retrograde transcription is nitrogen source-dependent even though the regulon has no requirement for the NCR-sensitive transcription factors (Tate and Cooper 2003).

The common denominator of these observations, and likely key to explaining them, is ammonia. **(i)** Both rapamycin-treatment and *ure2Δ* increase retrograde gene expression only with ammonia as nitrogen source (Tate and Cooper 2003). **(ii)** Ammonia is also reported to be the nutrient to which the retrograde regulon responds. **(iii)** Both rapamycin and *ure2Δ* increase NCR-sensitive transcription. Since expression of the ammonia permease *(MEP)* genes is NCR-sensitive, it is reasonable to argue that both perturbants increase *MEP* expression, ammonia permease activity, and intracellular ammonia levels. Measurements of all three parameters support this interpretation. From these experiments it was concluded that Torl/2 participation in the regulation of retrograde transcription is an indirect consequence of alterations that occur in nitrogen metabolism (Tate and Cooper 2003).

9.14.2 Strain variation is an important variable in the interpretation of retrograde expression data

One of the serious challenges to interpreting genome-wide or traditional transcription data from one laboratory to another are the effects of strain and culture condition differences, especially if they are unknown and hence cannot be taken into account. It is, for example, well known that ammonia is not nearly as repressive a nitrogen source for NCR-sensitive transcription in S288C-derived strains compared with those in a Σl278b genetic background (Bossinger and Cooper 1976). Knowledge of this strain difference markedly alters expectations and interpretation of data when using ammonia as a nitrogen source. Similar strain differences have complicated studies of retrograde gene expression as well. For example, *CIT2* expression does not occur in proline-grown Σ1278b strains, but does in those used by Wickner (Tate and Cooper 2003). Strain differences are also reported to account for Shamji's conclusion that Mksl is a positive regulator of retrograde transcription (Shamji et al. 2000), whereas others found it to a negative regulator (Tate et al. 2002; Sekito et al. 2002; Dilova et al. 2002). Given such variations, there will always be some risk involved in extrapolating data from one strain to another unless there is an independent way of detecting these variations and appropriately accounting for them.

9.14.3 The connection between Ure2 and Mks1

An understanding of the connection between Mks1 and Ure2 requires appreciation of the genetic selection used for the initial isolation of *ure2* mutants. Ure2 mutants derived unexpectedly from an investigation of the pyrimidine biosynthetic genes. Ureidosuccinate (USA), the first unique intermediate in uracil biosynthesis can fulfill the auxotrophy generated by *ura2* mutations *(URA2* encodes aspartate transcarbamylase, the first enzyme in uracil biosynthesis as well as uracil (Lacroute 1986; Drillien and Lacroute 1972; Drillen et al. 1973). However, USA successfully substitutes for uracil only when proline, but not ammonia, is provided as the sole nitrogen source. Although unknown until later, the ability of USA to substitute for uracil was NCR-sensitive. To investigate this somewhat surprising finding, mutants, in which USA could fulfill the auxotrophy of *ura2* mutations with ammonia as nitrogen source, were selected. These mutants, designated *ure2* (**URE**idosuccinate), acquired the ability to fulfill the *ura2* auxotrophy and were shown to do so by permitting USA transport with ammonia as nitrogen source. In other words, the *ure2* mutation abrogates "ammonia repression" (ammonia repression was replaced by the term NCR when the phenomenon was found to occur with other nitrogen sources as well). Moreover, *ure2* mutations were highly pleiotropic in that NCR was abrogated for multiple nitrogen catabolic genes (Hoffman-Bang 1999; ter Schure et al. 2000; Cooper 1996; Grenson et al. 1974). The physiological connection between uracil biosynthesis and NCR-sensitivity became clear when *UREP* (encoding USA permease) was found to be identical to *DAL5,* encoding the catabolic NCR-sensitive allantoate permease (Turoscy and Cooper 1987).

Identification of Mks1 as a negative rather than positive regulator of retrograde gene expression was key to explaining the observation which led to the conclusion that Mksl negatively regulates Ure2 (Edskes et al. 1999), i.e., over expression of *MKS1* suppresses the NCR-sensitivity of USA uptake in ammonia-grown cells (Tate et al. 2002). The primary observation can be explained in two ways: **(i)** Mkslp negatively regulates Ure2p function, or **(ii)** Mkslp diminishes the ability of ammonia to elicit NCR (Fig. 6). The latter best explains the published data. Since Mks1 is a strong negative-regulator of retrograde gene expression, production of the retrograde gene products, citrate synthase, aconitase, and isocitrate dehydrogenase, increases when *MKS1* is deleted. In contrast, Mksl over-production excessively down-regulates production of the same proteins, which are required to synthesize α-ketoglutarate (Fig. 6). The resulting limitation of α-ketoglutarate decreases the rate of ammonia assimilation to glutamate thereby decreasing its ability to elicit NCR. This loss permits sufficient *DAL5/UREP* expression to meet the requirements of the selection scheme (ureidosuccinate permease function) that identified a high copy plasmid carrying *MKS1* as a suppressor of NCR-sensitivity (Edskes et al. 1999). This reasoning also explains the recent report that *DAL5-lacZ* expression increases in an *rtg2Δ* (Pierce et al. 2001). Since Rtg2p is required for nuclear localization of the retrograde transcriptional activators, Rtgl/3p, its loss diminishes retrograde gene expression as occurs with over expression of *MKS1.*

9.15 Nuclear localization of Gln3 during glucose-starvation

Control of retrograde gene expression and the influence of Mks1 on NCR-sensitive transcription demonstrate that carbon and nitrogen metabolism are highly integrated in *S. cerevisiae.* Understanding this integration facilitated elucidation of the mechanism responsible for carbon starvation-induced NCR-sensitive gene expression. The critical and potentially exciting observation was that carbon starvation, like nitrogen starvation, resulted in nuclear localization of Gln3 and increased transcription of NCR-sensitive *GAP1, GDH2* and *PUT1* (Bertram et al. 2002). Carbon starvation-triggered nuclear localization of Gln3 was Snf1-dependent and correlated with hyper-phosphorylation of Gln3, the latter being just opposite of what occurs during nitrogen starvation, where phosphorylation levels decrease (Bertram et al. 2002). These observations implied that: **(i)** Gln3 localizes to the nucleus both during times of nitrogen limitation and excess, the latter occurring during carbon-starvation. **(ii)** Both phosphorylated and dephosphorylated forms of Gln3 are transported into the nucleus.

These paradoxical implications were resolved by experiments demonstrating that whether or not carbon starvation triggers nuclear localization of Gln3 does not derive from carbon starvation, but the nitrogen source being utilized during starvation (Cox et al. 2002). Carbon-starvation of cells growing in ammonia- but not glutamine-medium causes nuclear localization of Gln3 (Cox et al. 2002). These observations are straightforwardly explained by the fact that ammonia is assimilated into glutamate and glutamine, thereby repressing NCR-sensitive transcription, only if α-ketoglutarate is available to provide the carbon skeletons required for amino acid synthesis. Following the onset of glucose starvation in ammonia medium, carbon-containing compounds (including α-ketoglutarate) become increasingly limiting, thereby slowing ammonia assimilation and eventually starving the cells for nitrogen. This, in turn, relieves NCR and triggers nuclear localization of Gln3 (Cox et al. 2002). In contrast, when glutamine is provided as nitrogen source, the requirement for α-ketoglutarate is largely eliminated and the fact that it becomes limiting following the onset of carbon starvation no longer diminishes NCR.

9.16 Rapamycin and Tor protein regulation of transporter protein stability

Tor1/2 influence not only NCR-sensitive transcription, but also nitrogen-regulated membrane protein (Mep2, Gap1, and Tat2) trafficking and stability (Fig. 7). Nitrogen catabolic enzyme activities disappear from wild type yeast cells with a half-life that equals the cell's doubling time, following transfer from a poor to a good nitrogen source, i.e., they are lost by simple dilution rather than active proteolysis (Cooper 1980). NCR-sensitive permease activities required for transport of poor

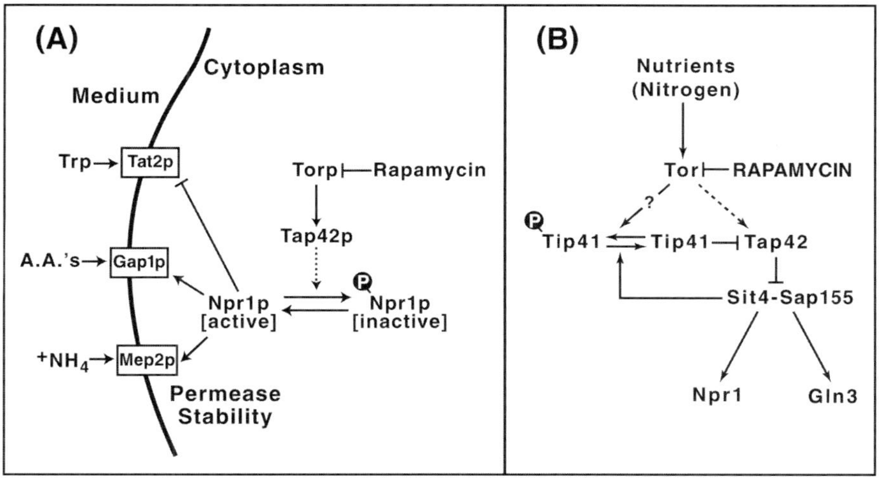

Fig. 7. Panel A. Regulation of nitrogen permease (Tat2, Gapl, and Mep2) stability by Torp, Tap42, and Nprl. Panel B. Model describing the interactions of Tor signal transduction pathway components, Tap42, Tip4l, Sit4, and Sapl55. Model taken from Jacinto et al. (2001).

nitrogen sources, on the other hand, disappear much more rapidly, 3-5 minutes for allantoin transport, following transfer to rich medium (Cooper 1980). Catabolic Gapl (general amino acid permease) responds similarly. Grenson and colleagues identified two proteins that participate in stabilizing and degrading the NCR-sensitive Gapl: Nprl, a protein kinase-related protein and Npil/Rsp5, an ubiquitin-protein ligase (E3 enzyme) (Grenson 1983). When grown with a poor nitrogen source, Gapl activity is high in wild type and low in an *nprl* mutant (Grenson 1983; Vandenbol et at. 1987, 1990; Stanbrough and Magasanik *1995;* Springael et al. 19996). In contrast, when cells are grown with a good nitrogen source, Gapl activity is low in wild type and high in an *npil* mutant (Grenson 1983; Vandenbol et al. 1987, 1990; Stanbrough and Magasanik 1995; Springael et al. 1999). Finally, Gapl activity is high an *nprlnpil* double mutant irrespective of the nitrogen source. Since *npil* mutations are epistatic to those in *nprl,* Nprl was concluded to protect Gapl from Npil-mediated inactivation (Grenson 1983; Vandenbol et al. 1987, 1990; Stanbrough and Magasanik 1995; Springael et al. 1999). Gapl is a phosphoprotein, the phosphorylated form of which correlates with high transport activity. In an *nprl* mutant, Gapl is dephosphorylated, ubiquinated, transported to the vacuole and degraded (Stanbrough and Magasanik 1995; Springael et al. 1999; Soetens et al. 2001; De Craeno et al. 2001). Newly synthesized Gapl is also directly sorted to the vacuole in an *npr1Δ* (Soetens et al. 2001; De Craeno et al. 2001). Mep2, the NCR-sensitive ammonia permease, appears to be similarly regulated (Fig. 7A) (Lorenz and Heitman 1998).

The response of Tat2, a tryptophan permease, whose cognate gene is not NCR-sensitive, to *npr1Δ* is just the opposite of Gapl (Fig. 7A). Overproduction of Nprl, rather than its loss, results in disappearance of Tat2 activity, as does treating cells with rapamycm (Schmidt et al. 1998), establishing a link between Tor, Nprl and Tat2 stability. The link was extended by the findings that Nprl is: **(i)** itself a phosphoprotein, and **(ii)** phosphorylated with ammonia as nitrogen source and dephosphorylated by rapamycin treatment or with proline as nitrogen source (Schmidt et al. 1998). Finally, Tap42, as noted above, is a regulatory protein that is phosphorylated by Torl/2 and is necessary for the maintenance of rapamycin-induced Nprl phosphorylation and Tat2 stability, i.e., neither occurs in a *tap42-11* mutant following treatment with rapamycin (Schmidt et al. 1998).

Two new members of the TOR signal transduction pathway have been identified that influence Nprl phosphorylation, Tip4l, identified in a two-hybrid screen of proteins interacting with Tap42, and Sapl55, a Sit4-associated protein (Jacinto et al. 2001). Investigation Tap42, Tip4l and Sapl55 yielded the following observations (Fig. 7B) (Jacinto et al. 2001): **(i)** Nprl is dephosphorylated, in a Sit4-dependent manner, when cells are treated with rapamycin. **(ii)** Nprl is partially dephosphorylated when *tip4l* or *sap155* mutants are treated with rapamycin. No dephosphorylation occurs, however, in a *tip4lsap155* double mutant. **(iii)** Tip4l is required for rapamycin-induced Tap42-Sit4 dissociation. **(iv)** Tip4l is required for rapamycin-induced nuclear localization of Gln3. **(v)** Tip4l is a phosphoprotein, which is dephosphorylated in wild type cells, and partially dephosphorylated in *sit4* cells following rapamycin treatment. **(vi)** Rapamycin-treatment increases the amount of Tip4l in a Gst-Tap42 immunoprecipitate in a Sit4-dependent manner. According to current explanations of these observations, Torl/2 directly or indirectly, phosphorylates Tip4l (Fig. 7B) (Jacinto et al. 2001). For the model to be tenable, the authors must argue that either Tor does not phosphorylate Tap42, as previously demonstrated *in vivo* and *in vitro* (Jiang and Broach 1999), or conclude that such phosphorylation is of only minor importance (Jacinto et al. 2001). Tip4l is proposed to bind to and inhibit Tap42. Tip4l-Tap42 binding is stimulated by rapamycin-induced, Sit4-dependent dephosphorylation of Tip4l. Finally, it is Tip4l-mediated inhibition of Tap42 that accounts for the Tip4l requirement for events downstream of Sit4, i.e., phosphorylation of Nprl and nuclear localization of Gln3.

The model generates two interesting questions: **(i)** since Tap42 has been shown to bind to Sit4 (62), are Tip4l and Sit4 competitive inhibitors of one another's binding to Tap42? **(ii)** Since evidence demonstrating Tor-dependent Tap42 phosphorylation (Jiang and Broach 1999), is as strong as the evidence presented for Tip4l and Nprl phosphorylation, how does Tap42 phosphorylation/dephosphorylation affect the Tip4l-Tap42 interaction?

The most striking data concerning Tip41, Tap42, and their regulation of Npr1 is the effect of mutation and growth condition on the quantity of Nprl. Nprl protein levels are enormously higher in *tip4l*, *sap155,* and *tip4lsap155* double mutants than in wild type or the *sit4* mutant. This occurs even though the loss of Nprl dephosphorylation is far more complete in the *sit4* mutant than in either a *tip4l* or *sap 155* mutant. Moreover, a very robust Nprl protein level is seen in wild type ammonia-grown cells, while in proline-grown cells Nprl is barely if at all detectable. The significance of this observation is that it is in wild type proline-grown cells that active, dephosphorylated Nprl functions to stabilize Gapl (Jacinto et al.

2001). Why do Nprl levels decrease so severely at just the time its kinase activity is most required? It could be argued that very little kinase protein is needed. However, from that vantage point, why does Nprl levels increase so much in ammonia grown cells where Nprl is hypothesized to be inactive? What is the function of so much phosphorylated, putatively inactive protein?

The Nprl model outlined above has recently been questioned by a report that rapamycin, and hence the Tor pathway, does not directly affect Gap1 sorting to the vacuole (Chen and Kaiser 2002). The signal directing Gap1 to sort into the vacuole is argued to be amino acids themselves. Here, as observed throughout this chapter, knowledge of the regulatory mechanisms potentially regulating nitrogen uptake and utilization has dramatically increased, but much remains to be done before the mechanisms will be fully understood. The preeminent challenge is no longer acquiring sufficient experimental data to accomplish the task of investigating regulatory networks, but that of effectively distinguishing the important from the trivial, direct relationships that elucidate the molecular mechanisms underlying regulation of a network from indirect relationships that derive from the subtle interplay between systems but in themselves do not identify the primary regulatory mechanisms.

References

Barbet NC, Schneider U, Helliwell SB, Stansfield I, Tuite MF, Hall MN (1996) TOR controls translation initiation and early Gl progression in yeast. Mol Biol Cell 7:25-42

Beck T, Hall MN (1999) The TOR signaling pathway controls nuclear localization of nutrient-regulated transcription factors. Nature 402:689-692

Benjamin PM, Wu JI, Mitchell AP, Magasanik B (1989) Three regulatory systems control expression of glutamine synthetase in *Saccharomyces cerevisiae* at the level of transcription. Mol Gen Genet 217:370-377

Bertram PG, Choi JH, Carvalho J, Chan TF, Ai W, Zheng XF (2002) Convergence of TOR-nitrogen and Snf1-glucose signaling pathways onto Gln3. Mol Cell Biol 22:1246-1252

Bertram PG, Choi JH, Carvalho J, Ai W, Zeng C, Chan T-F, Zheng XFS (2000) Tripartite regulation of Gln3 by TOR, Ure2, and phosphatases. J Biol Chem 275:35727-35733

Blinder D, Coschigano PW, Magasanik B (1996) Interaction of the GATA factor Gln3 with the nitrogen regulator Ure2 in *Saccharomyces cerevisiae.* J Bacteriol 178:4734-4736

Bossinger J, Cooper TG (1976) Sequence of molecular events involved in induction of allophanate hydrolase. J Bacteriol 126:198-204

Bysani N, Daugherty JR, Cooper TG (1991) Saturation mutagenesis of the UAS_{NTR} (GATAA) responsible for nitrogen catabolite repression-sensitive transcriptional activation of the allantoin pathway genes in *Saccharomyces cerevisiae*. J Bacteriol 173:4977-4982

Cardenas ME, Heitman J (1995) FKBP12-rapamycin target TOR2 is a vacuolar protein with an associated phosphatidylinositol-4 kinase activity. EMBO J 14:5892-5907

Cardenas ME, Cutler NS, Lorenz MC, Di Como CJ, Heitman J (1999) The TOR signaling cascade regulates gene expression in response to nutrients. Genes Dev 13:3271-3279

Carvaiho J, Zheng SXE (2003) Domains of Gln3p interacting with karyopherins, ure2p and the target of rapamycin protein (TOR). J Cell Biol (in press)

Carvalho J, Bertram PG, Wente SR, Zheng XFS (2001) Phosphorylation regulates the interaction between Gln3 and the nuclear import factor Srplp. J Biol Chem 276:25359-25365

Chelstowska A, Butow RA (1995) RTG genes in yeast that function in communication between mitochondria and the nucleus are also required for expression of genes encoding peroxisomal proteins. J Biol Chem 270:18141-18146

Chisholm G, Cooper TG (1982) Isolation and characterization of mutants that produce the allantoin-degrading enzymes constitutively in *Saccharomyces cerevisiae*. Mol Cell Biol 2:1088-1095

Coffman JA, Cooper TG (1997) Nitrogen GATA-factors participate in transcriptional regulation of vacuolar protease genes in *Saceharomyces cerevisiae.* J Bacteriol 179:5609-5613

Coffman JA, El Berry HM, Cooper TG (1994) *URE2* protein regulates nitrogen catabolic gene expression through the GATAA-containing UAS_{NTR} element in *Saccharomyces cerevisiae*. J Bacteriol 176:7476

Coffman JA, Rai R, Cooper TG (1995) Genetic evidence for Gln3-independent, nitrogen catabolite repression-sensitive gene expression in *Saceharomyces cerevisiae.* J Bacteriol 177:6910-6918

Coffman JA, Rai R, Cunningham T, Svetlov V, Cooper TG (1996) Gati, a GATA family protein whose production is sensitive to nitrogen catabolite repression, participates in transcriptional activation of nitrogen-catabolic genes in *Saccharomyces cerevisiae*. Mol Cell Biol 16:847-858

Coffman JA, Rai R, Loprete D, Cunningham T, Svetlov V, Cooper IG (1997) Cross regulation of four GATA factors that control nitrogen catabolic gene expression in *Saccharomyces cerevisiae*. J Bacteriol 179:3416-3429

Cooper AJ, Stephani RA, Meister A (1976) Enzymatic reactions of methionine sulfoximine. Conversion to the corresponding alpha-imino and alpha-keto acids and to alpha-ketobutyrate and methane sulfinimide. J Biol Chem 251:6674-6682

Cooper TG (1980) Selective gene expression and intracellular compartmentation: two means of regulating nitrogen metabolism in yeast. Trends Biochem Sci 5:332-334

Cooper TG (1996) Allantion degradative system - an integrated transcriptional response to multiple signals, in Mycota III, G Marzluf and R Bambrl, eds. Springer Verlag, Berlin, Heidelberg, pp 139-169

Cooper TG (2002) Transmitting the signal of excess nitrogen in *Saccharomyces cerevisiae* from the Tor proteins to the GATA factors: connecting the dots. FEMS Microbiol Rev 26:223-238

Cooper TG, Chisholm GE, Genbauffe FS (1983) Genetic control of gene expression in *S. cerevisiae* In Gene Expression Alan R. Liss, Inc New York, NY pp. 145-157

Cooper TG, Ferguson D, Rai R, Bysani N (1990) The *GLN3* gene product is required for transcriptional activation of allantoin system gene expression in *Saccharomyces cerevisiae*. J Bacteriol 172:1014-1018

Cooper TG, Rai R, Yoo HS (1989) Requirement of upstream activation sequences for nitrogen catabolite repression of the allantoin system genes in *Saccharomyces cerevisiae*. Mol Cell Biol 9:5440-5444

Corbett AH, Silver PA (1997) Nucleocytoplasmic transport of macromolecules. Microbiol Mol Biol Rev 61:193-211

Coshigano PW, Magasanik B (1991) The *URE2* gene product of *Saccharomyces cerevisiae* plays an important role in the cellular response to the nitrogen source and has homology to glutathione s-transferases. Mol Cell Biol 11:822-832

Courchesne WE, Magasanik B (1988) Regulation of nitrogen assimilation in *Saccharomyces cerevisiae:* roles of the *URE2* and *GLN3* genes. J Bacteriol 170:708-713

Cox KH, Pinchak AB, Cooper TG (1999) Genome-wide transcriptional analysis in *S. cerevisiae* by Mini-array membrane hybridization. Yeast 15:703-713

Cox KH, Rai R, Distler M, Daugherty JR, Coffman JA, Cooper TG (2000) GATA sequences function as TATA elements during nitrogen catabolite repression and when Gln3 is excluded from the nucleus by overproduction of Ure2. J Biol Chem 275:17611-1768

Cox KH, Tate JJ, Cooper TG (2002) Cytoplasmic compartmentation of Gln3 during nitrogen catabolite repression and the mechanism of its nuclear localization during carbon starvation in *Saccharomyces cerevisiae*. J Biol Chem 277:37559-37566

Cox KH, Kulkami A, Tate JJ, Cooper TG (2004) Submitted

Crespo JL, Hall MN (2002) Elucidating TOR signaling and rapamycin action: lessons from *Saccharomyces cerevisiae*. Microbiol Mol Biol Rev 66:579-591

Crespo JL, Powers T, Fowler B, Hall MN (2002) The TOR-controlled transcription activators GLN3, RTG1, and RTG3 are regulated in response to intracellular levels of glutamine. Proc Natl Acad Sci USA 99:6784-6789

Cunningham TS, Cooper TG (1993) The *Saccharomyces cerevisiae* DAL80 repressor protein binds to multiple copies of GATAA-containing sequences. J Bacteriol 175:5851-5861

Cunningham TS, Andhare R, Cooper TG (2000) Nitrogen catabolite repression of *DAL80* expression depends on the relative levels of Gat1 and Ure2 production in *Saccharomyces cerevisiae*. J Biol Chem 275:14408-14414

Cunningham TS, Dorrington RA, Cooper TG (1994) The *UGA4* UAS_{NTR} site required for GLN3-dependent transcriptional activation also mediates DAL80-responsive regulation and DAL80 protein binding in *Saccharomyces cerevisiae*. J Bacteriol 176:4718-4725

Cunningham TS, Rai R, Cooper TG (2000) The level of *DAL80* expression down-regulates GATA factor-mediated transcription in *Saccharomyces cerevisiae*. J Bacteriol 182:6584-6591

Cunningham TS, Svetlov VV, Rai R, Smart W, Cooper TG (1996) Gln3 is capable of binding to UAS_{NTR} elements and activating transcription in *Saccharomyces cerevisiae*. J Bacteriol 178:3470-3479

Cyert MS (2001) Regulation of nuclear localization during signaling. J Biol Chem 276:20805-20808

Daugherty JR, Rai R, El Berry HM, Cooper TG (1993) Regulatory circuit for responses of nitrogen catabolic gene expression to the Gln3 and Dal80 proteins and nitrogen catabolite repression in *Saccharomyces cerevisiae*. J Bacteriol 175:64-73

De Craene JO, Soetens O, Andre B (2001) The Nprl kinase controls biosynthetic and endocytic sorting of the yeast Gap1 permease. J Biol Chem 276:43939-43948

Dennis PB, Fumagalli S, Thomas G (1999) Target of rapamycin (TOR): balancing the opposing forces of protein synthesis and degradation. Curr Opin Genet Dev 9:49-54

Di Como CJ, Amdt KT (1996) Nutrients, via the Tor proteins, stimulated the association of Tap42 with type 2A phosphatases. Genes Dev 10:1904-1916

Dilova I, Chen CY, Powers T (2002) Mksl in concert with TOR signaling negatively regulates RTG target gene expression in S. *cerevisiae*. Curr Opin Biol 12:389-395

Drillien R, Lacroute F (1972) Ureidosuccinic acid uptake in yeast and some aspects of its regulation. J Bacteriol 109:203-208

Drillen RM, Aigle M, Lacroute F (1973) Yeast mutants pleiotropically impaired in the regulation of two glutamate dehydrogenases. Biochem Biophys Res Commun 53:367-372

Edskes HK, Wickner RB (2000) Protein required for prion generation: (URE3) induction requires the Ras-regulated Mksl protein. Proc Natl Acad Sci USA 97:6625-6629

Edskes HK, Hanover JA, Wickner RB (1999) Mkslp is a regulator of nitrogen catabolism upstream of Ure2 in *Saccharomyces cerevisiae*. Genetics 153:585-594

Feller A, Ramos F, Pierard A, Dujbois E (1997) Lys80p of *Saccharomyces cerevisiae,* previously proposed as a specific repressor of *LYS* genes, is a pleiotropic regulatory factor identical to Mksl. Yeast 13:1337-1346

Gorlich D, Kutay U (1999) Transport between the cell nucleus and the cytoplasm. Ann Rev Cell Dev Biol 15:607-660

Grenson M (1983) Study of the positive control of the general amino-acid permease and other ammonia-sensitive uptake systems by the product of the *NPR1* gene in the yeast *Saccharomyces cerevisiae*. Eur J Biochem 133:141-144

Grenson M, Dubois E, Piotrowska M (1974) Ammonia assimilation in *Saccharomyces cerevisiae* as mediated by the two glutamate dehydrogenases. Mol Gen Genet 128:73-85

Griffith OW, Meister A (1978) Differential inhibition of glutamine and gammaglutamylcysteine synthetases by aipha-alkyl analogs of methionine sulfoximine that induce convulsions. J Biol Chem 1978 253:2333-2338

Hardwick JS, Kuruvilla FG, Tong JF, Shamji AF, Schreiber SL (1999) Rapamycm-modulated transcription defines the subset of nutrient-sensitive signaling pathways directly controlled by the Tor proteins. Proc Natl Acad Sci USA 96:14866-14870

Heitman J, Movva NR, Hall MN (1991) Targets for cell cycle arrest by the immunosuppressant rapamycin in yeast. Science 253:905-909

Hoffman-Bang J (1999) Nitrogen catabolite repression in *Saceharomyces cerevisiae.* Mol Biotechnol 12:35-73

Ito T, Chiba T, Ozawa R, Yoshida M, Hottori M, Sakaki Y (2001) A comprehensive two-hybrid analysis to explore the yeast protein interactome. Proc Natl Acad Sci USA 98:4277-4278

Jiang Y, Broach JR (1999) Tor proteins and protein phosphatase 2A reciprocally regulate Tap42 in controlling cell growth in yeast. EMBO J 18:2782-2792

Jacinto E, Guo B, Arndt KT, Schmelzle T, Hall MN (2001) TIP41 interacts with TAP42 and negatively regulates the TOR signaling pathway. Mol Cell 8:1017-1026

Komeili A, Wedaman KP, O'Shea BK, Powers T (2000) Mechanism of metabolic control: target of rapamycin signaling links nitrogen quality to the activity of the Rtgl and Rtg3 transcription factors. J Cell Biol 15 1:863-878

Kulkarni AA, Abul-Hamd AT, Rai R, El Berry H, Cooper TG (2001) Gln3 nuclear localization and interaction with Ure2 *Saccharomyces cerevisiae*. J Biol Chem 276:32136-32144

Kunz J, Schneider U, Howald I, Schmidt A, Hall MN (2000) HEAT repeats mediate plasma membrane localization of Tor2 in Yeast. J Biol Chem 275:37011-37020

Lacroute F (1968) Regulation of pyrimidine biosynthesis in *Saccharomyces cerevisiae*. J Bacteriol 95:824-832

Lawther RP, Cooper TG (1973) Effects of inducer addition and removal upon the level of allophanate hydrolase in *Saccharomyces cerevisiae*. Biochem Biophys Res Commun 55:1100-1104

Lawther RP, Cooper TG (1975) Kinetics of induced and repressed enzyme synthesis in *Saccharomyces cerevisiae*. J Bacteriol 121:1064-1073

Lawther RP, Phillips SL, Cooper TG (1975) Lomofungin inhibition of allophanate hydrolase synthesis in *Saccharomyces cerevisiae*. Mol Gen Genet 137:89-99

Liao X, Butow RA (1993) *RTGJ* and *RTG2:* two yeast genes required for a novel path of communication from mitochondria to the nucleus. Cell 72:61-71

Liao XS, Small WC, Srere PA, Butow RA (1991) Intramitochondrial functions regulate nonmitochondrial citrate synthase *(CIT2)* expression in *Saceharomyces cerevisiae*. Mol Cell Biol 11:38-46

Liu Z, Sekito T, Epstein CB, Butow RA (2001) RTG-dependent mitochondria to nucleus signaling is negatively regulated by the seven WD-repeat protein Lst8p. EMBO J 20:7209-7219

Loewith R, Jacinto E, Wullschleger S, Lorberg A, Crespo JL, Bonenfant D, Oppliger W, Jenoe P, Hall MN (2002) Two TOR complexes, only one of which is rapamycin sensitive, have distinct roles in cell growth control. Mol Cell 10:457-468

Lorenz MC, Heitman J (1998) The *MEP2* ammonium permease regulates pseudohyphal differentiation in *Saccharomyces cerevisiae*. EMBO J 17:1236-1247

Luke MM, Seta FD, Di Como CJ, Kobayashi R, Arndt KT (1996) The SAPs, a new family of proteins, associate and function positively with S1T4 phosphatase. Mol Cell Biol 16:2744-2755

Matsuura A, Anraku Y (1993) Characterization of the *MKS1* gene, a new negative regulator of the RAS-cyclic AMP pathway in *Saccharomyces cerevisiae*. Mol Gen Genet 238:6-16

Mattaj IW, Englmeier L (1998) Nucleocytoplasmic transport: the soluble phase. Ann Rev Biochem 67:265-306

Mitchell AP, Magasanik B (1984) Regulation of glutamine-repressible gene products by the *GLN3* function in *Saccharomyces cerevisiae*. Mol Cell Biol 4:2758-2766

Omichinski JG, Close GM, Schaad O, Felsenfeld G, Trainor C, Appella E, Stahl SJ, Gronenbom AM (1993) NMR structure of a specific DNA complex of Zn-containing DNA binding domain of GATA-1. Science 261:438-446

Pierce MM, Maddelein ML, Roberts BT, Wickner RB (2001) A novel Rtg2p activity regulates nitrogen catabolism in yeast. Proc Natl Acad Sci USA 98:13213-13218

Rai R, Daugherty JR, Cunningham TS, Cooper TG (1999) Overlapping positive and negative GATA factor binding sites mediate inducible *DAL7* gene expression in *Saccharomyces cerevisiae*. J Biol Chem 274:28026-28034

Rai R, Genbauffe FS, Sumrada RA, Cooper TG (1989) Identification of sequences responsible for transcriptional activation of the allantoate permease gene in *Saccharomyces cerevisiae*. Mol Cell Biol 9:602-608

Rai R, Tate JJ, Cooper TG (2003) Ure2, a prion precursor with homology to glutathione S-transferase, protects *Saccharomyces cerevisiae* cells from heavy metal ion and oxidant toxicity. J Biol Chem 278(15):12826-12833

Ramos F, Wiame J-M (1985) Mutation affecting the specific regulatory control of lysine biosynthetic enzymes in *Saccharomyces cerevisiae*. Mol Gen Genet 200:291-294

Raught B, Gingras AC, Sonenberg N (2001) The target of rapamycin (TOR) proteins. Proc Natl Acad Sci USA 98:7037-7044

Richman PG, Orlowski M, Meister A (1973) Inhibition of gamma-glutamylcysteine synthetase by L-methionine-S-sulfoximine. J Biol Chem 248:6684-90

Rohde J, Heitman J, Cardenas ME (2001) The TOR kinases link nutrient sensing to cell growth. J Biol Chem 276: 9583-9586

Ronne H, Carlberg M, Hu GZ, Neblin JO (1991) Protein phosphatase 2A in *Saccharomyces cerevisiae:* effects on cell growth and bud morphogenesis. Mol Cell Biol 11:4876-4884

Rowen DW, Esiobu N, Magasanik B (1997) Role of GATA factor Nil2p in nitrogen regulation of gene expression in *Saccharomyces cerevisiae.* J Bacteriol 179:3761-3766

Schmelzle T, Hall MN (2000) TOR, a central controller of cell growth. Cell 103:253-262

Schmidt A, Beck T, Koller A, Junz J, Hall MN (1998) The TOR nutrient signaling pathway phosphorylates NPR1 and inhibits turnover of the tryptophan permease. EMBO J 17:6924-6931

Sekito T, Liu Z, Thornton J, Butow RA (2002) RTG-dependent mitochondria-to-nucleus signaling is regulated by MKS1 and is linked to formation of yeast prion (URE3). Mol Biol Cell 13:795-804

Sekito T, Thornton J, Butow RA (2000) Mitochondria-to-nuclear signaling is regulated by the subcellular localization of the transcription factors Rtglp and Rtg3p. Mol Biol Cell 11:2103-2115

Shamji AF, Kuruvilla FG, Schreiber SL (2000) Partitioning the transcriptional program induced by rapamycin among the effectors of the Tor proteins. Curr Biol 10:1574-1581

Soetens O, De Craene JO, Andre B (2001) Ubiquitin is required for sorting to the vacuole of the yeast general amino acid permease, Gap1. J Biol Chem 276:43949-43957

Springael J-Y, De Craene JO, Andre B (1999) The yeast Npil/Rsp5 ubiquitin ligase lacking its N-terminal C2 domain is competent for ubiquination but not for subsequent endocytosis of the Gaplp permease. Biochem Biophys Res Commun 257:561-566

Stanbrough M, Magasanik B *(1995)* Transcriptional and posttranslational regulation of the general amino acid permease of *Saccharomyces cerevisiae.* J Bacteriol 177:94-102

Sutton A, Immanuel D, Arndt KT (1991) The Sit4 protein phosphatase functions in late G1 for progression into S phase. Mol Cell Biol 11:2133-2148

Svetlov V, Cooper TG (1997) The minimal transactivation region of *Saccharomyces cerevisiae* Gln3 is localized to 13 amino acids. J Bacteriol 179:7644-7652

Svetlov VV, Cooper TG (1998) *The Saccharomyces cerevisiae* GATA factors Dal80 and Deh1 can form homo- and heterodimeric complexes. J Bacteriol 180:5682-5688

Tate JJ, Cooper TG (2003) Torl/2 regulation of retrograde gene expression in *S. cerevisiae* is an indirect consequence of alterations in nitrogen catabolism. J Biol Chem (submitted)

Tate JJ, Cox KH, Rai R, Cooper TG (2002) Mkslp is required for negative regulation of retrograde gene expression in *Saccharomyces cerevisiae* but does not affect nitrogen catabolite repression-sensitive gene expression. J Biol Chem 277:20477-20482

ter Schure EG, van Riel NA, Verrips CT (2000) The role of ammonia metabolism in nitrogen catabolite repression in *Saccharomyces cerevisiae.* FEMS Microbiol Rev 24:67-83

Turoscy V, Cooper TG (1987) Ureidosuccinate is transported by the allantoate transport system in *Saccharomyces cerevisiae.* J Bacteriol 169:2598-2600

Uetz P, Giot L, Cagney G, Mansfield TA, Judson RS, Knight JR, Lockshon D, Narayan V, Srinivasan M, Pochart P, Qureshi-Emili A, Li Y, Godwin B, Conover D, Kalbfleisch T, Vijayadamodar G, Yang M, Johnston M, Fields S, Rothberg JM (2000) A comprehensive analysis of protein-protein interactions in *Saccharomyces cerevisiae.* Nature 403:623-627

Vandenbol M, Jauniaux J-C, Grenson M (1990) The *Saceharomyces cerevisiae NPRJ* gene required for the activity of the ammonia-sensitive amino acid permeases encodes a protein kinase homologue. Mol Gen Genet 222:393-399

Vandenbol M, Jauniaux J-C, Vissers S, Grenson M (1987) Isolation of the NPR1 gene responsible for the reactivation of ammonia-sensitive amino-acid permeases in *Saccharomyces cerevisiae.* Eur J Biochem 164:607-612

van Vuuren HJJ, Daugherty JR, Rai R, Cooper TG (1991) Upstream induction sequence, the *cis*-acting element required for response to the allantoin pathway inducer and enhancement of operation of the nitrogen-regulated upstream activation sequence in *Saccharomyces cerevisiae.* J Bacteriol 173:7186-7195

Velot C, Haviernik P, Lauquin GJ (1996) The *Saccharomyces cerevisiae RTG2* gene is a regulator of aconitase expression under catabolite repression conditions. Genetics 144:893-903

Wedaman KP, Reinke A, Anderson S, Yates J, McCaffery JM, Powers T (2003) Tor Kinases are in distinct membrane-associated protein complexes in *Saceharomyces cerevisiae.* Mol Biol Cell 14:1204-1220

Wickner RB, Edskes HK, Maddelein ML, Taylor KL, Moriyama H (1999) Prions of yeast and fungi. Proteins as genetic material. J Biol Chem 274:555-558

Wu J, Tolstykh T, Lee J, Boyd K, Stock JB, Broach JR (2000) Carboxyl methylation of the phosphoprotein phosphatase 2A catalytic subunit promotes its functional association with regulatory subunits *in vivo*. EMBO J 19:5672-5681

Zhegchang L, Butow RA (1999) A transcriptional switch in the expression of yeast tricarboxylic acid cycle genes in response to a reduction or loss of respiratory function. Mol Cell Biol 19:6720-6728

Zheng XF, Florentino D, Chen J, Crabtree GR, Schreiber SL (1995) TOR kinase domains are required for two distinct functions, only one of which is inhibited by rapamycin. Cell 82:121-130

10 Glucose regulation of *HXT* gene expression in the yeast *Saccharomyces cerevisiae*

Amber L. Mosley, Megan L. Sampley, and Sabire Özcan

Abstract

Glucose, the most preferred carbon source, also functions as a signaling molecule by stimulating the expression of metabolic genes in eukaryotes. In the yeast *S. cerevisiae*, transcription of metabolic genes such as those encoding glucose transporters, glycolytic and ribosomal enzymes are induced in the presence of extracellular glucose. Induction of these genes by glucose enables its efficient utilization by the cell. This review will mainly focus on the regulation of hexose transporter (*HXT*) gene expression by glucose. Glucose transport is the first and rate-limiting step in the metabolism of glucose and the *HXT* genes that are upregulated by glucose encode glucose transporters.

In the absence of glucose, the *HXT* genes are expressed at very low levels due to repression by the transcription factor Rgt1. In the presence of glucose, repression of *HXT* gene expression by Rgt1 is abolished. Several components of the glucose induction pathway that lead to upregulation of *HXT* gene expression have been identified, which will be discussed in detail in this review. The glucose sensors Snf3 and Rgt2 and the ubiquitin ligase Grr1 are absolutely essential for induction of *HXT* gene expression by glucose. Repression of the *HXT* genes by Rgt1 in the absence of glucose requires the general repressor complex Ssn6-Tup1 and two homologous proteins known as Mth1 and Std1. Glucose upregulates *HXT* gene expression by causing the phosphorylation of Rgt1, which inhibits its binding to the *HXT* gene promoters.

10.1 Introduction

In addition to being the major energy and carbon source, glucose also regulates the expression of several metabolic genes in many organisms. In the yeast *Saccharomyces cerevisiae*, the presence of glucose causes repression of genes that are not required for its own utilization such as genes encoding galactose, maltose, or sucrose metabolizing enzymes (Gancedo 1998; Carlson 1999; Johnston 1999). On the other hand, glucose also induces the expression of genes important for its efficient metabolism such as genes encoding glucose transporters (Bisson et al. 1993; Kruckeberg 1996; Boles and Hollenberg 1997; Özcan and Johnston 1999). This review will focus on the positive regulatory effects of glucose on gene expression

Topics in Current Genetics, Vol. 7
J. Winderickx, P.M. Taylor (Eds) Nutrient-Induced Responses in Eukaryotic Cells

by discussing the induction of hexose transporter (*HXT*) gene expression by glucose.

In the yeast *S. cerevisiae*, the expression of hexose transporter genes *HXT1* through *HXT4* is upregulated by glucose by at least ten-fold (Özcan and Johnston 1995, 1996, 1999; Wendell and Bisson 1993, 1994; Theodoris et al. 1994). In the absence of glucose, the expression of the *HXT* genes is kept low due to a repression mechanism (Özcan and Johnston 1995; Özcan et al. 1996a). The target of the glucose induction pathway that regulates *HXT* gene expression is the transcription factor Rgt1 that binds to the *HXT* gene promoters and represses their transcription when glucose is absent (Özcan and Johnston 1995, 1996; Özcan et al. 1996a). Glucose induces *HXT* gene expression by abolishing repression by Rgt1 (Marshall-Carlson et al. 1991; Erickson and Johnston 1994; Özcan et al. 1996a). The presence and the concentration of extracellular glucose is sensed by two membrane-bound proteins Snf3 and Rgt2, which resemble glucose transporters (Celenza et al. 1998; Coons et al. 1995, 1997; Liang and Gaber 1996; Özcan et al. 1996, 1998; Vagnoli et al. 1998). While Snf3 is able to sense low levels of glucose, Rgt2 functions as a sensor of high concentrations of glucose (Özcan et al. 1996). Induction of *HXT* gene expression is completely abolished in a strain lacking both glucose sensors, indicating the requirement of Snf3 and Rgt2 for generation of the glucose signal that upregulates *HXT* gene expression (Özcan et al. 1998; Schmidt et al. 1999). The mechanisms by which Snf3 and Rgt2 sense extracellular glucose, and the nature of the intracellular glucose signal and how it is transmitted from the cytoplasm into the nucleus are unknown.

Induction of *HXT* gene expression by glucose is also dependent on the function of the protein Grr1, which encodes an ubiquitin ligase (Özcan et al. 1994; Vallier et al. 1994; Özcan and Johnston 1995). In a *grr1* mutant, Rgt1 behaves as a constitutive repressor and inhibits *HXT* gene expression in a carbon source independent manner (Vallier et al. 1994). Although, Grr1 has been implicated in targeting proteins such as the G1-cyclins for degradation by causing their ubiquitination (Barral et al. 1995), Rgt1 does not appear to be targeted for degradation by Grr1. Recent data indicate that Grr1 is required for degradation of a negative regulator of *HXT* gene expression (Flick et al. 2003).

Repression of *HXT* gene transcription by Rgt1 in the absence of glucose requires several proteins including the Ssn6-Tup1 repressor complex (Özcan and Johnston 1995; Özcan et al. 1996a) and Std1 and Mth1 (Özcan et al. 1993; Schmidt et al. 1999; Lafuente et al. 2000; Schulte et al. 2000; Flick et al. 2003; Lakshmanan et al. 2003). In a strain deleted for both *MTH1* and *STD1* genes (Schmidt et al. 1999; Lafuente et al. 2000; Flick et al. 2003; Lakshmanan et al. 2003) or Ssn6 or Tup1 (Özcan and Johnston 1995; Özcan et al. 1996a), repression by Rgt1 is abolished, and the expression of the *HXT* genes occurs constitutively. While Std1 and Mth1 cause repression of the *HXT* genes by directly interacting with Rgt1 (Ito et al. 2001; Tomas-Cobos et al. 2002; Lakshmanan et al. 2003), it remains to be determined whether Ssn6 and Tup1 cause repression by Rgt1 through a direct or indirect mechanism. Recent data indicate that the binding of the transcriptional regulator Rgt1 to the *HXT* gene promoters is regulated in a carbon source dependent manner (Mosley et al. 2003; Flick et al. 2003; Kim et al.

2003). Rgt1 is able to bind to the *HXT* gene promoters only when glucose is absent and becomes phosphorylated in response to glucose (Mosley et al. 2003; Flick et al. 2003; Kim et al. 2003). The modification of Rgt1 by phosphorylation inhibits Rgt1 binding to the *HXT* gene promoters (Kim et al. 2003).

10.2 The transcription factor Rgt1 is the ultimate target of the glucose induction pathway

The ultimate target of the glucose induction pathway that regulates *HXT* gene expression is the transcriptional regulator Rgt1 (Özcan and Johnston 1995; Özcan et al. 1996a). In the absence of glucose, Rgt1 binds to the *HXT* gene promoters (Mosley et al. 2003; Flick et al. 2003; Kim et al. 2003) and inhibits their expression (Özcan and Johnston 1995, 1996; Özcan et al. 1996a). However, in the presence of glucose, Rgt1 becomes phosphorylated (Mosley et al. 2003; Flick et al. 2003; Kim et al. 2003) and is unable to bind to the *HXT* gene promoters (Kim et al. 2003), which leads to induction of *HXT* gene expression.

Rgt1 belongs to the Zn_2Cys_6 family of transcription factors, which includes the transcriptional regulator Gal4 (Özcan et al. 1996a). Like Gal4, Rgt1 contains an amino-terminal zinc cluster as the DNA binding domain (Özcan et al. 1996a). Gal4 and other members of this transcription factor family bind as a dimer to two CGG repeats separated by a specific number of nucleotides that is different for each transcription factor (Marmorstein et al. 1992; Akache et al. 2001). In contrast, the binding site of Rgt1 contains only one CGG repeat indicating that it binds to DNA as a monomer (Özcan and Johnston 1996; Özcan et al. 1996a). Dimerization of Gal4 and other transcription factors of this family is mediated by a coiled-coil region following the DNA binding domain. Rgt1 lacks the coiled-coil domain required for dimerization, consistent with the idea that Rgt1 binds to its sequence as a monomer. Recent data indicate that the *HXT* gene promoters contain multiple binding sites for Rgt1 with the consensus sequence 5-CGGANNA-3', and these Rgt1 binding sites appear to act synergistically in repression of *HXT* gene transcription (Kim et al. 2003). Moreover, the presence of at least five copies of the Rgt1 binding site is required for repression by Rgt1 (Kim et al. 2003).

In addition to being a transcriptional repressor in the absence of glucose, Rgt1 is also able to activate transcription when fused to the lexA DNA binding domain and recruited to a LacZ reporter via this domain (Özcan et al. 1996a). However, Rgt1 functions as an activator of transcription only at high concentrations of glucose (Özcan et al. 1996a). Since deletion of *RGT1* reduces the induction of *HXT1* gene expression by five- to six-fold (Özcan and Johnston 1995, Özcan et al. 1996a), it was assumed, until recently, that Rgt1 was also involved in activating the expression of the *HXT1* gene at high concentrations of glucose by direct binding to the *HXT1* promoter (Özcan and Johnston 1995, Özcan et al. 1996a). However, recent data indicate that Rgt1 binds to the *HXT1* promoter only when glucose is absent (Mosley et al. 2003; Flick et al. 2003; Kim et al. 2003), suggesting that

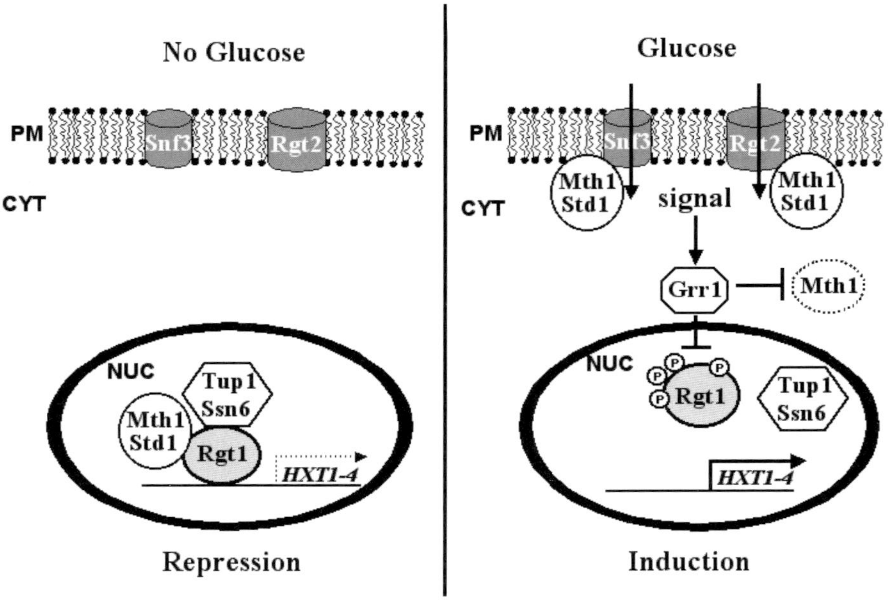

Fig. 1. Proposed model to explain the glucose induction of *HXT* gene expression. In the absence of glucose the transcriptional repressor Rgt1 is associated with the Mth1, Std1, Tup1, and Ssn6 proteins and is bound to the *HXT* gene promoters to repress their transcription (Tomas-Cobos and Sanz 2002; Mosley et al. 2003; Flick et al. 2003; Kim et al. 2003; Lakshmanan et al. 2003). In the presence of glucose, Rgt1 dissociates from Mth1, Std1, Tup1, and Ssn6 and becomes hyperphosphorylated. Phosphorylated Rgt1 is unable to bind to the *HXT* gene promoters, which leads to induction of *HXT* gene expression. It is likely that Std1 and Mth1 are translocated into the cytoplasm, because they interact with the carboxyl-terminal tails of the glucose sensors Snf3 and Rgt2 (Schmidt et al. 1999; Lafuente et al. 2000) when glucose is present. Recent data suggest that Grr1 mediates the degradation of Mth1 at high levels of glucose (Flick et al 2003).

the requirement of Rgt1 for maximal expression of the *HXT1* gene at high concentrations of glucose is mediated by an indirect mechanism (Mosley et al. 2003).

It has been recently demonstrated that Rgt1 binds to the *HXT* gene promoters only when glucose is absent and that it becomes hyperphosphorylated when glucose is present (Mosley et al. 2003; Flick et al. 2003; Kim et al. 2003). There is a correlation between phosphorylation of Rgt1 and its ability to bind to DNA. While hypophosphorylated Rgt1 is able to bind to the *HXT* gene promoters, hyperphosphorylated Rgt1 is unable to bind to the *HXT* gene promoters *in vivo* as well as *in vitro* (Kim et al. 2003). According to these data, glucose induces *HXT* gene expression by inhibiting the binding of Rgt1 to the *HXT* gene promoters (Fig.1). Furthermore, the ability of Rgt1 to activate transcription depends on its phosphorylation status (Fig. 1). Namely, only phosphorylated Rgt1 is able to activate transcription at high concentrations of glucose. In summary, Rgt1 binds to the *HXT* gene promoters and represses their transcription only when glucose is absent.

The presence of glucose induces the hyperphosphorylation of Rgt1 and inhibits its binding to the *HXT* gene promoters and thereby abolishes their repression (Fig. 1).

Rgt1 has been recently demonstrated to bind to the SUC2-B promoter region using a biochemical genomics approach where a nearly complete set of protein glutathione S-transferase fusions were used to identify proteins that bind to the *SUC2* promoter (Hazbun and Fields 2002). Deletion of *RGT1* reduces the induction of *SUC2* gene expression by low levels of glucose, suggesting that Rgt1 may function as an activator of *SUC2* expression (Hazbun and Fields 2002; Özcan et al. 1997).

10.3 Proteins that positively affect the glucose induction of *HXT* gene expression

Several proteins have been shown to be required for induction of *HXT* gene expression by glucose. These include the glucose-transporter like proteins Snf3 and Rgt2, which function as sensors of extracellular glucose and are required to generate the glucose signal for induction of *HXT* gene expression. The ubiquitin ligase Grr1 is also essential for upregulation of *HXT* gene transcription in response to glucose, and Grr1 is likely to be required for degradation of a negative regulator of *HXT* gene expression.

10.3.1 The glucose transporter-like proteins Snf3 and Rgt2 are required for sensing of extracellular glucose

SNF3 and *RGT2* encode membrane proteins with similarity to glucose transporters. They are 60% similar to each other and about 30% similar to other glucose transporters including the *HXT* genes (Fig. 2) (Özcan et al. 1996). Snf3 and Rgt2 function as receptors of extracellular glucose and generate a signal inside the cell that is required for induction of *HXT* gene expression (Özcan et al. 1996, 1998). While *snf3* mutants are defective in low glucose induction of *HXT2* gene expression, Rgt2 is required for maximal expression of the *HXT1* gene at high concentrations of glucose (Özcan et al. 1996). These data suggest that Snf3 may sense low levels of glucose, while Rgt2 may sense high concentrations of glucose. Consistent with a role for Snf3 and Rgt2 as sensors of glucose, dominant mutations in the *RGT2* and *SNF3* genes cause constitutive expression of the *HXT* genes (in the absence of glucose). The dominant mutations in Rgt2 (*RGT2-1*) or Snf3 (*SNF3-1*) change an arginine residue at position 231 or 229, respectively to a lysine residue (Özcan et al. 1996). This suggests that glucose signaling by Snf3 and Rgt2 is independent of glucose metabolism and thereby resembles the receptor-mediated hormone signaling in mammalian cells.

The expression levels of the *SNF3* and *RGT2* genes are very low compared to the *HXT* genes (Özcan et al. 1996). Furthermore, overexpression of *SNF3* and *RGT2* in a *hxt* null mutant (deleted for *HXT1-HXT7*), which is defective for

Fig. 2. Alignment of the glucose transporter Hxt1 with the glucose sensors Snf3, Rgt2 and Rag4 (from *K. lactis*). Residues that are identical are boxed in black and the gray shading depicts residues that are similar. The conserved motif within the carboxyl-terminal tails of the glucose sensors is boxed. Snf3 has two conserved motifs.

growth on glucose-containing media, does not restore the growth defect of this mutant compared to overexpression of the *HXT1* gene, suggesting that the glucose sensors Snf3 and Rgt2 are unable to transport sufficient amount of glucose (Özcan et al. 1998).

One unique feature of the Snf3 and Rgt2 glucose sensors is that they have a long carboxyl-terminal tail that is cytoplasmic. Normally, the carboxyl-terminal tail of glucose transporters is around sixty amino acids, while the carboxyl-terminal tails of Snf3 and Rgt2 are more than 200 amino acids long. There is no similarity within the carboxyl-terminal tails of Snf3 and Rgt2, except a stretch of 25 amino acids that is present twice in the carboxyl-terminal tail of Snf3 and once in Rgt2 (Fig. 2) (Özcan et al. 1998). Deletion of the carboxyl-terminal tails or the conserved motifs abolishes the ability of Snf3 and Rgt2 to sense extracellular glucose (Vagnoli et al. 1998; Özcan et al. 1998). The hexose transporters *HXT1* and *HXT2* do not display any sensor function by themselves and lack the carboxyl-terminal tail that is present in Snf3 and Rgt2. However, the attachment of the carboxyl-terminal tail of Snf3 or Rgt2 to *HXT1* or *HXT2* converts these metabolic transporters to sensors of glucose (Özcan et al. 1998). Recent data indicate that Snf3 is phosphorylated at the carboxyl-terminal tail (Dlugai et al. 2001). However

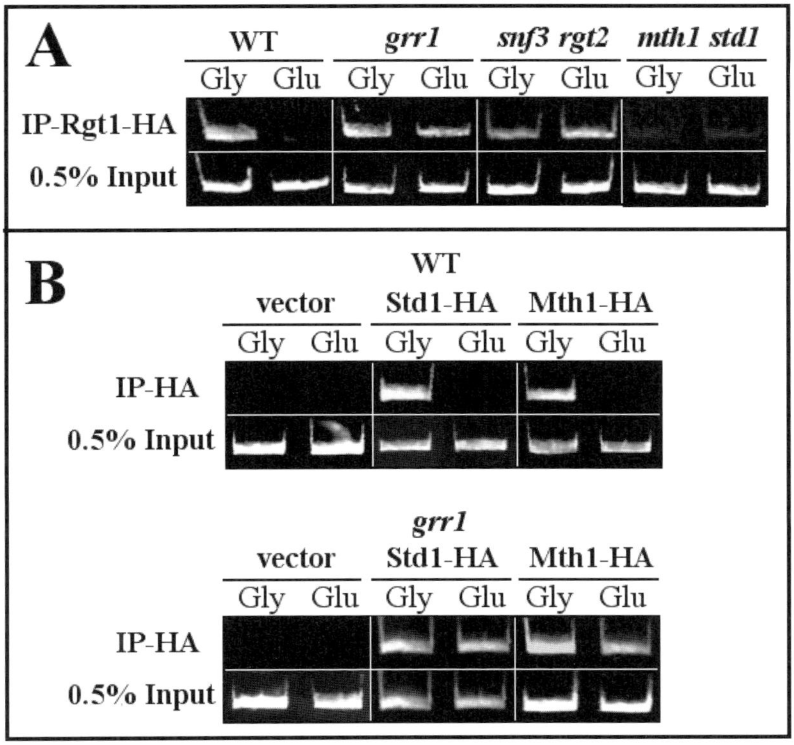

Fig. 3. Panel A: Rgt1 binds to the *HXT1* promoter constitutively in the *grr1* and *snf3 rgt2* mutants. In the *mth1 std1* double mutant, binding of Rgt1 to the *HXT1* promoter is abolished. Binding of Rgt1 to the *HXT1* gene promoter on 5% glycerol (Gly) or 4% glucose (Glc) was determined in *grr1*, *snf3 rgt2*, and *mth1 std1* mutants containing the *RGT1-HA* construct using the ChIP assay with HA antibodies (Mosley et al. 2003). Mouse IgG was used as a negative control for the ChIP assay and no PCR products were obtained with IgG precipitated samples as template (data not shown). Panel B: Mth1 and Std1 associate with the *HXT1* gene promoter only in the absence of glucose in wild type cells. In the *grr1* mutant, Mth1 and Std1 are recruited to the *HXT1* promoter independent of carbon source. Association of Std1-HA or Mth1-HA in wild type (upper panel) or *grr1* (lower panel) mutants incubated for 2 hours with 5% glycerol (Gly) or 4% glucose (Glc) was determined using the ChIP assay with HA antibodies. Wild type or *grr1* cells transformed with vector alone were used as a negative control for the ChIP assay. The *STD1-HA* and *MTH1-HA* constructs have been described previously (Schmidt et al. 1999).

the kinase that phosphorylates the carboxyl-terminal tail of Snf3 and the protein(s) that interact with Snf3 and Rgt2 to transmit the glucose signal remain to be identified. Two homologues proteins, Std1 and Mth1 have been shown to interact with the carboxyl-terminal tails of Snf3 and Rgt2 (Schmidt et al. 1999; Lafuente et al. 2000) and may be involved in the transmission of the glucose signal from the sensors to the nucleus.

In an *snf3 rgt2* double mutant, Rgt1 always behaves as a constitutive repressor of transcription and causes repression of *HXT* gene expression independent of carbon source (Özcan et al. 1998). Consistent with this finding, Rgt1 lacks the glucose-mediated hyperphosphorylation in the *snf3 rgt2* double mutant (Mosley et al. 2003) and binds to the *HXT* gene promoters constitutively (Fig. 3A). Interestingly, the *snf3 rgt2* double mutant is also defective for glucose repression of *GAL1* and *SUC2* expression (Özcan et al. 1998; Schmidt et al. 1999; Özcan 2002). This is caused by a lack of glucose uptake in the *snf3 rgt2* mutant, which is defective for glucose induction of *HXT* gene expression. Since the signal for glucose repression depends on the uptake and metabolism of glucose, this explains the glucose repression defect of the *snf3 rgt2* double mutant (Özcan 2002).

A glucose sensor like Snf3 or Rgt2 also exists in the milk yeast *Kluyveromyces lactis*. The Rag4 protein of *K. lactis* has about 50% identity to Snf3 and Rgt2 and contains an extended carboxyl-terminal tail that has the conserved motif present in Snf3 and Rgt2 (Fig. 2) (Betina et al. 2001). The glucose sensors Snf3, Rgt2 and Rag4 share a high degree of homology also outside the conserved carboxyl-terminal motif compared to the Hxt1 protein (Fig. 2). These conserved residues in the glucose sensors may be involved in the sensing and signaling of the glucose signal. This suggests the existence of Snf3- or Rgt2-like glucose sensors in other yeasts. However, there are no mammalian homologues of Snf3 and Rgt2 present in the available databases.

After the discovery of Snf3 and Rgt2, several other membrane proteins in yeast have been demonstrated to function as sensors of nutrients. While Ssy1 is involved in sensing of extracellular amino acid availability (Didion et al. 1998; Jorgensen et al. 1998; Klasson et al. 1999; Iraqui et al. 1999), the G-protein-coupled receptor Gpr1 (Xue et al. 1998; Kraakman et al. 1999; Lorenz et al. 2000) and the high-affinity ammonium transporter Mep2 (Lorenz et al. 1998) are sensors of fermentable sugars and of ammonium, respectively; and are both involved in the regulation of pseudohyphal growth.

10.3.2 The ubiquitin ligase Grr1 inhibits the repressor function of Rgt1 when glucose is present

Mutations in *GRR1* have pleiotropic effects, including elongated cell morphology, decreased glucose and aromatic amino acid uptake, and insensitivity to glucose repression and inactivation (Bailey and Woodward 1984; Flick and Johnston 1991; Jiang et al. 1997; Iraqui et al. 1999). Mutations in *GRR1* abolish the glucose-induced expression of the *HXT* genes (Özcan and Johnston 1995) and therefore, *grr1* mutants are defective in glucose transport (Özcan et al. 1994; Vallier et al. 1994). The glucose repression defect of the *grr1* mutant is due to its inability to transport glucose, since the generation of the glucose repression signal requires the transport and metabolism of glucose (Özcan 2002). Like the *snf3 rgt2* double mutant, the *grr1* mutant is also defective for growth on glucose-containing media caused by a lack of induction of *HXT* gene expression by glucose (Özcan et al. 1994, 1998).

Mutations in *RGT1* restore the defects in glucose uptake and glucose repression in a *grr1* mutant, suggesting that Grr1 is required to inhibit the repressor function of Rgt1 (Vallier et al. 1994; Özcan and Johnston 1995). Grr1 is an ubiquitin ligase and part of an ubiquitin-conjugating complex called SCF^{Grr1} (Skp1-Cullin-F-box) that targets proteins for degradation (Bai et al. 1996; Hochstrasser 1996; Li and Johnston 1997; Skowyra et al. 1997; Kishi et al. 1998). Interestingly, temperature-sensitive mutations in the components of the SCF^{Grr1} complex (such as *SKP1* and *CDC53*) have been shown to abolish *HXT* gene expression, suggesting that the SCF complex regulates both gene expression and the cell cycle in response to glucose (Li and Johnston 1997).

Although the exact role of Grr1 in glucose induction of *HXT* gene expression is not known, Grr1 appears to be required to inhibit repression of *HXT* gene transcription by Rgt1. Binding of Rgt1 to the *HXT* gene promoters is abolished in the presence of glucose due to phosphorylation of Rgt1 (Mosley et al. 2003; Flick et al. 2003; Kim et al. 2003). However in a *grr1* mutant, Rgt1 lacks the glucose-mediated phosphorylation (Mosley et al. 2003; Flick et al. 2003) and is constitutively bound to the *HXT* gene promoters (Fig. 3B). This explains why Rgt1 functions as a constitutive repressor of the *HXT* genes in the *grr1* mutant. Recent data indicate that Grr1 mediates the degradation of the Mth1 protein (Flick et al. 2003), which is a negative regulator of *HXT* gene expression (see section 10.3.1).

Like the glucose induction pathway that leads to upregulation of *HXT* gene expression, the amino acid sensing and signaling pathway also requires the ubiquitin ligase Grr1 (Forsberg and Ljungdahl 2001). However, the proteins that are degraded in a Grr1-dependent manner in the amino acid signaling pathway are not known. The target of the Grr1-dependent degradation pathway in glucose induction of *HXT* gene expression appears to be the negative regulator Mth1 (Flick et al. 2003); however, it remains to be established whether Std1 and Mth1 play a role in amino acid signaling.

10.4 Proteins that negatively regulate the glucose induction of *HXT* gene expression

Repression of *HXT* gene expression in the absence of glucose requires the two homologous proteins Std1 and Mth1. Furthermore, the general repressor complex Ssn6-Tup1 is also required for repression of *HXT* gene expression by Rgt1. It is not clear if Ssn6-Tup1 cause repression of *HXT* gene expression by directly interacting with Rgt1 or by an indirect mechanism.

10.4.1 Std1 and Mth1 are required for repression of *HXT* gene expression by Rgt1

Several lines of evidence indicate a role for Std1 (Msn3) and Mth1 in repression of *HXT* gene transcription (Özcan et al. 1993; Gamo et al. 1994; Schmidt et al.

1999; Schulte et al. 2000; Lafuente et al. 2000). Both proteins have been shown to interact with the carboxyl-terminal tails of the glucose sensors Snf3 and Rgt2 (Schmidt et al. 1999; Lafuente et al. 2000). A dominant mutation in *MTH1* isolated as *HTR1-23* or as *DGT1-1* causes constitutive repression of the *HXT* genes independent of the presence of glucose (Özcan et al 1994; Gamo et al. 1994; Schulte et al. 2000). Furthermore, Std1 has been shown to interact with Rgt1 *in vivo* (Tomas-Cobos and Sanz 2002). The interaction of Rgt1 with Std1 or Mth1 is regulated by glucose and occurs only in the absence of glucose (Lakshmanan et al. 2003). Consistent with this finding, Std1 and Mth1 become recruited to the *HXT* gene promoters only when glucose is absent (Fig. 3B). However in a *grr1* mutant, Std1 and Mth1 are associated with the *HXT* gene promoters independent of carbon source (Fig. 3B), which explains the constitutive repression of *HXT* gene expression as observed in the *grr1* mutant (Özcan and Johnston 1995; Özcan et al. 1996a).

Repression of *HXT* gene transcription by Rgt1 in the absence of glucose is completely abolished in a strain lacking both *STD1* and *MTH1* genes (Schmidt et al. 1999; Lafuente et al. 2000; Schulte et al. 2000; Flick et al. 2003; Lakshmanan et al. 2003). Std1 and Mth1 appear to function downstream of Grr1, since deletion of *MTH1* and *STD1* in a *grr1* mutant abolishes the constitutive repression of *HXT* gene expression observed in this mutant (Flick et al. 2003).

Previous data indicate that Rgt1 binds to *HXT* gene promoters and represses their transcription only when glucose is absent (Mosley et al. 2003; Flick et al. 2003; Kim et al. 2003). In the presence of glucose, Rgt1 becomes phosphorylated and is unable to bind to the *HXT* gene promoters. In an *std1 mth1* double mutant, Rgt1 is always phosphorylated and is unable to bind to the *HXT* gene promoters independent of carbon source (Fig. 3A) (Flick et al. 2003; Lakshmanan et al. 2003). This explains why Rgt1 is unable to repress the transcription of *HXT* gene expression in the absence of the *STD1* and *MTH1* genes.

Interestingly, Std1 also interacts with the Snf1 kinase and stimulates its kinase activity (Hubbard et al. 1994; Tomas-Cobos and Sanz 2002; Kuchin et al. 2003). However, deletion of the *SNF1* gene has no effect on Rgt1 phosphorylation and binding to the *HXT* gene promoters (data not shown). This suggests that the kinase that phosphorylates Rgt1 is not Snf1. Although Std1 and Mth1 are 61% identical, they have no homology to any of the known proteins from yeast and other organisms. The exact mechanism(s) of how Std1 and Mth1 repress *HXT* gene expression in the absence of glucose remains to be identified.

Recent data indicate that the Mth1 becomes degraded in the presence of glucose (Flick et al. 2003). This degradation event appears to be dependent of the ubiquitin ligase Grr1 (Flick et al. 2003). In this case, Mth1 associates with Rgt1 only in the absence of glucose to enable Rgt1 to bind to the *HXT* gene promoters and to repress transcription. In the presence of glucose, Mth1 is degraded and thereby the interaction between Mth1 and Rgt1 is disrupted, which causes the phosphorylation of Rgt1 and relieve of *HXT* gene repression (Fig. 1). Although this is a very attractive model to explain the function of Mth1 in causing repression of *HXT* gene expression, studies on Mth1 degradation are complicated by the fact that the transcription of the *MTH1* gene itself is under the control of glucose

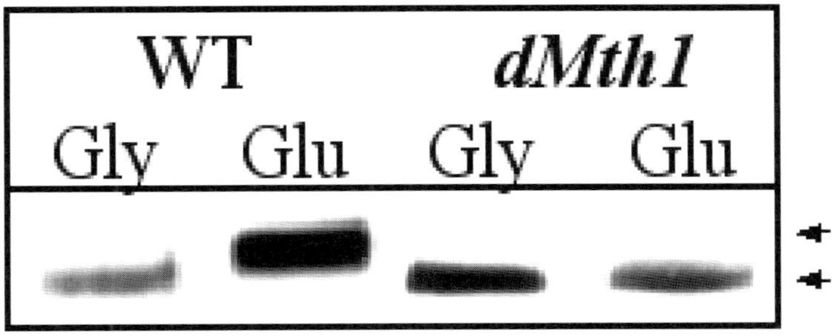

Fig. 4. The dominant mutation in Mth1, dMth1 (Htr1-23) prevents the glucose-mediated phosphorylation of Rgt1. Western blot analysis of Rgt1 phosphorylation in wild type and dMth1 (Htr1-23) strain transformed with the *lexA-RGT1* construct using lexA antibodies for immunoblotting of cell extracts prepared from 5% glycerol (Gly) or 4% glucose (Glc) incubated cells as described previously (Mosley et al. 2003). The arrows indicate Rgt1 and Phospho-Rgt1.

repression (Schmidt et al. 1999; Schulte et al. 2000; Ren et al. 2000). Furthermore, the *grr1* mutants are defective in glucose repression and thereby in a *grr1* mutant repression of *MTH1* gene transcription would be abolished. Therefore, analysis of Mth1 degradation in other glucose repression or proteosome-deficient mutants will be valuable in understanding the regulation of Mth1 function by glucose. Although, the transcription of the *STD1* gene is not regulated by glucose (Schmidt et al. 1999; Flick et al. 2003), it is not clear whether glucose regulates Std1 function by causing its degradation.

Htr1-23, a dominant allele of the *MTH1* gene causes constitutive repression of *HXT* gene expression, which is abolished by deletion of *RGT1* (Özcan et al. 1993; Schulte et al. 2000). This suggests that the interaction of Htr1-23 with the repressor protein Rgt1 occurs independent of glucose in a constitutive manner. Consistent with this idea, Rgt1 lacks the glucose-mediated phosphorylation in the Htr1-23 mutant (Fig. 4); therefore, it is likely that Rgt1 is always associated with the *HXT* gene promoters and represses their transcription independent of carbon source. The mutation in Htr1-23 changes isoleucine 85 of Mth1 to either an aspartate or serine (Schulte et al. 2000). There are several possibilities to explain the function of the dominant Mth1 mutant (Htr1-23): while wild type Mth1 may be normally degraded by glucose (Flick et al. 2003), Htr1-23 could be stable and present independent of carbon source. Alternatively, while the presence of glucose causes the translocation of Mth1 from the nucleus to the cytoplasm, the Htr1-23 mutant may always be present in the nucleus. Another possibility is that glucose may disrupt the interaction of Mth1 with Rgt1 by causing the modification of Mth1; the mutation in Htr1-23 may inhibit this modification and cause constitutive association of Htr1-23 with Rgt1.

10.4.2 The Ssn6-Tup1 repressor complex is required for repression of *HXT* gene expression in the absence of glucose

The Ssn6-Tup1 complex has been shown to repress the expression of several different genes (Gancedo 1998; Smith and Johnston 2000). Previous data indicate that Ssn6 and Tup1 are also required for repression of *HXT* gene expression by Rgt1 when glucose is absent (Özcan and Johnston 1995; Özcan et al. 1996a). Recent data suggest that Ssn6 interacts with Rgt1 *in vivo* (Tomas-Cobos and Sanz 2002) and that Ssn6 is associated with *HXT* gene promoters (Kim et al. 2003). Association of Ssn6 with the *HXT* gene promoters is dependent on Rgt1, because in an *rgt1* mutant, Ssn6 is unable to associate with the *HXT* gene promoters (Kim et al. 2003). Since the Ssn6-Tup1 complex, like Std1 and Mth1 is also required for repression of *HXT* gene transcription by Rgt1, it is possible that they exist in a complex. Alternatively, Std1 and Mth1 may be required to target the Ssn6-Tup1 complex to the *HXT* gene promoter to cause repression of *HXT* gene expression.

Tup1 has been proposed to repress gene expression by two different mechanisms, which involve the interaction of Tup1 with the mediator complex (Gromoller and Lehming 2000; Papamichos-Chronakis et al. 2000; Zaman et al. 2001) and the deacetylation of histones at specific promoters (Watson et al. 2000; Wu et al. 2001). However, the exact role of Tup1 and Ssn6 in repression of the *HXT* gene expression by Rgt1 remains to be determined.

10.5 Concluding remarks

The glucose induction pathway enables *Saccharomyces cerevisiae* to adapt to different extracellular concentrations of glucose and allows the efficient metabolism of the extracellular sugar. At high concentrations of extracellular glucose, *S. cerevisiae* induces the expression of the *HXT1* and *HXT3* genes, that function as low-affinity and high-capacity glucose transporters (Özcan and Johnston 1995; Reifenberger et al. 1997). Low levels of extracellular glucose stimulate the expression of *HXT2* (Özcan and Johnston 1995), *HXT6* and *HXT7* genes (Liang and Gaber 1996; Reifenberger et al. 1997), which encode for high-affinity glucose transporters (Reifenberger et al. 1997). Interestingly, the different concentrations of extracellular glucose are sensed by glucose-transporter like membrane proteins such as Snf3 and Rgt2, which appear to have different affinities for extracellular glucose (Özcan and Johnston 1999). In addition to glucose, *HXT* gene expression is also likely to be modulated by other environmental signals. For instance, transcription of the *HXT1* and *HXT5* genes is also regulated by hyperosmolarity (Hirayama et al. 1995; Rep et al. 2000). While it is very likely that the affinity of the different glucose transporters is modulated by posttranslational modification, the transcriptional regulation of the *HXT* genes is a major determinant of the kinetics of glucose transport in response to the glucose availability.

Acknowledgements

We would like to thank Mark Johnston and the members of his laboratory for sharing data prior to publication. We are grateful to all our colleagues who have shared information, ideas, and reagents with us in the past.

References

Akache B, Wu K, Turcotte B (2001) Phenotypic analysis of genes encoding yeast zinc cluster proteins. Nucleic Acids Res 29:2181-2190

Bai C, Sen P, Hofmann K, Ma L, Goebl M, Harper JW, Elledge SJ (1996) SKP1 connects cell cycle regulators to the ubiquitin proteolysis machinery through a novel motif, the F-box. Cell 86:263-274

Bailey RB, Woodward A (1984) Isolation and characterization of a pleiotropic glucose repression resistant mutant of *Saccharomyces cerevisiae*. Mol Gen Genet 193:507-512

Barral Y, Jentsch S, Mann C (1995) G1 cyclin turnover and nutrient uptake are controlled by a common pathway in yeast. Genes Dev 9:399-409

Betina S, Goffrini P, Ferrero I, Wesolowski-Louvel M (2001) *RAG4* gene encodes a glucose sensor in *Kluyveromyces lactis*. Genetics 156:541-548

Bisson LF, Coons DM, Kruckeberg AL, Lewis DA (1993) Yeast sugar transporters. Crit Rev Biochem Mol Biol 28:259-308

Boles E, Hollenberg CP (1997) The molecular genetics of hexose transport in yeasts. FEMS Microbiol Rev 21:85-111

Carlson M (1999) Glucose repression in yeast. Curr Opin Microbiol 2:202-207

Celenza JL, Marshall-Carlson L, Carlson M (1998) The yeast *SNF3* gene encodes a glucose transporter homologous to the mammalian protein. Proc Natl Acad Sci USA 85:2130-2134

Coons DM, Boulton RB, Bisson LF (1995) Computer-assisted nonlinear regression analysis of the multicomponent glucose uptake kinetics of *Saccharomyces cerevisiae*. J Bacteriol 177:3251-3258

Coons DM, Vagnoli P, Bisson LF (1997) The c-terminal domain of Snf3p is sufficient to complement the growth defect of *snf3* null mutations in *Saccharomyces cerevisiae*: *SNF3* functions in glucose recognition. Yeast 13:9-20

Didion T, Regenberg B, Jørgensen MU, Kielland-Brandt MC, Andersen HA (1998) The permease homologue Ssy1p controls the expression of amino acid and peptide transporter genes in *Saccharomyces cerevisiae.* Mol Microbiol 27:643–650

Dlugai S, Hippler S, Wieczorke R, Boles E (2001) Glucose-dependent and –independent functions of the glucose sensor Snf3. FEBS Lett 505:389-392

Erickson JR, Johnston M (1994) Suppressors reveal two classes of glucose repression genes in the yeast *Saccharomyces cerevisiae*. Genetics 136:1271-1278

Flick JS, Johnston M (1991) *GRR1* of *Saccharomyces cerevisiae* is required for glucose repression and encodes a protein with leucine-rich repeats. Mol Cell Biol 11:5101-5112

Flick KM, Spielewoy N, Guaderrama M, Kalishnikova T, Zhu QZ, Chang HC, Wittenberg C (2003) Grr1-dependent inactivation of Mth1 mediates glucose induced dissociation of Rgt1 from *HXT* gene promoters. Mol Biol Cell 14:3230-3241

Forsberg H, Ljungdahl PO (2001) Sensors of extracellular nutrients in *Saccharomyces cerevisiae.* Curr Genet 40:91-109

Gamo FJ, Lafuente MJ, Gancedo C (1994) The mutation in *DGT1-1* decreases glucose transport and alleviates carbon catabolite repression in *Saccharomyces cerevisiae.* J Bacteriol 176:7423-7429

Gancedo JM (1998) Yeast carbon catabolite repression. Microbiol Mol Biol Rev 62:334-361

Gromoller A, Lehming N (2000) Srb7p is a physical and physiological target of Tup1p. EMBO J 19:6845-6852

Hazbun TR, Fields S (2002) A Genome-wide screen for site-specific DNA-binding proteins. Mol Cell Proteom 1:538–543

Hirayama T, Maeda T, Saito H, Shinozaki 1995) Cloning and characterization of seven cDNAs for hyperosmolarity-responsive (HOR) genes of *Saccharomyces cerevisiae.* Mol Gen Genet 249:127-138

Hochstrasser M (1996) Protein degradation or regulation: Ub the judge. Cell 84:813-815

Hubbard R, Jiang R, Carlson M (1994) Dosage-dependent modulation of glucose repression by *MSN3* (*STD1*) in *Saccharomyces cerevisiae.* Mol Cell Biol 14:1972-1978

Iraqui I, Vissers S, Bernard F, de Craene JO, Boles E, Urrestarazu A, Andre B (1999) Amino acid signaling in *Saccharomyces cerevisiae*: a permease-like sensor of external amino acids and F-Box protein Grr1p are required for transcriptional induction of the *AGP1* gene, which encodes a broad-specificity amino acid permease. Mol Cell Biol 19:989-1001

Ito T, Chiba T, Ozawa R, Yoshida M, Hattori M, Sakaki Y (2001) A comprehensive two-hybrid analysis to explore the yeast protein interactome. Proc Natl Acad Sci USA 98:4569-4574

Jiang H, Medintz I, Michels CA (1997) Two glucose sensing/signaling pathways stimulate glucose-induced inactivation of maltose permease in *Saccharomyces.* Mol Biol Cell 8:1293-1304

Johnston M (1999) Fasting, feasting and fermenting: Glucose sensing in yeast and other cells. Trends Genet 15:29-33

Jorgensen MU, Brunn MB, Didion T, Kielland-Brandt MC (1998) Mutations in five loci affecting GAP1-independent uptake of neutral amino acids in yeast. Yeast 14:103–114

Kim JH, Polish J, Johnston M (2003) Specificity and regulation of DNA-binding by the yeast glucose transporter gene repressor Rgt1. Mol Cell Biol 23:5208-5216

Kishi T, Seno T, Yamao F (1998) Grr1 functions in the ubiquitin pathway in *Saccharomyces cerevisiae* through association with Skp1. Mol Gen Genet 257:143-148

Klasson H, Fink GR, Ljungdahl PO (1999) Ssy1p and Ptr3p are plasma membrane components of a yeast system that senses extracellular amino acids. Mol Cell Biol 19:5405–5416

Kuchin S, Vyas VK, Kanter ESP, Carlson M (2003) Std1p (Msn3p) positively regulates the Snf1 kinase in *Saccharomyces cerevisiae.* Genetics 163:507-514

Kraakman L, Lemaire K, Ma P, Teunissen AW, Donaton MC, Van Dijck P, Winderickx J, de Winde JH, Thevelein JM (1999) A *Saccharomyces cerevisiae* G-protein coupled receptor, Gpr1, is specifically required for glucose activation of the cAMP pathway during the transition to growth on glucose. Mol Microbiol 32:1002–1012

Kruckeberg AL (1996) The hexose transporter family of *Saccharomyces cerevisiae.* Arch Microbiol 166:283-292

Lafuente MJ, Gancedo C, Jauniaux JC, Gancedo JM (2000) Mth1 receives the signal given by the glucose sensors Snf3 and Rgt2 in *Saccharomyces cerevisiae*. Mol Microbiol 35:161-172

Lakshmanan J, Mosley AL, Özcan S (2003) Repression of transcription by Rgt1 in the absence of glucose requires Std1 and Mth1. Curr Genet 44:19-25

Liang H, Gaber RF (1996) A novel signal transduction pathway in *Saccharomyces cerevisiae* defined by Snf3-regulated expression of *HXT6*. Mol Biol Cell 7:1953-1966

Li FN, Johnston M (1997) Grr1 of *Saccharomyces cerevisiae* is connected to the ubiquitin proteolysis machinery through Skp1: coupling glucose sensing to gene expression and the cell cycle. EMBO J 16:5629-5638

Lorenz MC, Heitman J (1998) The MEP2 ammonium permease regulates pseudohyphal differentiation in *Saccharomyces cerevisiae*. EMBO J 17:1236–1247

Lorenz MC, Pan X, Harashima T, Cardenas ME, Xue Y, Hirsch JP, Heitman J (2000) The G protein-coupled receptor Gpr1 is a nutrient sensor that regulates pseudohyphal differentiation in *Saccharomyces cerevisiae*. Genetics 154:609–622

Marmorstein R, Carey M, Ptashne M, Harrison SC (1992) DNA recognition by GAL4: structure of a protein-DNA complex. Nature 356:408-414

Marshall-Carlson L, Neigeborn L, Coons D, Bisson L, Carlson M (1991) Dominant and recessive suppressors that restore glucose transport in a yeast *snf3* mutant. Genetics 128:505-511

Mosley AL, Lakshmanan J, Aryal BK, Özcan S (2003) Glucose-mediated phosphorylation converts the transcription factor Rgt1 from a repressor to an activator. J Biol Chem 278:10322-10327

Özcan S (2002) Two different signals regulate repression and induction of gene expression by glucose. J Biol Chem 277:46993-46997

Özcan S, Dover J, Johnston M (1998) Glucose sensing and signaling by two glucose receptors in the yeast *Saccharomyces cerevisiae*. EMBO J 17:2566-2573

Özcan S, Dover J, Rosenwald AG, Wölfl S, Johnston M (1996) Two glucose transporters in *Saccharomyces cerevisiae* are glucose sensors that generate a signal for induction of gene expression. Proc Natl Acad Sci USA 93:12428-12432

Özcan S, Freidel K, Leuker A, Ciriacy M (1993) Glucose uptake and catabolite repression in dominant *HTR1* mutants of *Saccharomyces cerevisiae*. J Bacteriol 175:5520-5528

Özcan S, Johnston M (1995) Three different regulatory mechanisms enable yeast hexose transporter (*HXT*) genes to be induced by different levels of glucose. Mol Cell Biol 15:1564-1572

Özcan S, Johnston M (1996) Two different repressors collaborate to restrict expression of the yeast glucose transporter genes *HXT2* and *HXT4* to low levels of glucose. Mol Cell Biol 16:5536-5545

Özcan S, Johnston M (1999) Function and regulation of yeast hexose transporters. Microbiol Mol Biol Rev 63:554-569

Özcan S, Leong T, Johnston M (1996a) Rgt1p of *Saccharomyces cerevisiae*, a key regulator of glucose-induced genes, is both an activator and repressor of transcription. Mol Cell Biol 16:6419-6426

Özcan S, Schulte F, Freidel K, Weber A, Ciriacy M (1994) Glucose uptake and metabolism in *grr1/cat80* mutants of *Saccharomyces cerevisiae*. Eur J Biochem 224:605-611

Özcan S, Vallier, LG, Flick JS, Carlson M, Johnston M (1997) Expression of the *SUC2* gene of *Saccharomyces cerevisiae* is induced by low levels of glucose. Yeast 13:127–137

Papamichos-Chronakis M, Conlan RS, Gounalaki N, Copf T, Tzamarias D (2000) Hrs1/Med3 is a Cyc8-Tup1 corepressor target in the RNA polymerase II holoenzyme. J Biol Chem 275:8397-8403

Reifenberger E, Boles E, Ciriacy M (1997) Kinetic characterization of individual hexose transporters of *Saccharomyces cerevisiae* and their relation to the triggering mechanisms of glucose repression. Eur J Biochem 245:324–333

Ren B, Robert F, Wyrick JJ, Aparicio O, Jennings EG, Simon I, Zeitlinger J, Schreiber J, Hannett N, Kanin E, Volkert TL, Wilson CJ, Bell SP, Young RA (2000) Genome-wide location and function of DNA binding proteins. Science 290:2306-2309

Rep M, Krantz M, Thevelein JM, Hohmann S (2000) The transcriptional response of *Saccharomyces cerevisiae* to osmotic shock. Hot1p and Msn2p/Msn4p are required for the induction of subsets of high osmolarity glycerol pathway dependent genes. J Biol Chem 275:8290–300

Schmidt MC, McCartney RR, Zhang X, Tillman TS, Solimeo H, Wölfl S, Almonte C, Watkins SC (1999) Std1 and Mth1 proteins interact with the glucose sensors to control glucose-regulated gene expression in *Saccharomyces cerevisiae*. Mol Cell Biol 19:4561-4571

Schulte F, Wieczorke R, Hollenberg CP, Boles E (2000) The *HTR1* gene is a dominant negative mutant allele of *MTH1* and blocks Snf3- and Rgt2-dependent glucose signaling in yeast. J Bacteriol 182:540-542

Skowyra D, Craig KL, Tyers M, Elledge SJ, Harper JW (1997) F-box proteins are receptors that recruit phosphorylated substrates to the SCF ubiquitin-ligase complex. Cell 91:209-219

Smith RL, Johnston AD (2000) Turning genes off by Ssn6-Tup1: a conserved system of transcriptional repression in eukaryotes. Trends Biochem Sci 25:325-330

Theodoris G, Fong NM, Coons DM, Bisson L (1994) High-copy suppression of glucose transport defects by *HXT4* and regulatory elements in the promoters of the *HXT* genes in *Saccharomyces cerevisiae*. Genetics 137:957-966

Tomas-Cobos L, Sanz P (2002) Active Snf1 protein kinase inhibits expression of the *Saccharomyces cerevisiae HXT1* glucose transporter gene. Biochem J 368:657-663

Vagnoli P, Coons DM, Bisson L (1998) The c-terminal domain of Snf3p mediates glucose-responsive signal transduction in *Saccharomyces cerevisiae*. FEMS Microbiol Lett 160:31-36

Vallier LG, CoonsD, Bisson LF, Carlson M (1994) Altered regulatory responses to glucose are associated with a glucose transport defect in *grr1* mutants of *Saccharomyces cerevisiae*. Genetics 136:1279-1285

Watson AD, Edmondson DG, Bone JR, Mukai Y, Yu Y, Du W, Stillman, DJ, Roth S (2000) Ssn6-Tup1 interacts with class I histone deacetylases required for repression. Genes Dev 14:2737-2744

Wendell DL, Bisson LF (1993) Physiological characterization of putative high-affinity glucose transport protein Hxt2 of *Saccharomyces cerevisiae* by use of anti-synthetic peptide antibodies. J Bacteriol 175:7689-7696

Wendell DL, Bisson LF (1994) Expression of high-affinity glucose transport protein Hxt2p of *Saccharomyces cerevisiae* is both repressed and induced by glucose and appears to be regulated posttranslationally. J Bacteriol 176:3730-3737

Wu J, Suk N, Carlson M, Grunstein M (2001) TUP1 utilizes histone H3/H2B-specific HDA1 deacetylase to repress gene activity in yeast. Mol Cell 7:117-126

Xue Y, Batlle M, Hirsch JP (1998) GPR1 encodes a putative G protein-coupled receptor that associates with the Gpa2p Galpha subunit and functions in a Ras-independent pathway. EMBO J 17:1996–2007

Zaman Z, Ansari AZ, Koh SS, Young R, Ptashne M (2001) Interaction of a transcriptional repressor with the RNA polymerase II holoenzyme plays a crucial role in repression. Proc Natl Acad Sci USA 98:2550-2554

11 Integration of nutrient signalling pathways in the yeast *Saccharomyces cerevisiae*

Johnny Roosen, Christine Oesterhelt, Katrien Pardons, Erwin Swinnen, and Joris Winderickx

Abstract

The ability to sense and to respond to changes in the nutrient availability is an essential feature for the survival of every organism. *Saccharomyces cerevisiae* has several signal transduction cascades to optimally adapt its metabolism to the availability of nutrients in the environment. In this chapter, we focus on the convergence of signal transduction pathways for nutrient sensing in budding yeast. In the first part, we will give an overview of the glucose-induced signal transduction pathways, focusing in particular on the Ras/cAMP pathway and its pleiotropic characteristics. Secondly, the current knowledge of the protein kinases Sch9 and Pho85 in nutrient-induced signal transduction is reviewed. Finally, the interconnectivity between these multiple pathways in glycogen biosynthesis, control of Msn2 activity and pseudohyphal growth will be discussed. We conclude that complex networks are involved in the integration of nutrient signals, regulating the complete transcriptome and metabolome in an intriguingly dynamical manner.

11.1 Glucose-induced signalling

Carbon sources are of central importance to living organisms. For *Saccharomyces cerevisiae*, glucose is the preferred carbon and energy source. Addition of a rapidly fermentable sugar, such as glucose, to cells grown on a non-fermentable carbon source, e.g. ethanol, causes a rapid metabolic transition from respiration to fermentation. This transition involves a variety of regulatory pathways, all directed to the optimal and exclusive use of the rich carbon source (Jiang et al. 1998). Two of the best-described glucose-induced signal transduction pathways are the main glucose repression pathway (*a.k.a.* the catabolite repression pathway) and the Ras/cAMP pathway. Activation of these pathways is dependent on both transport and phosphorylation of the fermentable sugar. Interestingly, no further metabolism beyond the step of phosphorylation appears to be required. Following transport into the cell, fermentable sugars are phosphorylated and enter glycolysis. Yeast contains three hexose kinases responsible for this phosphorylation: hexokinase PI (Hxk1), hexokinase PII (Hxk2), and glucokinase (Glk1). The two isoenzymes, Hxk1 and Hxk2, are both able to phosphorylate glucose, fructose and mannose, whereas Glk1 is specific for glucose and mannose (Lobo and Maitra

Topics in Current Genetics, Vol. 7
J. Winderickx, P.M. Taylor (Eds) Nutrient-Induced Responses in Eukaryotic Cells

1977). Hxk2 is clearly the main sugar-phosphorylating enzyme during growth on glucose (Herrero et al. 1995; de Winde et al. 1996). This is also reflected in the expressional regulation of the hexose kinases. Indeed, Hxk2 is induced (de Winde et al. 1996; Martinez-Campa et al. 1996), while Hxk1 and Glk1 are repressed in the presence of fermentable sugars, such as glucose, fructose and mannose (Herrero et al. 1995; de Winde et al. 1996). Interestingly, this regulation is mediated by the main glucose-repression and the Ras/cAMP pathway, the two signal transduction cascades triggered by sugar phosphorylation (Herrero et al. 1995; de Winde et al. 1996).

11.1.1 Main glucose repression pathway

The glucose repression pathway is a rapid response system that, when activated by glucose, represses genes encoding for proteins involved in uptake and utilisation of alternative carbon sources (Ronne 1995). Central components of glucose repression are the Mig1 transcriptional repressor complex, the Snf1 protein kinase complex and protein phosphatase 1. The three components of the Snf1 kinase complex (Snf1, Snf4, and the Sip proteins) show homology to the subunits of the related AMP activated protein kinase (AMPK) in mammalian cells (Stapleton et al. 1994; Woods et al. 1994). Although the Snf1 kinase is not directly activated by AMP, its activity correlates well to the AMP/ATP ratio (Wilson et al. 1996). In cells growing on glucose, the AMP content decreases due to the generation of ATP by glycolysis. Under these conditions, Snf1 is inactive, allowing repression of glucose-repressible genes. When glucose is exhausted, AMP levels are replenished, resulting in a high AMP/ATP ratio. Activation of Snf1 and subsequent relief of glucose repression are the result. Although an increase in the AMP/ATP ratio was observed upon removal of glucose from repressed cells, AMP and ATP levels appear however to be very similar during growth on glycerol and on glucose (Hardie and Carling 1997) and are therefore unlikely to be the main signal for glucose repression.

Alternatively, Hxk2 may harbour the glucose-sensing function. Hxk2 was identified as one of the first proteins involved in glucose repression (Zimmermann and Scheel 1977; Entian 1980) and a special role in signal transduction has always been attributed to this enzyme (Entian and Fröhlich 1984; Entian et al. 1984). Even though each one of the yeast sugar kinases (Hxk1, Hxk2, Glk1) is able to supply the cell with adequate levels of sugar phosphates (Beullens et al. 1988), only Hxk2 can sustain a long-term glucose-repression (de Winde et al. 1996; Sanz et al. 1996). Initially, a good correlation was observed between the extent of glucose repression and the catalytic activity of Hxk2 (Ma et al. 1989b; Rose et al. 1991). Recently, however, mutant alleles of Hxk2 have been isolated in which sugar kinase activity and glucose repression are differentially affected (Hohmann et al. 1999; Kraakman et al. 1999b; Mayordomo and Sanz 2001). This has been supported by structure-function analysis of Hxk2, suggesting that the establishment of glucose repression depends on the phosphoryl transfer reaction and is probably related to the formation of a stable transition intermediate (Kraakman et al. 1999b). Possibly, the conformational change within the enzyme triggers catabolite repression via protein-protein interaction and protein modification but, as yet, no protein has been identified that would directly interact with Hxk2. Another putative trigger for signal transduction is the phosphorylation status of Hxk2. It has been reported that the enzyme is mainly phosphorylated under glucose-derepressed conditions (Vojtek and Fraenkel 1990; Kriegel et al. 1994). Addition of glucose to cells causes dephosphorylation of Hxk2 by the Reg1/Glc7 protein phosphatase (Randez-Gil et al. 1998b; Alms et al. 1999). Furthermore, protein phosphorylation appears to affect the monomer-dimer equilibrium of Hxk2 (Furman and Neet 1983; Randez-Gil et al. 1998b). Phosphorylation shifts the equilibrium to the monomeric state and increases the glucose affinity of the enzyme (Behlke et al. 1998; Herrero et al. 1998). Interestingly, Hxk2 has also been shown to display protein kinase activity itself (Herrero et al. 1989). The latter is regulated by the availability of glucose in the medium. Under conditions of derepression, protein kinase activity apparently decreases significantly, although the sugar-phosphorylating activity remains unchanged. The target of protein phosphorylation by Hxk2 has not yet been identified. However, the enzyme could have a role in regulating the phosphorylation status of the regulatory subunit of protein phosphatase 1 (Reg1/Hex2), since Reg1 is phosphorylated in response to glucose limitation and dephosphorylated by Glc7 when glucose is present (Sanz et al. 2000). The protein kinase activity of Hxk2 has been reported to correlate with the ATP-dependent auto-phosphorylation at Ser-158 (Fernandez et al. 1988; Heidrich et al. 1997; Kraakman et al. 1999b) but the functional significance of Hxk2 auto-phosphorylation on glucose repression is still unclear (Ma et al. 1989a; Herrero et al. 1998; Randez-Gil et al. 1998a).

11.1.2 The Ras/cAMP pathway

The Ras/cAMP pathway (Fig. 1) plays a major role in the control of stress resistance, metabolism, and proliferation. Synthesis of cAMP from ATP is catalysed by adenylate cyclase *(CYR1)* (Matsumoto et al. 1984; Kataoka et al. 1985). Adenylate cyclase is activated by the two small G-class proteins Ras1 and Ras2 (Toda et al. 1985; Field et al. 1988). Exchange of GDP for GTP on the Ras proteins is stimulated by the activating 'guanine nucleotide exchange factor' (GEF), encoded by *CDC25/SDC25* (Camonis et al. 1986; Broek et al. 1987; Jones et al. 1991; Camus et al. 1994). Ira1 and Ira2, on the other hand, act as inhibitory regulators of Ras by stimulating the intrinsic GTPase-activity (Tanaka et al. 1990).

Only two stimuli are known to activate cAMP synthesis in yeast: intracellular acidification and addition of glucose to derepressed cells. It has been shown that intracellular acidification activates the pathway through the increase in the GTP/GDP ratio on Ras, possibly via inhibition of the GTPases Ira1 and Ira2 and independent of Cdc25/Sdc25 (Colombo et al. 1998). For glucose, it was originally proposed that Cdc25 and Ras would be involved in the transduction of the signal (Mbonyi et al. 1988; Munder and Küntzel 1989; Van Aelst et al. 1991), but later studies reported that Cdc25 was not the receiver of the glucose signal (Goldberg et al. 1994) and that glucose did not alter the GDP/GTP ratio bound on Ras (Colombo et al. 1998). Most recent data now indicate that an increased level of Ras-GTP could, at least partially, be involved in glucose-induced activation of the pathway (Rudoni et al. 2001).

A second G-protein involved in the glucose-induced activation of the Ras/cAMP pathway is Gpa2. *GPA2* originally has been cloned as a yeast homologue of mammalian heterotrimeric Gα proteins (Nakafuku et al. 1988). Remarkably, no accompanying Gβγ subunits have yet been identified and hence the question remains whether the Gpa2 protein acts alone or with a yet unidentified Gβγ partner. Compared to other Gα proteins, Gpa2 contains an unusually long N-terminal extension, which might take over the function of the Gβγ dimer. The recent identification of two Kelch-repeat proteins, Gpb1 and Gpb2, suggests that signalling could possibly occur via these proteins which structurally mimic typical β subunits (Harashima and Heitman 2002).

Together with the upstream receptor-like protein, Gpr1, Gpa2 constitutes a G-protein coupled receptor system (GPCR) (Yun et al. 1997; Xue et al. 1998; Kraakman et al. 1999a). This G-protein coupled receptor has the same characteristics as its mammalian counterparts, i.e. 7 transmembrane domains, an extracellular N-terminus, a large third intracellular loop and a long cytoplasmic C-terminus (Baldwin 1994; Bockaert and Pin 1999). Gpa2 is negatively regulated by Rgs2, originally isolated as a multicopy suppressor of glucose-induced loss of stress resistance in stationary phase cells (Versele et al. 1999). Rgs2 belongs to the family of regulators of heterotrimeric G-protein signalling (RGS) and acts as a GTPase for Gpa2. Another yeast RGS protein, Sst2, is involved in regulating Gα protein activity in the pheromone pathway. Even though the two RGS proteins fall into the same class of regulatory proteins, they both exert specific, non-overlapping functions (Versele et al. 1999).

Glucose-induced activation of the Ras/cAMP pathway requires two sensing systems (Fig. 1) (Campbell-Burk et al. 1987; Rolland et al. 2001, 2002). The first system is intracellular sugar phosphorylation (Beullens et al. 1988) and appears to act as a prerequisite for activation of the Ras/cAMP pathway by glucose (Campbell-Burk et al. 1987; Rolland et al. 2001, 2002). In contrast to the glucose repression pathway, any of the three sugar phosphorylating enzymes (Hxk1, Hxk2, and Glk1) can activate the Ras/cAMP pathway. Activation of cAMP synthesis beyond the basal level is a transient effect, limited to the short period of the onset of fermentation, when all three kinases are functional (de Winde et al. 1996). The complete absence of a glucose- or fructose-induced cAMP peak in a *hxk1Δ hxk2Δ glk1Δ* strain shows that sugar phosphorylation is an absolute requirement for activation of the pathway (Hohmann et al. 1999; Kraakman et al. 1999b). Neither glucose-6-P (Glc-6-P) nor ATP, the substrate of adenylate cyclase, are the (sole) triggers for cAMP synthesis (Campbell-Burk et al. 1987; Rolland et al. 2001, 2002).

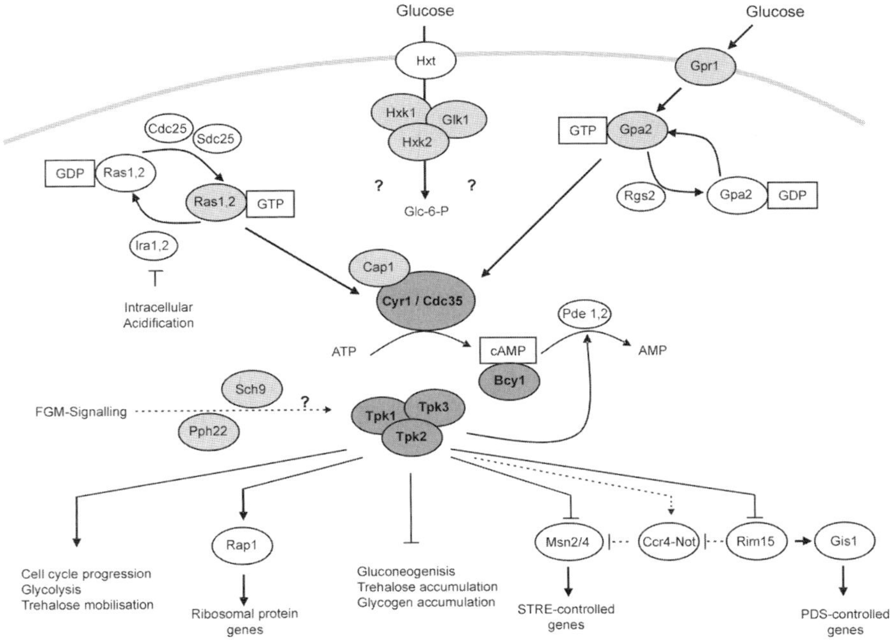

Fig. 1. The Ras/cAMP pathway in *S. cerevisiae*. In derepressed yeast cells, the addition of glucose triggers a boost in the second messenger cAMP and, subsequently, the activation of protein kinase A. PKA phosphorylates a number of downstream targets and thereby initiates the adaptation of cells to the new nutritional situation. The presence of glucose has to be sensed in two ways in order to activate the pathway: binding of external glucose to the receptor Gpr1 activates the GPCR system, while internal glucose is phosphorylated by any one of the three hexose kinases (Hxk1, Hxk2, and Glk1). The FGM pathway also acts on PKA, independently of cAMP. (For details, see sections 11.1.2. and 11.2.1.). Arrows and bars represent positive and negative interactions respectively. Dashed lines represent putative interactions.

Secondly, extracellular glucose is detected by the GPCR system, consisting of Gpr1 and Gpa2. Addition of glucose to derepressed cells activates the low affinity Gpr1 receptor, which in turn stimulates the exchange of GDP for GTP on Gpa2 (Kraakman et al. 1999a). Gpr1 displays a high specificity for D-glucose and sucrose, with an apparent Km for activation of 75 mM and 1 mM, respectively (Rolland et al. 2001). The low affinity of Gpr1 for glucose explains the need for a relatively high concentration of D-glucose (100 mM) to achieve an optimal cAMP peak. Addition of other sugars, such as fructose, mannose or glucose analogs, i.e. L-glucose, does not trigger a cAMP response.

An intriguing item that remains to be resolved is how both glucose-sensing systems are connected. As described above, only when there is intracellular glucose phosphorylation a further activation of the Ras/cAMP pathway may be obtained via the GPCR system. We already proposed a possible link between the two G-protein systems that control adenylate cyclase activity, i.e. Gpr1-Gpa2-Rgs2 and Cdc25-Ras-Ira. Early studies have already proposed a role as glucose-signal transducers for the Ras proteins (Mbonyi et al. 1988) and Cdc25 (Munder and Küntzel 1989; Van Aelst et al. 1990, 1991). Despite many efforts, no upstream activator of Cdc25 has yet been identified. Possibly, the sugar kinases are required for the activation of Cdc25. In this case, activated Cdc25 would increase the Ras-GTP content and stimulate adenylate cyclase activity. Cdc25 was already shown to bind directly to adenylate cyclase (Mintzer and Field 1999). This interaction could promote an efficient assembly of the Ras-GTP-adenylate cyclase complex. Interestingly, both Ras and Cdc25 are known to be involved in membrane targeting of adenylate cyclase (Pardo et al. 1993; Bhattacharya et al. 1995). The role of Ras in the cAMP pathway might thus be to increase the responsiveness of adenylate cyclase to stimulation by the GPCR system. Other results indicate that the Ras proteins are also involved in the feedback-mechanism that quickly downregulates cAMP levels, following activation of the Ras/cAMP pathway (Gross et al. 1992, 1999). Addition of glucose to derepressed cells causes a hyperphosphorylation of Cdc25 resulting in the relocalization of the protein from the membrane to the cytoplasm. Cytosolic localization reduces the interaction between Cdc25 and membrane-bound Ras. Activation of Ras is thus prevented, ultimately resulting in an extinction of the cAMP signal (Gross et al. 1992). Hence, the role of Ras proteins can be regarded as a molecular buffer system to control adenylate cyclase activity.

Since the glucose activation pathway ultimately acts on adenylate cyclase, the sugar kinases may also act directly on this protein. The N-terminal part of adenylate cyclase has been reported to act as an inhibitor of the catalytic activity of the protein (Heideman et al. 1987; Uno et al. 1987). Hence, a model in which sugar-induced activation of cAMP synthesis leads to a relief of auto-inhibition on the catalytic part of adenylate cyclase is feasible. Possibly, this allows for interaction with and stimulates activation by the Gα protein Gpa2.

What is remarkable about the involvement of the sugar kinases in glucose-induced signalling is that for the main glucose repression pathway, Hxk2 clearly has a predominant role in activation and regulation, whereas any functional hexose kinase is sufficient for triggering cAMP synthesis. Still, the glucose-repressible aspect of the Ras/cAMP pathway suggests that there is a connection with the main glucose repression pathway. Indeed, no glucose-induced cAMP increase can be observed in *snf1Δ* mutants, which are unable to derepress (Arguelles et al. 1990).

Cyclic AMP controls a number of processes in yeast, such as responses to changes in the nutrient availability and passage of cells through G1 of the cell cycle (Gibbs and Marshall 1989; Broach and Deschenes 1990). All of these effects are mediated through protein phosphorylation by protein kinase A (PKA). The sole function of cAMP in yeast appears to be activation of PKA (Matsumoto et al. 1985). The protein kinase is a hetero-tetramer consisting of two catalytic subunits, encoded redundantly by *TPK1*, *TPK2*, and *TPK3* (Toda et al. 1987b), and two regulatory subunits encoded by *BCY1* (Toda et al. 1987a; Gibbs and Marshall 1989; Taylor et al. 1990). Cyclic AMP activates PKA by binding to its regulatory subunit, thereby releasing and activating the catalytic subunits (Kuret et al. 1988). The phosphotransferase activity of PKA, as well as the binding site for Bcy1, have been mapped to the N-terminus of the Tpk proteins (Gibbs et al. 1992). Since Bcy1 is an inhibitor of PKA activity, its deletion can rescue the growth defect of a *cyr1Δ* (adenylate cyclase) mutant (Matsumoto et al. 1982) as well as the lethal phenotype of a *ras1Δ ras2Δ* double deletion, by rendering PKA active (Toda et al. 1985). Cyclic AMP in turn is hydrolysed by the low- and high-affinity phosphodiesterases, encoded by *PDE1* and *PDE2*, respectively (Sass et al. 1986; Nikawa et al. 1987; Wilson and Tatchell 1988). The nature of the cAMP signal and the subsequent boost in PKA activity are transient due to a strong and fast feedback inhibition on cAMP levels (Mbonyi et al. 1990). This feedback mechanism involves PKA itself and Pde1 (Ma et al. 1999) but might also affect adenylate cyclase directly or indirectly through the Ras proteins or Cdc25.

Another mode of regulation of PKA activity, apart from feedback-control, involves the regulation of the subcellular localization. In glucose-derepressed cells, Bcy1 is equally distributed throughout the cytosol and the nucleus (Griffioen et al. 2000). When cells are transferred to glucose, cytosolic Bcy1 is targeted towards the nucleus. This change in intracellular localization can only be achieved in response to glucose-availability. Nitrogen starvation in the presence of glucose has no effect. The intracellular localization of Bcy1 also strictly depends on its phosphorylation status: phosphorylated Bcy1 localizes to the cytosol, non-phosphorylated Bcy1 to the nucleus (Griffioen et al. 2001). Bcy1 has multiple phosphorylation sites (Werner-Washburne et al. 1991) but only the N-terminal domain of the protein (residues 1-124) appears to be essential and sufficient for proper nutrient-regulated localization of the protein (Griffioen et al. 2000). The localization of the catalytic subunit Tpk1 follows that of Bcy1 and strictly depends on it (Griffioen et al. 2000).

Inactivation of any of the PKA catalytic subunits is viable, whereas deletion of all three genes is lethal, suggesting a largely overlapping function for Tpk1-3, which is also reflected by a high degree of homology (Toda et al. 1987b). Despite the high redundancy for viability, the catalytic subunits also exhibit separate functions. For instance, Tpk2 is required for the induction of pseudohyphal differentiation, whereas Tpk1 and Tpk3 have a repressive effect on this phenotype (Robertson and Fink 1998; Pan and Heitman 1999) (see also 11.3.2). Recent genome-wide transcriptional profiling indicates that Tpk2 is specifically involved in the repression of genes involved in iron uptake and inhibition of respiratory growth, whereas Tpk1 regulates genes involved in the branched-chain amino acid pathway, mitochondrial DNA stability and the exit from stationary phase (Robertson et al. 2000).

The main function of the Ras/cAMP pathway is to signal the presence of glucose in the medium and, in turn, to switch cells from respiration to fermentation (Boy-Marcotte et al. 1996; Jiang et al. 1998). It is important to note that this signal transduction cascade can only be activated in glucose-derepressed cells. Once the switch from respiration to fermentation has been achieved and cells are glucose-repressed, they exhibit a 'high PKA phenotype' and the Ras/cAMP pathway is no longer responsive. In general, a 'high PKA phenotype' includes an upregulation of glycolytic enzymes and a downregulation of gluconeogenic enzymes, a mobilization of storage carbohydrates, such as trehalose and glycogen (Thevelein 1992,

1994), a repression of stress-responsive genes (Ruis and Schuller 1995) and an induction of ribosomal protein genes (Kraakman et al. 1993; Griffioen et al. 1996). As a consequence of these changes, cells lose their stress resistance (Thevelein 1994). In the absence of a fermentable carbon source, cells exhibit a 'low PKA phenotype' with opposite phenotypic characteristics.

The multiple stress responses have been correlated to the activity of the transcriptional activators Msn2 and Msn4 (Schmitt and McEntee 1996). (For details about transcriptional control by Msn2/4: see 11.3.1). Activity of these transcription factors is required for the expression of STRE (general Stress Response Element)-controlled genes and is inhibited by PKA (Boy-Marcotte et al. 1998; Moskvina et al. 1998; Smith et al. 1998) (Fig. 1). Under stress conditions and during growth on non-fermentable carbon sources, when PKA activity is low, Msn2 and Msn4 are targeted towards the nucleus. This nuclear localization is strictly dependent on the phosphorylation status of the two proteins. During growth on glucose, Msn2 and Msn4 are phosphorylated and retained in the cytosol. When glucose becomes exhausted, cells undergo a diauxic shift, which precedes respiration. Upon this diauxic transition, the proteins become hyperphosphorylated, translocated to the nucleus and thus activated. This hyperphosphorylation is prevented by high PKA activity (Gorner et al. 1998; Garreau et al. 2000). *In vitro* phosphorylation of Msn2 and Msn4 by PKA occurs within their nuclear localization domain (residues 576-704) and, most likely, affects protein transport into the nucleus (Gorner et al. 2002). Finally, deletion of *MSN2* and *MSN4* can rescue an otherwise lethal *tpk null* strain (Smith et al. 1998). Another effector of PKA in mediating transcriptional regulation of the STRE-controlled genes is the Ccr4-Not complex. It has been suggested that the complex contributes to repress transcription by Msn2/4 (Fig. 1), possibly via a direct interaction with Tpk2 (Lenssen et al. 2002). Proper entry into diauxic shift also requires the transcriptional activator Gis1. This factor binds the so-called post diauxic shift (PDS) elements. The activity of Gis1 as well as that of Msn2 and Msn4 depend on the protein kinase Rim15, which itself is negatively regulated by PKA (Reinders et al. 1998; Pedruzzi et al. 2000; Lenssen et al. 2002). Not surprisingly, Rim15 is also required for proper entry into the stationary phase and loss of Rim15 can rescue the lethality of a *tpk null* strain (Reinders et al. 1998).

While the Gis1- and Msn2/4-mediated expression of PDS- and STRE-controlled genes is repressed by PKA, cell proliferation and the expression of ribosomal protein (rp) genes are positively regulated by PKA (Herruer et al. 1987; Kraakman et al. 1993; Griffioen et al. 1994). The multifunctional transcriptional activator-repressor Rap1 (Shore et al. 1984; Sussel and Shore 1991) has been implicated in the regulation of rp gene expression (Teem et al. 1984; Moehle and Hinnebusch 1991; Lascaris et al. 1999). Regulation of the growth rate by rp genes strongly depends on PKA activity in response to environmental conditions. Constitutively activated PKA increases transcription of rp genes by increasing Rap1-dependent transcriptional activity (Klein and Struhl 1994).

Activation of the Ras/cAMP pathway also initiates transcription of *HIS3* and *HIS4*, mediated by the transcription factor Gcn4 (Engelberg et al. 1994). Since

Rap1 is involved in the Gcn4-mediated transcription of *HIS4*, a link between PKA activity and nitrogen metabolism can be established (Yu et al. 2001).

In addition to these transcriptional regulations, key enzymes of the biosynthesis of storage carbohydrates are subject to PKA phosphorylation in an apparent Rim15-dependent manner (Reinders et al. 1998; Pedruzzi et al. 2000). Glycogen synthase (*GSY2*) is inactivated *in vitro* by PKA (Peng et al. 1990) as well as the subunits of trehalose-6P-synthase, *TPS1* and *TPS2* (Panek et al. 1987; Fernandes et al. 1997). In addition, Tps1 is also subject to catabolite inactivation (François et al. 1991). Synthesis of glycogen and trehalose is, therefore, inhibited under conditions of high glucose availability and high PKA activity (François et al. 1988; Thevelein 1996; Hampsey 1997). At the same time, PKA-mediated phosphorylation of glycogen phosphorylase (*GPH1*) and trehalase (*NTH1*) leads to enzyme activation (Uno et al. 1983; Zahringer et al. 1998) and in turn to a breakdown of storage carbohydrates. PKA is also a central element in regulating glycolytic fluxes. Fructose-2.6-bisphosphatase (*FBP26*) (Kretschmer et al. 1987), fructose-1.6-bisphosphatase (*FBP1*) (Rittenhouse et al. 1987; Hoffman and Winston 1991), 6-phosphofructo-2-kinase (*PFK26)* (Vaseghi et al. 2001), and pyruvate kinase (*CDC19* and *PYK2*) (Cytrynska et al. 2001; Portela et al. 2001) are all targets of PKA phosphorylation. Eventually, these post-translational protein modifications result in the stimulation of glycolysis in the presence of glucose, while gluconeogenetic and respiratory metabolism are negatively regulated (Goncalves et al. 1997).

In general, PKA is one of the key switches in regulating metabolic adaptations to nutrient availability in yeast. Disruption of PKA activity such as deletion of the three catalytic subunits Tpk1-3 or uncoupling of the catalytic subunits from its regulatory elements is therefore highly detrimental to the cell. For instance, mutants with constitutively active PKA, such as in a *bcy1Δ* or in a RAS2^{Val19} strain rapidly acquire suppressor mutations, which alleviate the defect. The same is true for strains with a constitutively low or attenuated PKA activity such a *tpk1*^{w} *tpk2Δ tpk3Δ*. Other characteristic phenotypes of strains with overactive PKA (e.g. *bcy1Δ*) include no growth on ethanol or glycerol because of *ADH2* repression (Taylor and Young 1990), increased sensitivity to stress due to constitutive repression of STRE-controlled genes (see above) and low levels of storage compounds (see above). Proliferation is stimulated while the total life span is reduced (Lin et al. 2000; Fabrizio et al. 2001, 2003). Finally, sporulation is inhibited in cells with high PKA activity (Malone 1990; Matsuura et al. 1990), while cells with a low PKA activity (e.g. overexpression of *BCY1*) sporulate without the need for a nutrient starvation signal (Portela et al. 2001).

11.2 Nitrogen-, amino acid-, and phosphate-induced signalling

Besides glucose-induced signalling, multiple pathways are involved in signalling the presence of nutrients such as nitrogen, amino acids and phosphate in the environment. Here, we will focus in particular on two protein kinases, Sch9 and Pho85, which have an important integrative function because of their crosstalk with other pathways. First, we will focus on the integration of nitrogen- and amino acid-induced signalling through the protein kinase Sch9. Secondly, the phosphate-induced control of the Pho85 protein kinase and its pleiotropic features will be discussed.

11.2.1 The role of Sch9 in nutrient-signalling

The Ras/cAMP pathway triggers multiple responses upon the addition of glucose. However, the activation of PKA by cAMP is only transient (see above) and the active state of the Ras/cAMP pathway seems to be restricted to the transition from respiratory to fermentative growth (Jiang et al. 1998). Nevertheless, fermenting cells exhibit a typical 'high PKA phenotype' (Thevelein 1994). Readdition of any essential nutrient, such as nitrogen or amino acids, to cells starved for this nutrient on glucose-containing medium, triggers similar responses as activation of the Ras/cAMP pathway. Among the phenotypes are activation of trehalase (Hirimburegama et al. 1992), induction of ribosomal protein genes (Neuman-Silberberg et al. 1995; Griffioen et al. 1996; Winderickx et al. 1996) and the repression of STRE-controlled genes (Boy-Marcotte et al. 1998; Gorner et al. 1998; Tadi et al. 1999). These nutrient-induced effects appear to be independent of cAMP (Hirimburegama et al. 1992; Giots et al. 2003) and are also observed in mutants lacking Bcy1, the regulatory subunit of PKA. Moreover, in contrast to the Ras/cAMP pathway, sugar phosphorylation is not required (Pernambuco et al. 1996), while activity of the catalytic subunits of PKA is essential (Cameron et al. 1988; Durnez et al. 1994). Since this cAMP-independent control of PKA targets requires a fermentable carbon source and all other nutrients for growth, the pathway was called the 'Fermentable Growth Medium-induced (FGM) pathway'.

The protein kinase Sch9 has recently been implicated in nitrogen- and amino acid-induced FGM-signalling (Crauwels et al. 1997). The C-terminal part of Sch9 is highly homologous to the AGC serine/threonine protein kinase and is functionally related to mammalian PKB/Akt (Geyskens et al. 2001). Unlike PKB/Akt, which contains a PH-domain, able to bind PI(3,4,5)P_3, thereby allowing phosphorylation and subsequent activation of PKB, Sch9 contains a C2 phospholipid and a calcium-binding motif at the N-terminus (Geyskens et al. 2001). In yeast, phosphoinositides appear to act as second messengers in nitrogen-signalling (Schomerus and Kuntzel 1992; Bergsma et al. 2001; Wera et al. 2001). However, the correlation between phosphoinositide signalling and the activation of Sch9 still remains to be established.

Sch9 is structurally related to the catalytic subunits of the cAMP-dependent protein kinase. It was originally isolated as a multi-copy suppressor of strains deficient in PKA activity (Toda et al. 1988). An *sch9Δ* strain exhibits a slow growth and small colony phenotype which can be suppressed by activation of the Ras/cAMP pathway (Toda et al. 1988; Hartley et al. 1994). Deletion of *SCH9* is synthetically lethal with *gpr1Δ*, *gpa2Δ,* or *ras2Δ* mutations (Kraakman et al.

1999a; Lorenz et al. 2000). Given the structural homology between Sch9 and the catalytic subunits of PKA, it may not be surprising that both kinases share a large number of targets. Indeed, high throughput protein chip analysis showed that both PKA and Sch9 can phosphorylate the same substrates, including Hog1 and Pfk2, albeit with different relative specificities (Zhu et al. 2000). For the concept of FGM signalling, the precise relationship between PKA and Sch9 remains to be clarified. It has been suggested that Sch9 acts directly on the free catalytic subunits of PKA (Thevelein 1992; Durnez et al. 1994; Thevelein 1994) (Fig. 1). Although *in vitro* PKA activity is enhanced in *sch9Δ* strains, direct post-translational modification of PKA by Sch9 has not been shown (Crauwels et al. 1997). Other observations suggest that PKA and Sch9 may control parallel pathways which converge downstream of the kinases. For instance, expression of *ADH2* is negatively regulated by PKA via inactivation of the transcriptional activator Adr1 (Cherry et al. 1989), whereas Sch9 acts as a positive modulator of *ADH2* expression, independent of Adr1 (Denis and Audino 1991). In addition, our recent data confirm that the PKA-pathway and the Sch9 pathway act in parallel but that they converge on the protein kinase Rim15 (Roosen, manuscript in preparation).

One of the major questions that remains to be answered is how the activity of Sch9 is altered in response to the nutritional status and how its signal is transmitted to downstream targets. Recent research has started to unravel specific cellular functions of the Sch9 protein kinase and has shed new light onto the possible mechanisms involved.

It has been proposed that Sch9 might define a downstream effector branch of the Gpr1-Gpa2 GPCR system in parallel to the Ras/cAMP pathway (Xue et al. 1998; Lorenz et al. 2000). The integrity of this GPCR system depends also on an interaction with the phospholipase C, Plc1 (Ansari et al. 1999), which in turn interacts with Tor2, one of the two "targets of rapamycin" in yeast (Lin et al. 1998). This points to a direct or indirect involvement of Sch9 in Tor-dependent signalling, which would not only provide an explanation for the remarkable similarity of nitrogen-dependent targets of Sch9 and Tor but it could lead to a model that resembles the situation as described in higher eukaryotes for PKB and mTOR signalling.

Similar to the function of mammalian PKB (Scheid and Woodgett 2001; Kozma and Thomas 2002), Sch9 also appears to be involved in the regulation of cell size (Jorgensen et al. 2002). Deletion of *SCH9* (Toda et al. 1988) results in a slow growth phenotype (Blumberg and Silver 1991) and small cell size known as the 'Wee-phenotype' (Xu and Norris 1998; Jorgensen et al. 2002). The same phenotype has been described for a *sfp1Δ* strain. Further analysis uncovered a synthetic genetic interaction between Sfp1 and Sch9, indicating that Sch9 and the DNA-binding protein Sfp1 (Blumberg and Silver 1991) might converge on the same processes (Jorgensen et al. 2002). It has been proposed that Sfp1 acts in concert with Rap1 to regulate transcription of ribosomal protein (rp) genes by binding to the RPPE (Hughes et al. 2000) and PAC (Dequard-Chablat et al. 1991) promoter elements (Jorgensen et al. 2002). Consistently, nitrogen- and amino acid-induced expression of rp-genes is largely abolished in *sch9Δ* mutants (Crauwels et al. 1997; Geyskens et al. 2001).

There have also been a number of reports suggesting an involvement of Sch9 in the regulation of longevity and stress resistance of cells. Reducing the activity of the cAMP/PKA pathway or lowering the amounts of glucose (calorie-restriction) extends the life span of cells. Several pathways are controlling this phenotype. One involves the NAD-dependent histone deacetylase Sir2 (Lin et al. 2000), which mediates life span extension in response to calorie restriction or high external osmolarity via Hog1, Cdc25, Npt1, and Gpd1, but not Msn2, Msn4, or Tps1 (Kaeberlein et al. 2002). The other pathways involve adenylate cyclase (*CYR1*/*CDC35*) and Sch9. Mutations in *CYR1*/*CDC35* or *SCH9* increase the resistance to oxidants and extend the lifespan of yeast. In contrast to the Sir2-mediated pathway, this process is partially mediated by the protein kinase Rim15 and the transcriptional activators Msn2 and Msn4 (Fabrizio et al. 2001). Ageing and early cell death has been related to oxidative damage of macromolecules. Two antioxidant enzymes, CuZn superoxide dismutase (*SOD1*) and Mn superoxide dismutase (*SOD2*) have been identified as contributing to the long-term survival of yeast (Longo et al. 1999). Similar to loss of Rim15 function, deletion of *SOD2* abolishes life span extension in *sch9Δ* mutants (Fabrizio et al. 2003). This is probably due to the transcriptional regulation of *SOD2* via the transcription factors Msn2, Msn4, and Gis1, which appear to be regulated by Rim15 (Flattery-O'Brien et al. 1997; Reinders et al. 1998; Pedruzzi et al. 2000). Interestingly, however, deletion of Msn2 and Msn4 did not affect the longevity of an *sch9Δ* mutant, while it clearly reduced the life span of a *cyr1Δ* mutant (Fabrizio et al. 2001). This finding supports the idea that Sch9 and PKA act through parallel pathways.

In addition, similar to the deletion of *SCH9*, overexpression of *PPH22*, encoding the catalytic subunits of the protein phosphatase PP2A, leads to a 'high PKA phenotype' including defects in trehalose accumulation, high trehalase activity and constitutive expression of STRE-controlled genes. These effects are dependent on the presence of Sch9, suggesting that Pph22 dephosphorylates Sch9 or that it may interfere with a downstream component (Sugajska et al. 2001).

Finally, Sch9 regulates the *in vivo* activity of the Hsp90 chaperone complex (Morano and Thiele 1999). Hsp90 is an abundant and ubiquitous heat shock protein essential for viability of eukaryotic cells (Borkovich et al. 1989; Cutforth and Rubin 1994). Its function is required for Hsf1 (heat shock factor)-mediated transcriptional control in response to stress (Morano and Thiele 1999). Moreover, the Hsp90 chaperone complex is required for low basal activity of the pheromone signalling pathway and for pheromone induction using Ste11 as a substrate (Louvion et al. 1998). Consistently, loss of Sch9 function leads to derepression of Hsp90 signal transduction and hyperactivation of the pheromone response MAPK pathway in the absence of pheromone (Morano and Thiele 1999).

Although the precise mechanisms remain to be elucidated, Sch9 appears to function at the intersection of several pathways. Most likely, its role is to act as an integrator of the glucose and nitrogen signals so as to coordinate adequate responses to an ever-changing nutritional environment.

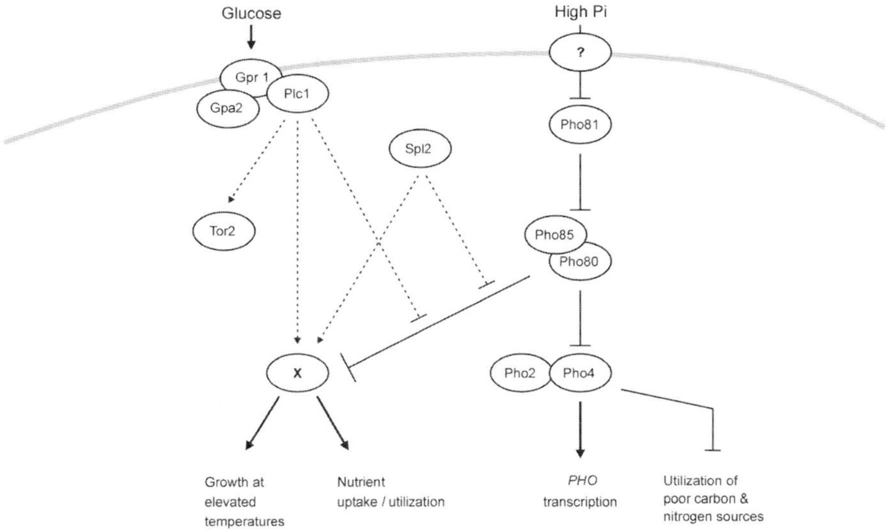

Fig. 2. Schematic depiction of the PHO pathway and its genetic interactions with other nutrient-regulated systems. The Pho85-Pho80 (CDK/cyclin) complex is involved in phosphate-dependent regulation of the transcription factor Pho4 through the action of the Pho81 CKI. Pho85-Pho80 also regulates another factor, designated as "X", which genetically interacts with the phospholipase C system and Spl2, a Pho81-homologue. Furthermore, the PHO pathway seems to be involved in the utilization of poor carbon and nitrogen sources in a Pho4-dependent manner. (For details, see section 11.2.2.). Arrows and bars represent positive and negative interactions respectively. Dashed lines represent putative interactions.

11.2.2 The role of Pho85 in nutrient-signalling

Under conditions of low inorganic phosphate (Pi) availability, yeast cells activate a regulatory mechanism known as the PHO pathway (Lenburg and O'Shea 1996), which leads to an increased expression of genes involved in the acquisition, uptake and storage of Pi (Ogawa et al. 2000). Three of these genes, *PHO5*, *PHO11*, and *PHO12*, encode for secreted acid phosphatases, which enable the cells to use organic phospho-substrates as a Pi-source. Induction of *PHO* genes in response to phosphate starvation depends on the transcription factors Pho2 and Pho4 (Fig. 2). The latter is phosphorylated at several Ser residues by the cyclin dependent protein kinase (CDK) Pho85, in association with the Pho80 cyclin (Kaffman et al. 1994). This phosphorylation of Pho4 leads to an inactivation of the transcription factor by disrupting Pho4-Pho2 interaction, stimulating nuclear export of Pho4 and inhibiting its nuclear import (O'Neill et al. 1996; Kaffman et al. 1998a, 1998b; Komeili and O'Shea 1999; Jeffery et al. 2001). Regulation of the kinase activity of the Pho85-Pho80 complex in response to phosphate levels occurs through the

Pho81 cyclin-dependent protein kinase inhibitor (CKI) protein (Schneider et al. 1994). Pho81 physically interacts with the kinase complex, mainly through interaction with Pho80, both under high and low phosphate conditions, although it only inhibits the kinase complex under low-Pi conditions. This indicates that Pho81 activity is controlled in response to phosphate by an as yet unknown post-translational mechanism. Expression of *PHO81* itself is also controlled by the PHO system, providing a positive feedback loop in which both enhanced expression of *PHO81* and a low-Pi signal from the medium are necessary for proper regulation of the PHO pathway (Creasy et al. 1993; Ogawa et al. 1995). The nature of the phosphate sensor that regulates the PHO pathway remains unclear. Recently, the holoenzyme casein kinase 2 (CK2), consisting of the regulatory subunits Ckb1 and Ckb2 and catalytic subunits Cka1 and Cka2, has been implicated in the regulation of PHO pathway genes in a subunit- and isoform-specific manner, affecting both *PHO4* expression and Pho4 function (Barz et al. 2003). Also, Pho2 seems to be regulated through phosphorylation in a Cdc28-dependent manner and this phosphorylation appears to be critical for the Pho2-Pho4 interaction (Liu et al. 2000). Interestingly, the PHO system seems to interact genetically with the phospholipase C system (Flick and Thorner 1998). The temperature sensitivity of a *plc1Δ* strain can be suppressed by overexpression of *PHO81* or a homologue, *SPL2*, by deletion of *PHO85* or *PHO80*, or by growing cells in a low phosphate medium. As mentioned above, Plc1 has been shown to interact with Gpr1 (Ansari et al. 1999) and Tor2 (Lin et al. 1998), and thus the activity of the Pho85-Pho80 complex appears to be linked to the nutrient-status of the cell. Yet, the precise relationship between these proteins remains to be elucidated. Surprisingly, the complementations of Plc1 are independent of Pho4, suggesting that the Pho85-Pho80 complex regulates additional targets required for growth on elevated temperatures and nutrient uptake and/or utilization.

Aside from the phosphate-sensitive regulation of Pho85 activity, via the Pho81 CKI, the Pho85 requirement is also linked to the quality of carbon and nitrogen sources, as shown by the inability of *pho85Δ* cells to grow on nonfermentable carbon sources or on medium with proline as the sole nitrogen source (Lee et al. 2000). Surprisingly, both these growth defects can be suppressed by an additional deletion of *PHO4*, suggesting an interplay between the PHO pathway and carbon/nitrogen utilization (Nishizawa et al. 1999; Popova Iu et al. 2000). A *pho80Δ* strain shows similar growth defects as a *pho85Δ* strain, and recently, the Pcl6 and Pcl7 cyclins have also been implicated to be involved in the use of alternative carbon sources and proline (Gilliquet and Berben 1993; Nishizawa et al. 1999; Lee et al. 2000; Dimmer et al. 2002). Interestingly, the activity of the Pho85-Pcl7 complex is regulated in response to phosphate levels in a Pho81-dependent manner (Lee et al. 2000). Pho85 may thus be essential for growth on poor carbon and nitrogen sources by mechanisms that use multiple Pcl proteins. Moreover, the expression of these Pcl proteins is also regulated by the carbon source, growth phase and the Pho85 kinase itself (Nishizawa et al. 2001). In this way, Pho85 resembles Cdc28, the essential CDK involved in the yeast cell cycle, which can also associate with multiple cyclins to perform a wide variety of functions (Andrews and

Measday 1998). Although Pho85 shows 51% identity with Cdc28, neither protein can substitute for the other *in vivo* (Toh-e et al. 1988; Santos et al. 1995).

Regarding nitrogen utilization, Pho85 is also involved in the control of the general amino acid control response. This pathway leads to an activation of the Gcn4 transcription factor in response to starvation for amino acids, and Gcn4, in turn, will activate the transcription of genes involved in amino acid biosynthesis (Hinnebusch and Natarajan 2002 and references therein). Apart from a control on the translational level, Gcn4 activity is also regulated at the level of protein stability (Wek et al. 1995; Albrecht et al. 1998; Chi et al. 2001; Valenzuela et al. 2001; Kubota et al. 2003). This post-translational regulation of Gcn4 stability occurs in the nucleus, indicating a spatial separation of protein synthesis and degradation for Gcn4 in yeast (Pries et al. 2002). Protein-degradation by the 26S proteasome is dependent on the ubiquitin-conjugation enzyme Cdc34 and the SCF^{CDC4} ubiquitin ligase complex (Kornitzer et al. 1994; Meimoun et al. 2000). Ubiquitination of Gcn4 requires a preceding phosphorylation of the protein (Meimoun et al. 2000). Two cyclin-dependent kinases, Srb10 and Pho85, have been found to phosphorylate Gcn4 and hereby affect its stability *in vivo* (Chi et al. 2001). Both kinases have additive effects on Gcn4 degradation *in vivo* and appear to be differentially regulated. While phosphorylation of Gcn4 by Srb10 is thought to be a constitutive process, Pho85-dependent phosphorylation is regulated by amino acid availability. Gcn4 stability is increased upon amino acid starvation due to a decrease in Pho85-dependent phosphorylation (Meimoun et al. 2000; Shemer et al. 2002). Pcl5 is specifically required for Pho85-mediated Gcn4 phosphorylation and degradation (Shemer et al. 2002). *PCL5* itself is induced by Gcn4 under amino acid starvation at the transcriptional level, but the amount of Pcl5 protein decreases due to a rapid constitutive turnover and generally reduced levels of protein biosynthesis under these conditions (Shemer et al. 2002). Because of its instability, Pcl5 may act as a sensor for the general capacity of protein biosynthesis in yeast. The fact that *PCL5* is induced in the presence of Gcn4 suggests that it is part of a homeostatic mechanism, which reduces Gcn4 levels upon recovery from starvation.

Another major function of Pho85 appears to be the regulation of transcription. A *pho85Δ* strain shows an increased expression of stationary phase-specific genes and those involved in amino acid biosynthesis, while the expression of translation-related genes is decreased relative to the wild type (Huang et al. 2002). The increased expression of amino acid biosynthesis genes could be explained by a stabilization of Gcn4 in a *pho85Δ* strain (see above). Other reports show that deletion of *PHO85* increases the expression of several stress response genes as well as genes involved in carbohydrate metabolism (Timblin et al. 1996; Timblin and Bergman 1997; Nishizawa et al. 2001). In line with this, *pho85Δ* cells hyperaccumulate glycogen, exhibit sporulation defects, show morphological abnormalities and are hypersensitive to a large number of stress conditions (Gilliquet and Berben 1993; Timblin et al. 1996; Measday et al. 1997; Huang et al. 2002). Except for UDP-glucose pyrophosphorylase (*UGP1*), all these increased expressions are independent of the Pho4 transcription factor (Timblin et al. 1996; Timblin and Bergman 1997; Carroll et al. 2001; Francois and Parrou 2001; Huang et al. 2002). The regulation of stress response genes by Pho85 may also be, at least in part, independent of the STRE promoter element (Timblin and Bergman 1997; Francois and Parrou 2001). Recently, a more general role for Pho85 in producing a generic response to many types of cellular stress has been suggested by studies using a chemical genetic approach (Carroll et al. 2001). Rapid loss of Pho85 activity (through inhibition of the *F82G* allele by the cell-permeable drug 1-Na PP1) causes the induction of a diverse set of genes, which have recently been characterized as comprising a generic response to stressful conditions, known as the environmental stress response (ESR) (Gasch et al. 2000; Carroll et al. 2001). This induction is also independent of the Pho4 transcription factor (Carroll et al. 2001). Most of the ESR genes do not appear to be constitutively expressed in a *pho85Δ* strain, suggesting that activation of the ESR is repressed under these conditions.

In addition to the phenomena mentioned above, several other functions and *in vivo* targets of Pho85 have been described in the literature. Pho85 contributes to G_1 progression by at least two mechanisms. First, Pho85 activity in G_1 decreases the stability of Sic1, an S-phase inhibitor (Espinoza et al. 1994; Measday et al. 1994; Nishizawa et al. 1998). Second, Pho85 seems to positively regulate morphological aspects of G_1 progression, probably by upregulating Cdc42 and Rvs167 activities (Lee et al. 1998; Colwill et al. 1999; Lenburg and O'Shea 2001). Swi5, a zinc finger transcription factor required for expression of a number of genes early in the cell cycle, is another potential *in vivo* substrate for Pho85 (Measday et al. 2000; Simon et al. 2001). Pho85 is also necessary for the asymmetric accumulation of Ash1, a daughter cell–specific repressor of *HO* gene transcription (McBride et al. 2001). It appears that Pho85 regulates the stability of Ash1 through direct phosphorylation. These multiple roles of Pho85 link the protein to cell cycle regulation. The connection to the nutritional status, however, remains obscure and therefore these functions will not be discussed further.

Finally, Pho85 has been extensively studied for its role in glycogen metabolism. Together with several cyclin partners, Pho85 is part of an elaborate network controlling glycogen accumulation, which will be discussed in part 11.3.3.

11.3 Integration of nutrient signals

Obviously, multiple pathways contribute to the pleiotropic effects of cells in a complex environment. Although considerable progress has been made in understanding the diverse signal transduction pathways, the precise nature of their interconnectivity remains unclear. In the following part, we will discuss the potential convergence of several pathways in yeast by means of three well established and generally used read-outs, i.e. STRE-regulated gene expression, pseudohyphal differentiation and glycogen content of cells.

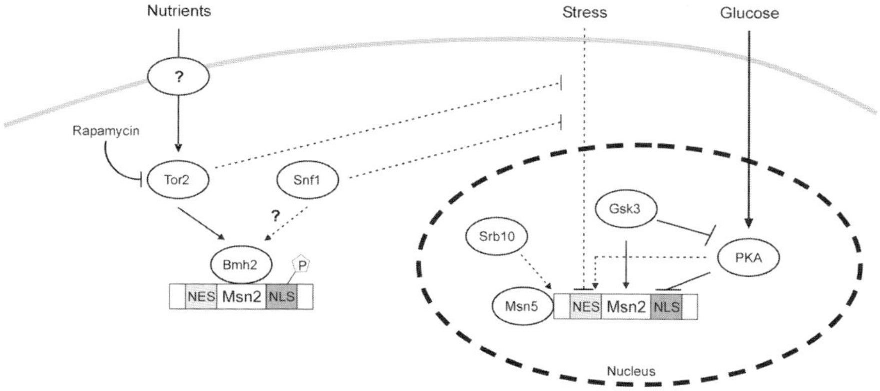

Fig. 3. Nucleocytoplasmic relocalization of Msn2 involves the Ras/cAMP and Tor pathway. Msn2 translocates to the nucleus upon stress and glucose-limitation. Msn2 can be phosphorylated at multiple sites. PKA specifically phosphorylates and inhibits the nuclear localization domain (NLS) of Msn2. Phosphorylated, cytoplasmic Msn2, is bound by the 14-3-3 proteins Bmh1 and Bmh2 in a rapamycin-sensitive manner. The nuclear export domain (NES) is presumably subjected to regulation via the Tor proteins, the Snf1 protein kinase and environmental stresses. Nuclear export is also mediated through the Srb10 protein kinase and the Msn5 importin (For details, see section 11.3.1) (Adapted from Görner et al. 2002). Arrows and bars represent positive and negative interactions respectively. Dashed lines represent putative interactions.

11.3.1 Msn2-mediated transcriptional control

Entry into diauxic shift as a consequence of nutrient limitation - i.e. when glucose becomes limiting and yeast cells start to respire the products of glycolysis such as ethanol - or exposure of cells to mild stress leads to increased expression of STRE (general Stress Response Element)-regulated genes (Boy-Marcotte et al. 1998; Moskvina et al. 1998; Treger et al. 1998). In contrast, PKA represses STRE-dependent gene expression (Boy-Marcotte et al. 1998; Smith et al. 1998; Tadi et al. 1999). The redundant *trans*-acting transcriptional activators Msn2 and Msn4 specifically bind to the STRE consensus sequence (AGGGG) and induce gene expression (Martinez-Pastor et al. 1996; Schmitt and McEntee 1996; Gasch et al. 2000).

Several studies have indicated that Msn2 is localized in the cytoplasm and translocates to the nucleus upon diauxic shift, nutrient starvation and mild osmotic or heat stress (Gorner et al. 1998; Garreau et al. 2000; Van Wuytswinkel et al. 2000; Chi et al. 2001). PKA directly phosphorylates and inhibits Msn2 on its nuclear localisation domain (NLS), thereby stimulating nuclear exclusion (Gorner et al. 1998, 2002; Garreau et al. 2000) (Fig. 3). Furthermore, the NLS phosphorylation status is highly sensitive to carbohydrate fluctuation during fermentative growth (Gorner et al. 2002). Once in the cytosol, phosphorylated Msn2 binds to Bmh2 and forms a stable complex (Fig. 3). The Tor proteins inhibit nuclear accumulation of Msn2 by regulating its interaction with Bmh2 (Beck and Hall 1999). The region responsible for nuclear export (NES) responds to signals generated by stress, the TOR pathway and PKA (Gorner et al. 2002). Moreover, the Snf1 protein kinase, for which Msn2 was first identified as a multi-copy suppressor (Estruch and Carlson 1993), probably converges on the same components as the TOR proteins, thereby inhibiting nuclear import of Msn2 (Mayordomo et al. 2002). Similarly, Srb10 has been implicated in the phosphorylation and nuclear exclusion of Msn2 in a process mediated by the nuclear export protein Msn5 (Chi et al. 2001). Finally, the yeast homologues of mammalian glycogen synthase kinase 3 (GSK-3), encoded by *MCK1*, *MDS1*, *MRK1* and *YOL128c*, are required for proper binding of Msn2 to its DNA-element without affecting the subcellular localisation of Msn2 (Hirata et al. 2003). Since, Mck1 directly binds and inhibits the catalytic subunits of PKA (Rayner et al. 2002) this protein may exhibit a dual function, i.e. Mck1 inhibits PKA-mediated nuclear exclusion of Msn2 and at the same time enhances DNA-binding of the transcription factor.

Thus, Msn2 might be considered as a point of convergence for multiple pathways in regulating a group of stress-responsive genes. Its function is strictly controlled by protein phosphorylation at distinct sites, resulting in nuclear or cytoplasmic retention dependent on the environmental stimuli and the cellular needs.

11.3.2 Pseudohyphal differentiation

In response to nutrient limitation, such as nitrogen depletion, yeast cells switch their morphology from the normal round shape to a filamentous invasive and pseudohyphal form. This morphological switch is generally considered to facilitate foraging for scarce nutrients in the environment and involves an integrated signalling network (Gancedo 2001; Gagiano et al. 2002; Palecek et al. 2002) (Fig. 4). Pseudohyphal differentiation requires Ras2-dependent activation of the pheromone responsive MAPK pathway (Mosch et al. 1996) through activation of the Ste20, Ste11, Ste7, and Kss1 protein kinases (Gancedo 2001 and references therein). Interestingly, the 14-3-3 proteins, Bmh1 and Bmh2, were found to interact with Ste20 and to fulfil an essential role in the MAPK signalling pathway in that they are required for Ras2- and Cdc42-mediated induction of filamentation and cell elongation (Roberts et al. 1997). Besides the MAPK cascade, the Ras/cAMP pathway is also extensively involved in the regulation of pseudohyphal differentiation (Pan et al. 2000). The G-protein coupled receptor Gpr1-Gpa2 regulates pseudohyphal differentiation by modulating the cAMP levels and thereby PKA activity (Kubler et al. 1997; Lorenz and Heitman 1997; Pan and Heitman 1999). Constitutive activation of the Ras/cAMP pathway leads to filamentous/pseudohyphal growth (Gimeno et al. 1992; Kubler et al. 1997; Lorenz and Heitman 1997), while overexpression of *PDE2* inhibits this morphology (Ward et al. 1995). However, the A kinase catalytic subunits, Tpk1, Tpk2, and Tpk3, appear

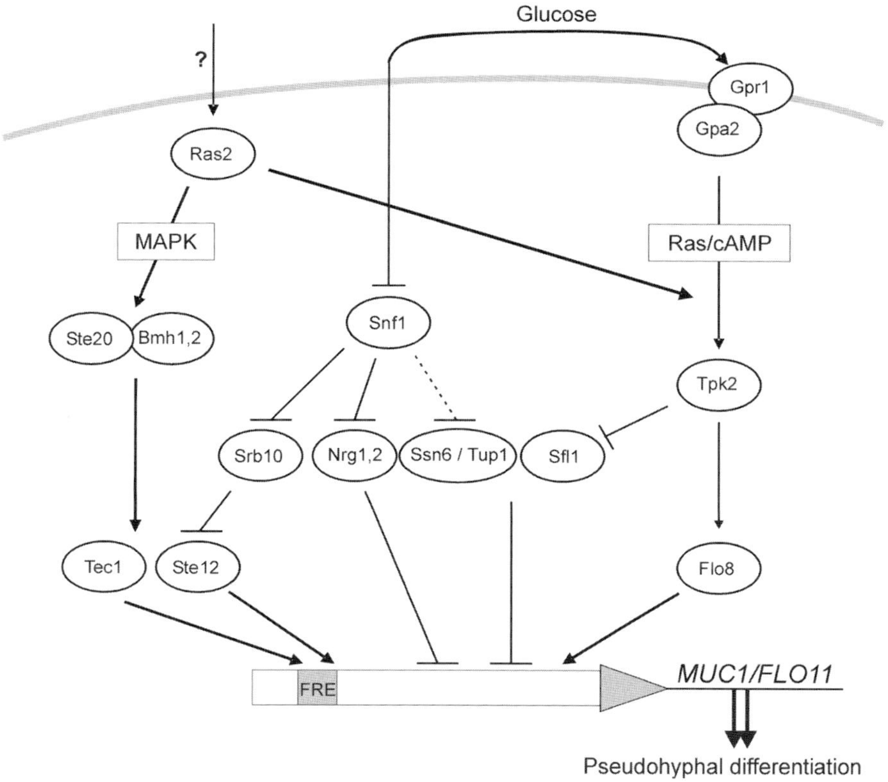

Fig. 4. Pseudohyphal differentiation is controlled through pathway-specific transcriptional regulators. Expression of the *MUC1/FLO11* gene is required for pseudohyphal growth. Activation of the MAPK pathway through the transcriptional activators Ste12 and Tec1 induces filamentation. The Ras/cAMP pathway controls *FLO11* expression mainly through Ste12-independent mechanisms. The Snf1 protein kinase renders *FLO11* expression more susceptible to other pathways in glucose-limiting conditions through inhibition of transcriptional repressors (For details, see section 11.3.2). Arrows and bars represent positive and negative interactions respectively. Dashed lines represent putative interactions.

to have distinct roles. Tpk2 induces filamentous growth, while Tpk1 and Tpk3 counteract this phenotype (Robertson and Fink 1998; Pan and Heitman 1999).

The *MUC1/FLO11* gene product, a cell surface flocculin, is an essential player required for filamentous/pseudohyphal growth. Transcription of *FLO11* is mediated by the MAPK-activated transcription factors Ste12 and Tec1 through its FRE- (Filamentation and invasive Response Element) element (Madhani and Fink 1997). In contrast, Tpk2-mediated *FLO11* transcription requires the transcriptional activator Flo8 (Rupp et al. 1999). In addition, Tpk2 specifically interacts with and inhibits the transcriptional repressor Sfl1 to positively regulate *FLO11* transcription (Robertson and Fink 1998). Apparently, Tpk2 operates as a developmental switch to control assembly of diverse transcription factors on the *FLO11* promoter in that it simultaneously stimulates an activator and inhibits a repressor (Pan and Heitman 2002). It was suggested that Sfl1 might recruit the Ssn6-Tup1 co-repressor for transcriptional repression of *FLO11*, *SUC2* and *HSP26* (Conlan and Tzamaris 2001) and thus that the Tpk2-mediated control of filamentation may resemble at least in part the mechanism used for Snf1-mediated control of glucose repression. Indeed, Snf1 inhibits Mig1-dependent recruitment of the Ssn6-Tup1 repression complex (Treitel and Carlson 1995; Ostling and Ronne 1998). Therefore, it may not be surprising that recent studies implicated also Snf1 in the regulation of filamentous and invasive growth (Cullen and Sprague 2000; Palecek et al. 2002). Snf1 physically interacts with the Srb/mediator proteins to regulate glucose-dependent transcription (Kuchin et al. 2000) and mutations in different components of this complex were found to induce invasive growth (Palecek et al. 2000). More recently, the RNA polymerase II holoenzyme-associated cyclin dependent kinase (CDK), Srb10/Cdk8, was shown to negatively regulate Ste12 activity (Nelson et al. 2003). Similarly, two other repressors of *FLO11* transcription, Nrg1 and Nrg2 (Negative Regulators of Glucose-controlled genes), are negatively regulated by Snf1 (Vyas et al. 2001; Kuchin et al. 2002) (Fig. 4).

Apart from the above-mentioned components of the MAPK, the Ras/cAMP and the glucose-repression pathways, a number of other proteins have been implicated in pseudohyphal differentiation (Lorenz and Heitman 1998) but their precise relationship to these pathways is not clear yet. For instance, a complex transcription factor cascade involving Sok2, Ash1, Phd1 and Swi5 has been proposed to regulate filamentation (Pan and Heitman 2000). Notably, the action of the Tpk's on pseudohyphal differentiation appears to be independent of Phd1 and Sok2 as well as the protein kinases Rim15 and Yak1 (Pan and Heitman 1999). The protein kinases, Sch9 and Yak1 are also involved in pseudohyphal differentiation (Lorenz et al. 2000; Zhang et al. 2001) but the underlying mechanism in Sch9-dependent pseudohyphal differentiation have not yet been elucidated completely. Finally, rapamycin blocks filamentous growth through inhibition of the Tor proteins and thus can be restored either by activation of the Ras/cAMP or MAPK pathways or by mutation of the Sok2 repressor suggesting that the TOR pathway acts in parallel to these known cascades (Cutler et al. 2001). Notably, the 14-3-3 proteins Bmh1 and Bmh2 have been demonstrated to be key components in the rapamycin-sensitive signalling (Bertram et al. 1998).

In conclusion, pseudohyphal differentiation involves a multitude of pathways each of which may control different transcription factors in response to nutrient depletion in order to control the expression of one single target gene (Fig. 4).

11.3.3 Regulation of glycogen biosynthesis

At the diauxic shift, when glucose becomes exhausted, yeast cells will start to accumulate the reserve carbohydrate glycogen. This occurs, on one hand, through transcriptional induction of genes involved in the synthesis of glycogen and on the other hand through posttranslational activation of glycogen synthase and inactivation of glycogen phosphorylase. *Saccharomyces cerevisiae* has two isoforms of glycogen synthase, encoded by the *GSY1* and *GSY2* genes. Glycogen synthase 2 (Gsy2) represents the predominant activity in vegetative growing cells and expression of *GSY2*, and not *GSY1*, is linked to nutrient limitation (Francois and Parrou 2001). Glycogen phosphorylase is encoded by the *GPH1* gene (Hwang et al. 1989). The activity of both glycogen synthase and glycogen phosphorylase is dependent on their phosphorylation state and thus on the relative activity of protein kinases and protein phosphatases. Gsy2 activity is regulated *in vivo* through phosphorylation/ dephosphorylation of 3 residues in its C-terminal part; Ser-650, Ser-654 and Thr-667 (Hardy and Roach 1993). Phosphorylation of Gsy2 impairs glycogen synthase activity, and consequently, Gsy2 activation is established through dephosphorylation of the enzyme (Francois and Parrou 2001). Gph1 is phosphorylated on a Thr residue at the N-terminus, and, in contrast with Gsy2, phosphorylation of Gph1 activates the enzyme (Lin et al. 1995). How the dynamic process of phosphorylation and dephosphorylation is regulated is not yet very clear, but several reports indicate a major role of the metabolite glucose-6-phosphate (Glc-6-P) (Francois and Parrou 2001). Glc-6-P, aside from being a substrate in glycogen biosynthesis, acts as a potent stimulator of the dephosphorylation and as an inhibitor of the phosphorylation processes (Francois and Hers 1988; Lin et al. 1996; Huang et al. 1997). Glc-6-P also non-competitively inhibits glycogen phosphorylase (Lin et al. 1995). Several protein kinases and phosphatases have been implicated in the control of glycogen biosynthesis (Fig. 5).

PKA and Snf1 exert antagonistic effects on the glycogen synthase activity and glycogen levels (Francois and Parrou 2001). A mutation in *SNF1* fails to accumulate glycogen in response to abrupt depletion of glucose (Thompson-Jaeger et al. 1991). Strains defective in Snf1 show a modest decrease in *GSY2* expression, and upon entry into stationary phase, the Gsy2 enzyme is blocked in the inactive, phosphorylated state (Hardy et al. 1994). Furthermore, *snf1Δ* cells have reduced levels of Glc-6-P compared to wild type cells (Huang et al. 1997). Additional mutations that elevate intracellular Glc-6-P, such as deletion of *PFK1*, *PFK2*, or *PGI1*, have been shown to restore glycogen accumulation (Corominas et al. 1992; Huang et al. 1997). In *bcy1Δ* cells, which have constitutively active protein kinase A, greatly reduced levels of *GSY2* mRNA were observed, as well as for *GAC1* and *GLC3*, two other genes necessary for glycogen accumulation (Hardy et al. 1994). All these genes contain STRE elements in their promoter and the reduction in expression is most probably caused by PKA-mediated inhibition of the Msn2/4 transcription factors (Gorner et al. 1998, 2002; Parrou et al. 1999). Although PKA can phosphorylate Gsy2 *in vitro*, this occurs at sites different from those that are phosphorylated *in vivo* (Hardy and Roach 1993; Hardy et al. 1994), suggesting that PKA influences glycogen synthase phosphorylation indirectly. Finally, PKA is shown to phosphorylate and activate Gph1 *in vitro* (Manhart and Holzer 1988).

The major enzyme responsible for dephosphorylation, and thus activation, of Gsy2 is the yeast protein phosphatase type I, encoded by the *GLC7* gene (Hardy and Roach 1993; Francois and Parrou 2001). Glc7 is targeted to the Gsy2 substrate by the regulatory subunit Gac1 (Francois et al. 1992). Another protein, Pig1, may also play an additional regulatory role in this process (Cheng et al. 1997). The phosphatase for inactivating Gph1 has not been identified yet.

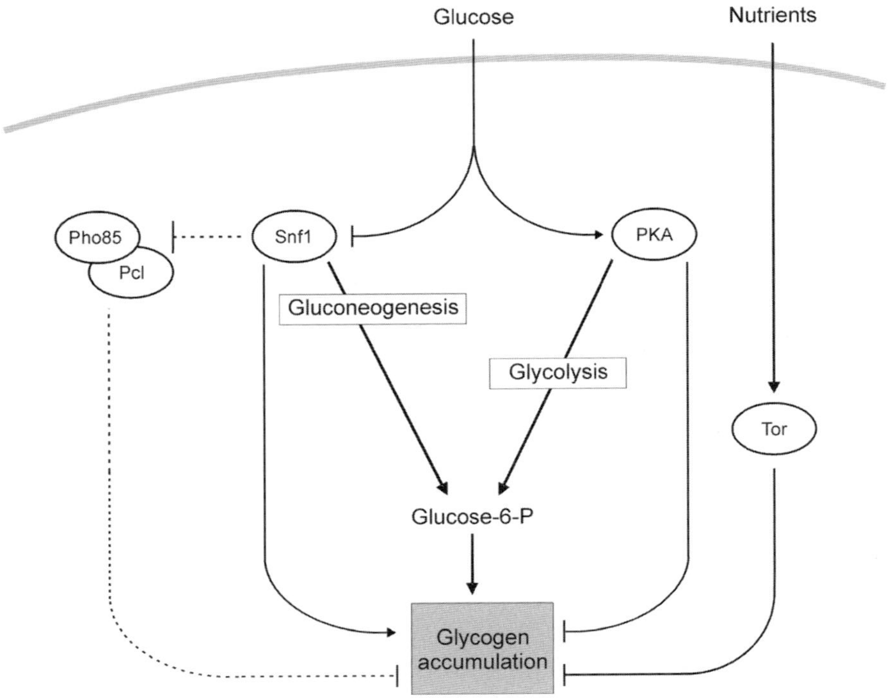

Fig. 5. Control of glycogen accumulation in *S. cerevisiae* through combined action of multiple signalling pathways. PKA and Snf1 antagonistically control glycogen levels in yeast. Snf1 appears, at least in part, to operate through Pho85 to perform this function. However, not depicted in this figure, Pho85 also seems to have positive inputs into glycogen accumulation (see section 11.3.3 for details). In addition, the Tor proteins negatively regulate glycogen levels. Finally, Glucose-6-P acts as a key metabolic intermediate, stimulating glycogen accumulation. Arrows and bars represent positive and negative interactions respectively. Dashed lines represent putative interactions.

The Pho85 CDK also has several inputs that control glycogen levels. First, deletion of *PHO85* leads to an increased expression of the *GSY2* gene, and this increase is, at least in part, independent of the STRE elements in the *GSY2* promoter (Timblin et al. 1996; Timblin and Bergman 1997). Second, Gsy2 activity is significantly increased in *pho85Δ* mutants (Huang et al. 1996). Pho85 specifically phosphorylates the Ser-654 and Thr-667 residues of Gsy2 *in vivo*, which are part of a CDK Ser/Thr-Pro consensus sequence (Huang et al. 1996, 1998). The cyclin partners responsible for directing Pho85 towards Gsy2 phosphorylation are Pcl8 and Pcl10 (Huang et al. 1998; Wilson et al. 1999). The kinase responsible for *in vivo* phosphorylation of the third residue, Ser-650, has not been characterized yet, although this residue would acquire the consensus recognition motif for the GSK-3-family enzymes, S-X-X-X-S/T, upon phosphorylation of Ser-654 by Pho85

(Harwood 2001). However, no functional link between yeast GSK-3 and Gsy2 phosphorylation has been reported to date. Third, besides elevated glycogen synthase activity, *pho85Δ* cells also display an increased glycogen phosphorylase activity (Wilson et al. 2002). Since Gph1 is activated through phosphorylation, Pho85-dependent regulation of Gph1 most likely occurs indirect. This function may require the Pcl6 and Pcl7 cyclins (Wang et al. 2001b). Fourth, Glc8, a glucose-repressible activator of the Glc7 protein phosphatase-1, has been shown to be another *in vivo* target of Pho85 (DeRisi et al. 1997; Nigavekar et al. 2002; Tan et al. 2003). *GLC8* was originally discovered in a screen for glycogen-deficient mutants (Cannon et al. 1994). Activity of Glc8 is regulated by at least two known mechanisms. First, *GLC8* expression is induced in stationary phase (DeRisi et al. 1997). Second, Glc8 requires phosphorylation on Thr-118 for *in vivo* function. Pho85 was identified as the sole Glc8 kinase *in vivo* (Tan et al. 2003). Pcl6 and Pcl7 comprise the major cyclins required by Pho85 to perform this phosphorylation, although Pcl8 and Pcl10 can also fulfil this function to some extent. Pho85-dependent activation of Glc7, mediated by the Glc8 protein, would represent a positive input into glycogen synthesis, since this leads to an activation of Gsy2 via Glc7-dependent dephosphorylation.

Finally, the Tor proteins have also been implicated as regulators of glycogen levels in yeast, indicated by the glycogen accumulation observed in rapamycin-treated cells (Loewith et al. 2002). However, very little is known about how the Tor proteins control glycogen accumulation. One input seems to comprise the inhibition of Msn2 and Msn4 function by Tor (Beck and Hall 1999), which has been discussed in part 11.3.1.

Deletion of *PHO85* in a *bcy1Δ* background does not suppress the glycogen deficiency (Timblin et al. 1996), suggesting that PKA exercises its effects on glycogen accumulation largely independent of Pho85. In contrast, epistasis analyses indicated that Pho85 may be acting downstream of Snf1 in the control of glycogen levels, as a *snf1Δ pho85Δ* strain restores glycogen synthase activity and glycogen levels back to wild type levels or more (Timblin et al. 1996). This suppression occurs without increasing the Glc-6-P levels (Huang et al. 1997). Furthermore, the opposite effects of STRE-independent transcriptional regulation by Snf1 and Pho85 also support the model of upstream control of Pho85 by Snf1 (Francois and Parrou 2001). However, Snf1 may still act in a pathway parallel of Pho85, for instance through the inhibition of other glycogen synthase kinases or activation of protein phosphatases such as Glc7-Gac1, or through modulation of Glc-6-P levels.

Although glycogen synthase activity is restored, a *snf1Δ pcl8Δ pcl10Δ* strain still fails to accumulate glycogen (Huang et al. 1998). Therefore, *snf1* mutants have a second deficiency that, apart from the lack of glycogen synthase activation, disables glycogen synthesis. This suggests the existence of additional control levels over glycogen synthesis that require Snf1 and most probably also involve Pho85, since glycogen accumulation and glycogen synthase activity are both restored in *snf1Δ pho85Δ* cells. Two genes, *APG1* and *APG13*, encoding for essential components in the regulation of autophagy, were recently isolated as multicopy suppressors of the glycogen-deficient phenotype of *snf1Δ pcl8Δ pcl10Δ* mutants, thereby functionally linking autophagy to glycogen accumulation

(Abeliovich and Klionsky 2001; Huang et al. 2001; Wang et al. 2001b). Apg1 is a protein kinase, which through association with the Apg13 protein regulates autophagy in response to nutrient starvation. Although autophagy has been reported to be negatively regulated by PKA (Noda and Ohsumi 1998), the yeast Tor proteins have been suggested to act as upstream nutrient sensors for autophagy, exerting their effect through regulation of the Apg1-Apg13 complex, whereby Tor-activity inhibits the autophagic process (Noda and Ohsumi 1998; Kamada et al. 2000; Abeliovich et al. 2003). This Tor-dependent modulation of Apg1-Apg13 function appears to be Tap42-independent. Interestingly, *snf1Δ pcl8Δ pcl10Δ* mutants behave similar to mutants defective for autophagy in that both are able to synthesize glycogen upon approaching the stationary phase, but are unable to maintain their glycogen stores, because subsequent synthesis is impaired and degradation by glycogen phosphorylase, Gph1, is enhanced. This indicates that it is the process of autophagy itself that influences glycogen storage rather than functions specifically related to Apg1 and/or Apg13. Induction of autophagy in *pho85Δ* mutants entering the stationary phase was exaggerated compared to the level in wild type cells, but was blocked in *pho85Δ apg1Δ* mutants. These data suggest that Snf1 and Pho85 are, respectively, positive and negative regulators of autophagy and they may also, like Tor, act via the Apg1-Apg13 complex (Wang et al. 2001a; Abeliovich et al. 2003). Defective glycogen storage in *snf1Δ* cells can thus be attributed to both defective synthesis upon entry into stationary phase and impaired maintenance of glycogen levels caused by the lack of autophagy (Wang et al. 2001a). Both these processes can be suppressed by deletion of *PHO85* in a *snf1Δ* strain. The cyclin partners directing Pho85 towards regulation of autophagy are not known. Although recently it has been shown that glycogen accumulation is restored in a *snf1Δ pcl6Δ pcl7Δ pcl8Δ pcl10Δ* strain (Wang et al. 2001b), Pcl6 and Pcl7 do not seem to be involved in autophagy (Wang et al. 2001a). This indicates that this suppression is likely mediated by changes in metabolic enzymes, possibly Gph1, since glycogen levels are also restored in a *snf1Δ pcl8Δ pcl10Δ gph1Δ* strain (Wang et al. 2001a, 2001b; Wilson et al. 2002). The exact nature of the input of the Pho85-Pcl6 and Pho85-Pcl7 kinase complexes in glycogen metabolism is still unknown.

11.4 Concluding Remarks

Most microorganisms, such as yeast, live under conditions of nutrient-limited growth. Therefore, it is of extreme importance that scarce nutrients can be used efficiently. To achieve this, microorganisms possess complex signal transduction cascades that enable them to integrate nutritional information and translate this information into adequate responses to coordinate their growth, proliferation, and metabolism. Although there remains still a lot to be elucidated about the triggers and sensing mechanisms involved in this nutritional response, recent research allowed to grasp an idea about the increasing complexity of these pathways. In the present era of genomics, proteomics, and metabolomics, a new picture is evolving where the individual pathways are just parts of integrative and converging signalling networks and where these networks are the units that assure a dynamical and balanced translation of extracellular signals.

References

Abeliovich H, Klionsky DJ (2001) Autophagy in yeast: mechanistic insights and physiological function. Microbiol Mol Biol Rev 65:463-479

Abeliovich H, Zhang C, Dunn WA Jr, Shokat KM, Klionsky DJ (2003) Chemical genetic analysis of Apg1 reveals a non-kinase role in the induction of autophagy. Mol Biol Cell 14:477-490

Albrecht G, Mosch HU, Hoffmann B, Reusser U, Braus GH (1998) Monitoring the Gcn4 protein-mediated response in the yeast *Saccharomyces cerevisiae*. J Biol Chem 273:12696-12702

Alms GR, Sanz P, Carlson M, Haystead TA (1999) Reg1p targets protein phosphatase 1 to dephosphorylate hexokinase II in *Saccharomyces cerevisiae*: characterizing the effects of a phosphatase subunit on the yeast proteome. EMBO J 18:4157-4168

Andrews B, Measday V (1998) The cyclin family of budding yeast: abundant use of a good idea. Trends Genet 14:66-72

Ansari K, Martin S, Farkasovsky M, Ehbrecht IM, Kuntzel H (1999) Phospholipase C binds to the receptor-like GPR1 protein and controls pseudohyphal differentiation in *Saccharomyces cerevisiae*. J Biol Chem 274:30052-30058

Arguelles JC, Mbonyi K, Van Aelst L, Vanhalewyn M, Jans AW, Thevelein JM (1990) Absence of glucose-induced cAMP signaling in the *Saccharomyces cerevisiae* mutants cat1 and cat3 which are deficient in derepression of glucose-repressible proteins. Arch Microbiol 154:199-205

Baldwin JM (1994) Structure and function of receptors coupled to G proteins. Curr Op Cell Biol 6:180-190

Barz T, Ackermann K, Pyerin W (2003) Perturbation of protein kinase CK2 uncouples executive part of phosphate maintenance pathway from cyclin-CDK control(1). FEBS Lett 537:210-214

Beck T, Hall MN (1999) The TOR signalling pathway controls nuclear localization of nutrient-regulated transcription factors. Nature 402:689-692

Behlke J, Heidrich K, Naumann M, Muller EC, Otto A, Reuter R, Kriegel T (1998) Hexokinase 2 from *Saccharomyces cerevisiae*: regulation of oligomeric structure by *in vivo* phosphorylation at serine-14. Biochemistry 37:11989-11995

Bergsma JC, Kasri NN, Donaton MC, De Wever V, Tisi R, de Winde JH, Martegani E, Thevelein JM, Wera S (2001) PtdIns(4,5)P(2) and phospholipase C-independent Ins(1,4,5)P(3) signals induced by a nitrogen source in nitrogen-starved yeast cells. Biochem J 359:517-523

Bertram PG, Zeng C, Thorson J, Shaw AS, Zheng XF (1998) The 14-3-3 proteins positively regulate rapamycin-sensitive signaling. Curr Biol 8:1259-1267

Beullens M, Mbonyi K, Geerts L, Gladines D, Detremerie K, Jans AW, Thevelein JM (1988) Studies on the mechanism of the glucose-induced cAMP signal in glycolysis and glucose repression mutants of the yeast *Saccharomyces cerevisiae*. Eur J Biochem 172:227-231

Bhattacharya S, Chen L, Broach JR, Powers S (1995) Ras membrane targeting is essential for glucose signaling but not for viability in yeast. Proc Natl Acad Sci (USA) 92:2984-2988

Blumberg H, Silver P (1991) A split zinc-finger protein is required for normal yeast growth. Gene 107:101-110

Bockaert J, Pin JP (1999) Molecular tinkering of G protein-coupled receptors: an evolutionary success. EMBO J 18:1723-1729

Borkovich KA, Farrelly FW, Finkelstein DB, Taulien J, Lindquist S (1989) hsp82 is an essential protein that is required in higher concentrations for growth of cells at higher temperatures. Mol Cell Biol 9:3919-3930

Boy-Marcotte E, Perrot M, Bussereau F, Boucherie H, Jacquet M (1998) Msn2p and Msn4p control a large number of genes induced at the diauxic transition which are repressed by cyclic AMP in *Saccharomyces cerevisiae*. J Bacteriol 180:1044-1052

Boy-Marcotte E, Tadi D, Perrot M, Boucherie H, Jacquet M (1996) High cAMP levels antagonize the reprogramming of gene expression that occurs at the diauxic shift in *Saccharomyces cerevisiae*. Microbiol 142:459-467

Broach JR, Deschenes RJ (1990) The function of *RAS* genes in *Saccharomyces cerevisiae*. Adv Cancer Res 54:79-139

Broek D, Toda T, Michaeli T, Levin L, Birchmeier C, Zoller M, Powers S, Wigler M (1987) The *S. cerevisiae CDC25* gene product regulates the RAS/adenylate cyclase pathway. Cell 48:789-799

Cameron S, Levin L, Zoller M, Wigler M (1988) cAMP-independent control of sporulation, glycogen metabolism and heat shock resistance in *S. cerevisiae*. Cell 53:555-566

Camonis JH, Kalékine M, Gondré B, Garreau H, Boy-Marcotte E, Jacquet M (1986) Characterization, cloning and sequence analysis of the *CDC25* gene which controls the cyclic AMP level of *Saccharomyces cerevisiae*. EMBO J 5:375-380

Campbell-Burk SL, Jones KA, Shulman RG (1987) 31P NMR Saturation-transfer measurements in *Saccharomyces cerevisiae*: characterization of phosphate exchange reactions by iodoacetate and antimycin A inhibition. Biochemistry 26:7483-7492

Camus C, Boymarcotte E, Jacquet M (1994) Two subclasses of guanine exchange factor (GEF) domains revealed by comparison of activities of chimeric genes constructed from CDC25, SDC25 and BUD5 in *Saccharomyces cerevisiae*. Mol Gen Genet 245:167-176

Cannon JF, Pringle JR, Fiechter A, Khalil M (1994) Characterization of glycogen-deficient *GLC* mutants of *Saccharomyces cerevisiae*. Genetics 136:485-503

Carroll AS, Bishop AC, DeRisi JL, Shokat KM, O'Shea EK (2001) Chemical inhibition of the Pho85 cyclin-dependent kinase reveals a role in the environmental stress response. Proc Natl Acad Sci USA 98:12578-12583

Cheng C, Huang D, Roach PJ (1997) Yeast PIG genes: PIG1 encodes a putative type 1 phosphatase subunit that interacts with the yeast glycogen synthase Gsy2p. Yeast 13:1-8

Cherry JR, Johnson TR, Dollard C, Shuster JR, Denis CL (1989) Cyclic AMP-dependent protein kinase phosphorylates and inactivates the yeast transcriptional activator ADR1. Cell 56:409-419

Chi Y, Huddleston MJ, Zhang X, Young RA, Annan RS, Carr SA, Deshaies RJ (2001) Negative regulation of Gcn4 and Msn2 transcription factors by Srb10 cyclin-dependent kinase. Genes Dev 15:1078-1092

Colombo S, Ma P, Cauwenberg L, Winderickx J, Crauwels M, Teunissen A, Nauwelaers D, de Winde JH, Gorwa MF, Colavizza D, Thevelein JM (1998) Involvement of distinct G-proteins, Gpa2 and Ras, in glucose- and intracellular acidification-induced cAMP signalling in the yeast *Saccharomyces cerevisiae*. EMBO J 17:3326-3341

Colwill K, Field D, Moore L, Friesen J, Andrews B (1999) *In vivo* analysis of the domains of yeast Rvs167p suggests Rvs167p function is mediated through multiple protein interactions. Genetics 152:881-893

Conlan RS, Tzamaris D (2001) Sfl1 functions via co-repressor Ssn6-Tup1 and the cAMP-dependent protein kinase Tpk2. J Mol Biol 309:1007-1015

Corominas J, Clotet J, Fernandez-Banares I, Boles E, Zimmermann FK, Guinovart JJ, Arino J (1992) Glycogen metabolism in a *Saccharomyces cerevisiae* phosphoglucose isomerase (pgil) disruption mutant. FEBS Lett 310:182-186

Crauwels M, Donaton MC, Pernambuco MB, Winderickx J, de Winde JH, Thevelein JM (1997) The Sch9 protein kinase in the yeast *Saccharomyces cerevisiae* controls cAPK activity and is required for nitrogen activation of the fermentable-growth-medium-induced (FGM) pathway. Microbiology 143:2627-2637

Creasy CL, Madden SL, Bergman LW (1993) Molecular analysis of the PHO81 gene of *Saccharomyces cerevisiae*. Nucleic Acids Res 21:1975-1982

Cullen PJ, Sprague GF Jr (2000) Glucose depletion causes haploid invasive growth in yeast. Proc Natl Acad Sci USA 97:13619-13624

Cutforth T, Rubin GM (1994) Mutations in Hsp83 and cdc37 impair signaling by the sevenless receptor tyrosine kinase in *Drosophila*. Cell 77:1027-1036

Cutler NS, Pan X, Heitman J, Cardenas ME (2001) The TOR signal transduction cascade controls cellular differentiation in response to nutrients. Mol Biol Cell 12:4103-4113

Cytrynska M, Frajnt M, Jakubowicz T (2001) *Saccharomyces cerevisiae* pyruvate kinase Pyk1 is PKA phosphorylation substrate *in vitro*. FEMS Microbiol Lett 203:223-227

de Winde JH, Crauwels M, Hohmann S, Thevelein JM, Winderickx J (1996) Differential requirement of the yeast sugar kinases for sugar sensing in establishing the catabolite-repressed state. Eur J Biochem 241:633-643

Denis CL, Audino DC (1991) The Ccr1 (Snf1) and Sch9 protein kinases act independently of cAMP-dependent protein kinase and the transcriptional activator Adr1 in controlling yeast *ADH2* expression. Mol Gen Genet 229:395-399

Dequard-Chablat M, Riva M, Carles C, Sentenac A (1991) RPC19, the gene for a subunit common to yeast RNA polymerases A (I) and C (III). J Biol Chem 266:15300-15307

DeRisi JL, Iyer VR, Brown PO (1997) Exploring the metabolic and genetic control of gene expression on a genomic scale. Science 278:680-686

Dimmer KS, Fritz S, Fuchs F, Messerschmitt M, Weinbach N, Neupert W, Westermann B (2002) Genetic basis of mitochondrial function and morphology in *Saccharomyces cerevisiae*. Mol Biol Cell 13:847-853

Durnez P, Pernambuco MB, Oris E, Arguelles JC, Mergelsberg H, Thevelein JM (1994) Activation of trehalase during growth induction by nitrogen sources in the yeast *Saccharomyces cerevisiae* depends on the free catalytic subunits of cAMP-dependent protein kinase, but not on functional ras proteins. Yeast 10:1049-1064

Engelberg D, Klein C, Martinetto H, Struhl K, Karin M (1994) The UV response involving the ras signaling pathway and ap-1 transcription factors is conserved between yeast and mammals. Cell 77:381-390

Entian KD (1980) Genetic and biochemical evidence for hexokinase PII as a key enzyme involved in carbon catabolite repression in yeast. Mol Gen Genet 178:633-637

Entian KD, Fröhlich KU (1984) *Saccharomyces cerevisiae* mutants provide evidence of hexokinase PII as a bifunctional enzyme with catalytic and regulatory domains for triggering carbon catabolite repression. J Bacteriol 158:29-35

Entian KD, Kopetzki E, Frohlich KU, Mecke D (1984) Cloning of hexokinase isoenzyme PI from *Saccharomyces cerevisiae*: PI transformants confirm the unique role of hexokinase isoenzyme PII for glucose repression in yeasts. Mol Gen Genet 198:50-54

Espinoza FH, Ogas J, Herskowitz I, Morgan DO (1994) Cell cycle control by a complex of the cyclin HCS26 (PCL1) and the kinase PHO85. Science 266:1388-1391

Estruch F, Carlson M (1993) Two homologous zinc finger genes identified by multicopy suppression in a SNF1 protein kinase mutant of *Saccharomyces cerevisiae*. Mol Cell Biol 13:3872-3881

Fabrizio P, Liou LL, Moy VN, Diaspro A, SelverstoneValentine J, Gralla EB, Longo VD (2003) SOD2 functions downstream of Sch9 to extend longevity in yeast. Genetics 163:35-46

Fabrizio P, Pozza F, Pletcher SD, Gendron CM, Longo VD (2001) Regulation of longevity and stress resistance by Sch9 in yeast. Science 292:288-290

Fernandes PM, Trugo LC, Panek AD (1997) Nitrogen starvation in a *Saccharomyces cerevisiae* strain deleted in the trehalose-6-phosphate synthase complex. Biochem Mol Biol Int 43:901-911

Fernandez R, Herrero P, Fernandez E, Fernandez T, Lopez-Boado YS, Moreno F (1988) Autophosphorylation of yeast hexokinase PII. J Gen Microbiol 134:2493-2498

Field J, Nikawa J-I, Broek D, MacDonald B, Rodgers L, Wilson IA, Lerner RA, Wigler M (1988) Purification of a RAS-responsive adenylate cyclase complex from *Saccharomyces cerevisiae* by use of an epitope addition method. Mol Cell Biol 8:2159-2165

Flattery-O'Brien JA, Grant CM, Dawes IW (1997) Stationary-phase regulation of the *Saccharomyces cerevisiae* SOD2 gene is dependent on additive effects of HAP2/3/4/5- and STRE-binding elements. Mol Microbiol 23:303-312

Flick JS, Thorner J (1998) An essential function of a phosphoinositide-specific phospholipase C is relieved by inhibition of a cyclin-dependent protein kinase in the yeast *Saccharomyces cerevisiae*. Genetics 148:33-47

Francois J, Hers HG (1988) The control of glycogen metabolism in yeast. 2. A kinetic study of the two forms of glycogen synthase and of glycogen phosphorylase and an investigation of their interconversion in a cell-free extract. Eur J Biochem 174:561-567

François J, Neves MJ, Hers HG (1991) The control of trehalose biosynthesis in *Saccharomyces cerevisiae* - evidence for a catabolite inactivation and repression of trehalose-6-phosphate synthase and trehalose-6-phosphate phosphatase. Yeast 7:575-587

Francois J, Parrou JL (2001) Reserve carbohydrates metabolism in the yeast *Saccharomyces cerevisiae*. FEMS Microbiol Rev 25:125-145

François J, Villanueva ME, Hers HG (1988) The control of glycogen metabolism in yeast. 1. Interconversion *in vivo* of glycogen synthase and glycogen phosphorylase induced by glucose, a nitrogen source or uncouplers. Eur J Biochem 174:551-559

Francois JM, Thompson-Jaeger S, Skroch J, Zellenka U, Spevak W, Tatchell K (1992) GAC1 may encode a regulatory subunit for protein phosphatase type 1 in *Saccharomyces cerevisiae*. EMBO J 11:87-96

Furman TC, Neet KE (1983) Association equilibria and reacting enzyme gel filtration of yeast hexokinase. J Biol Chem 258:4930-4936

Gagiano M, Bauer FF, Pretorius IS (2002) The sensing of nutritional status and the relationship to filamentous growth in *Saccharomyces cerevisiae*. FEM Yeast Res 2:433-470

Gancedo JM (2001) Control of pseudohyphae formation in *Saccharomyces cerevisiae*. FEMS Microbiol Rev 25:107-123

Garreau H, Hasan RN, Renault G, Estruch F, Boy-Marcotte E, Jacquet M (2000) Hyperphosphorylation of Msn2p and Msn4p in response to heat shock and the diauxic shift is inhibited by cAMP in *Saccharomyces cerevisiae*. Microbiology 146:2113-2120

Gasch AP, Spellman PT, Kao CM, Carmel-Harel O, Eisen MB, Storz G, Botstein D, Brown PO (2000) Genomic expression programs in the response of yeast cells to environmental changes. Mol Biol Cell 11:4241-4257

Geyskens I, Kumara SHMC, Donaton MCV, Bergsma JCT, Thevelein JM, Wera S (2001) Expression of mammalian PKB complements deletion of the yeast protein kinase Sch9. Nato Science Series A316:117-126

Gibbs CS, Knighton DR, Sowadski JM, Taylor SS, Zoller MJ (1992) Systematic mutational analysis of cAMP-dependent protein kinase identifies unregulated catalytic subunits and defines regions important for the recognition of the regulatory subunit. J Biol Chem 267:4806-4814

Gibbs JB, Marshall MS (1989) The *ras* oncogene - an important regulatory element in lower eucaryotic organisms. Microbiol Rev 53:171-185

Gilliquet V, Berben G (1993) Positive and negative regulators of the *Saccharomyces cerevisiae* 'PHO system' participate in several cell functions. FEMS Microbiol Lett 108:333-339

Gimeno CJ, Ljungdahl PO, Styles CA, Fink GR (1992) Unipolar cell divisions in the yeast *S. cerevisiae* lead to filamentous growth: regulation by starvation and *RAS*. Cell 68:1077-1090

Giots F, Donaton MC, Thevelein JM (2003) Inorganic phosphate is sensed by specific phosphate carriers and acts in concert with glucose as a nutrient signal for activation of the protein kinase A pathway in the yeast *Saccharomyces cerevisiae*. Mol Microbiol 47:1163-1181

Goldberg D, Segal M, Levitzki A (1994) Cdc25 is not the signal receiver for glucose induced cAMP response in *S. cerevisiae*. FEBS Lett 356:249-254

Goncalves PM, Griffioen G, Bebelman JP, Planta RJ (1997) Signalling pathways leading to transcriptional regulation of genes involved in the activation of glycolysis in yeast. Mol Microbiol 25:483-493

Gorner W, Durchschlag E, Martinez-Pastor MT, Estruch F, Ammerer G, Hamilton B, Ruis H, Schuller C (1998) Nuclear localization of the C2H2 zinc finger protein Msn2p is regulated by stress and protein kinase A activity. Genes Dev 12:586-597

Gorner W, Durchschlag E, Wolf J, Brown EL, Ammerer G, Ruis H, Schuller C (2002) Acute glucose starvation activates the nuclear localization signal of a stress-specific yeast transcription factor. EMBO J 21:135-144

Griffioen G, Anghileri P, Imre E, Baroni MD, Ruis H (2000) Nutritional control of nucleocytoplasmic localization of cAMP-dependent protein kinase catalytic and regulatory subunits in *Saccharomyces cerevisiae*. J Biol Chem 275:1449-1456

Griffioen G, Branduardi P, Ballarini A, Anghileri P, Norbeck J, Baroni MD, Ruis H (2001) Nucleocytoplasmic distribution of budding yeast protein kinase A regulatory subunit Bcy1 requires Zds1 and is regulated by Yak1- dependent phosphorylation of its targeting domain. Mol Cell Biol 21:511-523

Griffioen G, Laan RJ, Mager WH, Planta RJ (1996) Ribosomal protein gene transcription in *Saccharomyces cerevisiae* shows a biphasic response to nutritional changes. Microbiol 142:2279-2287

Griffioen G, Mager WH, Planta RJ (1994) Nutritional upshift response of ribosomal protein gene transcription in *Saccharomyces cerevisiae*. FEMS Microbiol Lett 123:137-144

Gross A, Winograd S, Marbach I, Levitzki A (1999) The N-terminal half of Cdc25 is essential for processing glucose signaling in *Saccharomyces cerevisiae*. Biochemistry 38:13252-13262

Gross E, Goldberg D, Levitzki A (1992) Phosphorylation of the *S. cerevisiae* Cdc25 in response to glucose results in its dissociation from Ras Nature 360:762-765

Hampsey M (1997) A review of phenotypes in *Saccharomyces cerevisiae*. Yeast 13:1099-1133

Harashima T, Heitman J (2002) The Galpha protein Gpa2 controls yeast differentiation by interacting with kelch repeat proteins that mimic Gbeta subunits. Mol Cell 10:163-173

Hardie DG, Carling D (1997) The AMP-activated protein kinase--fuel gauge of the mammalian cell? Eur J Biochem 246:259-273

Hardy TA, Huang D, Roach PJ (1994) Interactions between cAMP-dependent and SNF1 protein kinases in the control of glycogen accumulation in *Saccharomyces cerevisiae*. J Biol Chem 269:27907-27913

Hardy TA, Roach PJ (1993) Control of yeast glycogen synthase-2 by COOH-terminal phosphorylation. J Biol Chem 268:23799-23805

Hartley AD, Ward MP, Garrett S (1994) The yak1 protein kinase of *Saccharomyces cerevisiae* moderates thermotolerance and inhibits growth by an Sch9 protein kinase-independent mechanism. Genetics 136:465-474

Harwood AJ (2001) Regulation of GSK-3: a cellular multiprocessor. Cell 105:821-824

Heideman W, Casperson GF, Bourne HR (1987) Adenylyl cyclase in yeast. Hydrodynamic properties and activation by trypsin. J Biol Chem 262:7087-7091

Heidrich K, Otto A, Behlke J, Rush J, Wenzel KW, Kriegel T (1997) Autophosphorylation-inactivation site of hexokinase 2 in *Saccharomyces cerevisiae*. Biochemistry 36:1960-1964

Herrero P, Fernandez R, Moreno F (1989) The hexokinase isoenzyme PII of *Saccharomyces cerevisiae* is a protein kinase. J Gen Microbiol 135:1209-1216

Herrero P, Galindez J, Ruiz N, Martinezcampa C, Moreno F (1995) Transcriptional regulation of the *Saccharomyces cerevisiae HXK1*, *HXK2* and *GLK1* genes. Yeast 11:137-144

Herrero P, Martinez Campa C, Moreno F (1998) The hexokinase 2 protein participates in regulatory DNA-protein complexes necessary for glucose repression of the *SUC2* gene in *Saccharomyces cerevisiae*. FEBS Lett 434:71-76

Herruer MH, Mager WH, Woudt LP, Nieuwint RTM, Wassenaar GM, Groeneveld P, Planta RJ (1987) Transcriptional control of yeast ribosomal protein synthesis during carbon-source upshift. Nucl Acids Res 15:10133-10144

Hinnebusch AG, Natarajan K (2002) Gcn4p, a master regulator of gene expression, is controlled at multiple levels by diverse signals of starvation and stress. Eukaryot Cell 1:22-32

Hirata Y, Andoh T, Asahara T, Kikuchi A (2003) Yeast glycogen synthase kinase-3 activates Msn2p-dependent transcription of stress responsive genes. Mol Biol Cell 14:302-312

Hirimburegama K, Durnez P, Keleman J, Oris E, Vergauwen R, Mergelsberg H, Thevelein JM (1992) Nutrient-induced activation of trehalase in nutrient-starved cells of the yeast *Saccharomyces cerevisiae*: cAMP is not involved as second messenger. J Gen Microbiol 138:2035-2043

Hoffman CS, Winston F (1991) Glucose repression of transcription of the *Schizosaccharomyces pombe fbp1* gene occurs by a cAMP signaling pathway. Genes Dev 5:561-571

Hohmann S, Winderickx J, de Winde JH, Valckx D, Cobbaert P, Luyten K, de Meirsman C, Ramos J, Thevelein JM (1999) Novel alleles of yeast hexokinase PII with distinct effects on catalytic activity and catabolite repression of *SUC2*. Microbiol 145:703-714

Huang D, Farkas I, Roach PJ (1996) Pho85p, a cyclin-dependent protein kinase, and the Snf1p protein kinase act antagonistically to control glycogen accumulation in *Saccharomyces cerevisiae*. Mol Cell Biol 16:4357-4365

Huang D, Moffat J, Andrews B (2002) Dissection of a complex phenotype by functional genomics reveals roles for the yeast cyclin-dependent protein kinase Pho85 in stress adaptation and cell integrity. Mol Cell Biol 22:5076-5088

Huang D, Moffat J, Wilson WA, Moore L, Cheng C, Roach PJ, Andrews B (1998) Cyclin partners determine Pho85 protein kinase substrate specificity *in vitro* and *in vivo*: control of glycogen biosynthesis by Pcl8 and Pcl10. Mol Cell Biol 18:3289-3299

Huang D, Wilson WA, Roach PJ (1997) Glucose-6-P control of glycogen synthase phosphorylation in yeast. J Biol Chem 272:22495-22501

Huang S, Jeffery DA, Anthony MD, O'Shea EK (2001) Functional analysis of the cyclin-dependent kinase inhibitor Pho81 identifies a novel inhibitory domain. Mol Cell Biol 21:6695-6705

Hughes JD, Estep PW, Tavazoie S, Church GM (2000) Computational identification of cis-regulatory elements associated with groups of functionally related genes in *Saccharomyces cerevisiae*. J Mol Biol 296:1205-1214

Hwang PK, Tugendreich S, Fletterick RJ (1989) Molecular analysis of *GPH1*, the gene encoding glycogen phosphorylase in *Saccharomyces cerevisiae*. Mol Cell Biol 9:1659-1666

Jeffery DA, Springer M, King DS, O'Shea EK (2001) Multi-site phosphorylation of Pho4 by the cyclin-CDK Pho80-Pho85 is semi-processive with site preference. J Mol Biol 306:997-1010

Jiang Y, Davis C, Broach JR (1998) Efficient transition to growth on fermentable carbon sources in *Saccharomyces cerevisiae* requires signaling through the Ras pathway. EMBO J 17:6942-6951

Jones S, Vignais M-L, Broach JR (1991) The CDC25 protein of *Saccharomyces cerevisiae* promotes exchange of guanine nucleotides bound to RAS. Molecular and Cellular Biology 11:2641-2646

Jorgensen P, Nishikawa JL, Breitkreutz BJ, Tyers M (2002) Systematic identification of pathways that couple cell growth and division in yeast. Science 297:395-400

Kaeberlein M, Andalis AA, Fink GR, Guarente L (2002) High osmolarity extends life span in *Saccharomyces cerevisiae* by a mechanism related to calorie restriction. Mol Cell Biol 22:8056-8066

Kaffman A, Herskowitz I, Tjian R, O'Shea EK (1994) Phosphorylation of the transcription factor PHO4 by a cyclin-cdk complex, PHO80-PHO85. Science 263:1153-1156

Kaffman A, Rank NM, O'Neill EM, Huang LS, O'Shea EK (1998a) The receptor Msn5 exports the phosphorylated transcription factor Pho4 out of the nucleus. Nature 396:482-486

Kaffman A, Rank NM, O'Shea EK (1998b) Phosphorylation regulates association of the transcription factor Pho4 with its import receptor Pse1/Kap121. Genes Dev 12:2673-2683

Kamada Y, Funakoshi T, Shintani T, Nagano K, Ohsumi M, Ohsumi Y (2000) Tor-mediated induction of autophagy via an Apg1 protein kinase complex. J Cell Biol 150:1507-1513

Kataoka T, Broek D, Wigler M (1985) DNA sequence and characterization of the *S. cerevisiae* gene encoding adenylate cyclase. Cell 43:493-505

Klein C, Struhl K (1994) Protein kinase A mediates growth-regulated expression of yeast ribosomal protein genes by modulating RAP1 transcriptional activity. Mol Cell Biol 14:1920-1928

Komeili A, O'Shea EK (1999) Roles of phosphorylation sites in regulating activity of the transcription factor Pho4. Science 284:977-980

Kornitzer D, Raboy B, Kulka RG, Fink GR (1994) Regulated degradation of the transcription factor Gcn4. EMBO J 13:6021-6030

Kozma SC, Thomas G (2002) Regulation of cell size in growth, development and human disease: PI3K, PKB and S6K. Bioessays 24:65-71

Kraakman L, Lemaire K, Ma P, Teunissen AW, Donaton MC, Van Dijck P, Winderickx J, de Winde JH, Thevelein JM (1999a) A *Saccharomyces cerevisiae* G-protein coupled receptor, Gpr1, is specifically required for glucose activation of the cAMP pathway during the transition to growth on glucose. Mol Microbiol 32:1002-1012

Kraakman LS, Griffioen G, Zerp S, Groeneveld P, Thevelein JM, Mager WH, Planta RJ (1993) Growth-related expression of ribosomal protein genes in *Saccharomyces cerevisiae.* Mol Gen Genet 239:196-204

Kraakman LS, Winderickx J, Thevelein JM, de Winde JH (1999b) Structure-function analysis of yeast hexokinase: structural requirements for triggering cAMP signalling and catabolite repression. Biochem J 343:159-168

Kretschmer M, Schellenberger W, Otto A, Kessler R, Hofmann E (1987) Fructose-2,6-bisphosphatase and 6-phosphofructo-2-kinase are separable in yeast. Biochem J 246:755-759

Kriegel TM, Rush J, Vojtek AB, Clifton D, Fraenkel DG (1994) *In vivo* phosphorylation site of hexokinase 2 in *Saccharomyces cerevisiae*. Biochemistry 33:148-152

Kubler E, Mosch HU, Rupp S, Lisanti MP (1997) Gpa2p, a G-protein alpha-subunit, regulates growth and pseudohyphal development in *Saccharomyces cerevisiae* via a cAMP-dependent mechanism. J Biol Chem 272:20321-20323

Kubota H, Obata T, Ota K, Sasaki T, Ito T (2003) Rapamycin-induced translational derepression of GCN4 mRNA involves a novel mechanism for activation of the eIF2alpha kinase GCN2. J Biol Chem 278:20457-20460

Kuchin S, Treich I, Carlson M (2000) A regulatory shortcut between the Snf1 protein kinase and RNA polymerase II holoenzyme. Proc Natl Acad Sci USA 97:7916-7920

Kuchin S, Vyas VK, Carlson M (2002) Snf1 protein kinase and the repressors Nrg1 and Nrg2 regulate FLO11, haploid invasive growth, and diploid pseudohyphal differentiation. Mol Cell Biol 22:3994-4000

Kuret J, Johnson KE, Nicolette C, Zoller MJ (1988) Mutagenesis of the regulatory subunit of yeast cAMP-dependent protein kinase. Isolation of site-directed mutants with altered binding affinity for catalytic subunit. J Biol Chem 263:9149-9154

Lascaris RF, Mager WH, Planta RJ (1999) DNA-binding requirements of the yeast protein Rap1p as selected in silico from ribosomal protein gene promoter sequences. Bioinformatics 15:267-277

Lee J, Colwill K, Aneliunas V, Tennyson C, Moore L, Ho Y, Andrews B (1998) Interaction of yeast Rvs167 and Pho85 cyclin-dependent kinase complexes may link the cell cycle to the actin cytoskeleton. Curr Biol 8:1310-1321

Lee M, O'Regan S, Moreau JL, Johnson AL, Johnston LH, Goding CR (2000) Regulation of the Pcl7-Pho85 cyclin-cdk complex by Pho81. Mol Microbiol 38:411-422

Lenburg ME, O'Shea EK (1996) Signaling phosphate starvation. Trends Biochem Sci 21:383-387

Lenburg ME, O'Shea EK (2001) Genetic evidence for a morphogenetic function of the *Saccharomyces cerevisiae* Pho85 cyclin-dependent kinase. Genetics 157:39-51

Lenssen E, Oberholzer U, Labarre J, De Virgillio C, Collart MA (2002) Contributes to control of Msn2-dpendent transcription by the Ras/cAMP pathway. Mol Microbiol 43:1023-1037

Lin H, Choi JH, Vancura A (1998) Phosphoinositide-specific phospholipase C interacts with phosphatidylinositol kinase homolog TOR2. Biochem Biophys Res Commun 252:285-289

Lin K, Hwang PK, Fletterick RJ (1995) Mechanism of regulation in yeast glycogen phosphorylase. J Biol Chem 270:26833-26839

Lin K, Rath VL, Dai SC, Fletterick RJ, Hwang PK (1996) A protein phosphorylation switch at the conserved allosteric site in GP. Science 273:1539-1542

Lin SJ, Defossez PA, Guarente L (2000) Requirement of NAD and SIR2 for life-span extension by calorie restriction in *Saccharomyces cerevisiae*. Science 289:2126-2128

Liu C, Yang Z, Yang J, Xia Z, Ao S (2000) Regulation of the yeast transcriptional factor PHO2 activity by phosphorylation. J Biol Chem 275:31972-31978

Lobo Z, Maitra PK (1977) Genetics of yeast hexokinase. Genetics 86:727-744

Loewith R, Jacinto E, Wullschleger S, Lorberg A, Crespo JL, Bonenfant D, Oppliger W, Jenoe P, Hall MN (2002) Two TOR complexes, only one of which is rapamycin sensitive, have distinct roles in cell growth control. Mol Cell 10:457-468

Longo VD, Liou LL, Valentine JS, Gralla EB (1999) Mitochondrial superoxide decreases yeast survival in stationary phase. Arch Biochem Biophys 365:131-142

Lorenz MC, Heitman J (1997) Yeast pseudohyphal growth is regulated by GPA2, a G protein a homolog. EMBO J 16:7008-7018

Lorenz MC, Heitman J (1998) The MEP2 ammonium permease regulates pseudohyphal differentiation in *Saccharomyces cerevisiae*. EMBO J 17:1236-1247

Lorenz MC, Pan X, Harashima T, Cardenas ME, Xue Y, Hirsch JP, Heitman J (2000) The G protein-coupled receptor Gpr1 is a nutrient sensor that regulates pseudohyphal differentiation in *Saccharomyces cerevisiae*. Genetics 154:609-622

Louvion JF, Abbas-Terki T, Picard D (1998) Hsp90 is required for pheromone signaling in yeast. Mol Biol Cell 9:3071-3083

Ma H, Bloom LM, Dakin SE, Walsh CT, Botstein D (1989a) The 15 N-terminal amino acids of hexokinase II are not required for *in vivo* function: analysis of a truncated form of hexokinase II in *Saccharomyces cerevisiae*. Proteins 5:218-223

Ma H, Bloom LM, Walsh CT, Botstein D (1989b) The residual enzymatic phosphorylation activity of hexokinase II mutants is correlated with glucose repression in *Saccharomyces cerevisiae*. Mol Cell Biol 9:5643-5649

Ma P, Wera S, Van Dijck P, Thevelein JM (1999) The *PDE1* encoded low-affinity phosphodiesterase in the yeast *Saccharomyces cerevisiae* has a specific function in controlling agonist-induced cAMP signalling. Mol Biol Cell 10:91-104

Madhani HD, Fink GR (1997) Combinatorial control required for the specificity of yeast MAPK signaling. Science 275:1314-1317

Malone RE (1990) Dual regulation of meiosis in yeast. Cell 61:375-378

Manhart A, Holzer H (1988) Substrate specificity of the phosphorylated fructose-1,6-bisphosphatase dephosphorylating protein phosphatase from *Saccharomyces cerevisiae*. Yeast 4:227-232

Martinez-Campa C, Herrero P, Ramirez M, Moreno F (1996) Molecular analysis of the promoter region of the hexokinase 2 gene of *Saccharomyces cerevisiae*. FEMS Microbiol Lett 137:69-74

Martinez-Pastor MT, Marchler G, Schuller C, Marchler-Bauer A, Ruis H, Estruch F (1996) The *Saccharomyces cerevisiae* zinc finger proteins Msn2p and Msn4p are required for transcriptional induction through the stress response element (STRE). EMBO J 15:2227-2235

Matsumoto K, Uno I, Ishikawa T (1984) Identification of the structural gene and nonsense alleles for adenylate cyclase in *Saccharomyces cerevisiae*. J Bacteriol 157:277-282

Matsumoto K, Uno I, Ishikawa T (1985) Genetic analysis of the role of cAMP in yeast. Yeast 1:15-24

Matsumoto K, Uno I, Oshima Y, Ishikawa T (1982) Isolation and characterization of yeast mutants deficient in adenylate cyclase and cAMP-dependent protein kinase. Proc Natl Acad Sci USA 79:2355-2359

Matsuura A, Treinin M, Mitsuzawa H, Kassir Y, Uno I, Simchen G (1990) The adenylate cyclase/protein kinase cascade regulates entry into meiosis in *Saccharomyces cerevisiae* through the gene IME1. EMBO J 9:3225-3232

Mayordomo I, Estruch F, Sanz P (2002) Convergence of the target of rapamycin and the Snf1 protein kinase pathways in the regulation of the subcellular localization of Msn2, a transcriptional activator of STRE (Stress Response Element)-regulated genes. J Biol Chem 277:35650-35656

Mayordomo I, Sanz P (2001) Hexokinase PII: structural analysis and glucose signalling in the yeast *Saccharomyces cerevisiae*. Yeast 18:923-930

Mbonyi K, Beullens M, Detremerie K, Geerts L, Thevelein JM (1988) Requirement of one functional RAS gene and inability of an oncogenic ras variant to mediate the glucose-induced cyclic AMP signal in the yeast *Saccharomyces cerevisiae*. Mol Cell Biol 8:3051-3057

Mbonyi K, van Aelst L, Arguelles JC, Jans AW, Thevelein JM (1990) Glucose-induced hyperaccumulation of cyclic AMP and defective glucose repression in yeast strains with reduced activity of cyclic AMP- dependent protein kinase. Mol Cell Biol 10:4518-4523

McBride HJ, Sil A, Measday V, Yu Y, Moffat J, Maxon ME, Herskowitz I, Andrews B, Stillman DJ (2001) The protein kinase Pho85 is required for asymmetric accumulation of the Ash1 protein in *Saccharomyces cerevisiae*. Mol Microbiol 42:345-353

Measday V, McBride H, Moffat J, Stillman D, Andrews B (2000) Interactions between Pho85 cyclin-dependent kinase complexes and the Swi5 transcription factor in budding yeast. Mol Microbiol 35:825-834

Measday V, Moore L, Ogas J, Tyers M, Andrews B (1994) The PCL2 (ORFD)-PHO85 cyclin-dependent kinase complex: A cell cycle regulator in yeast. Science 266:1391-1395

Measday V, Moore L, Retnakaran R, Lee J, Donoviel M, Neiman AM, Andrews B (1997) A family of cyclin-like proteins that interact with the Pho85 cyclin-dependent kinase. Mol Cell Biol 17:1212-1223

Meimoun A, Holtzman T, Weissman Z, McBride HJ, Stillman DJ, Fink GR, Kornitzer D (2000) Degradation of the transcription factor Gcn4 requires the kinase Pho85 and the SCF(CDC4) ubiquitin-ligase complex. Mol Biol Cell 11:915-927

Mintzer KA, Field J (1999) The SH3 domain of the *S. cerevisiae* Cdc25p binds adenylyl cyclase and facilitates Ras regulation of cAMP signalling. Cell Signal 11:127-135

Moehle CM, Hinnebusch AG (1991) Association of RAP1 binding sites with stringent control of ribosomal protein gene transcription in *Saccharomyces cerevisiae*. Mol Cell Biol 11:2723-2735

Morano KA, Thiele DJ (1999) The Sch9 protein kinase regulates Hsp90 chaperone complex signal transduction activity *in vivo*. EMBO J 18:5953-5962

Mosch HU, Roberts RL, Fink GR (1996) Ras2 signals via the Cdc42/Ste20/mitogen-activated protein kinase module to induce filamentous growth in *Saccharomyces cerevisiae*. Proc Natl Acad Sci USA 93:5352-5356

Moskvina E, Schuller C, Maurer CT, Mager WH, Ruis H (1998) A search in the genome of *Saccharomyces cerevisiae* for genes regulated via stress response elements. Yeast 14:1041-1050

Munder T, Küntzel H (1989) Glucose-induced cAMP signaling in *Saccharomyces cerevisiae* is mediated by the CDC25 protein. FEBS Lett 242:341-345

Nakafuku M, Obara T, Kaibuchi K, Miyajima I, Miyajima A, Itoh H, Nakamura S, Arai K, Matsumoto K, Kaziro Y (1988) Isolation of a second yeast *Saccharomyces cerevisiae* gene (*GPA2*) coding for guanine nucleotide-binding regulatory protein: studies on its structure and possible functions. Proc Natl Acad Sci USA 85:1374-1378

Nelson C, Goto S, Lund K, Hung W, Sadowski I (2003) Srb10/Cdk8 regulates yeast filamentous growth by phosphorylating the transcription factor Ste12. Nature 421:187-190

Neuman-Silberberg FS, Bhattacharya S, Broach JR (1995) Nutrient availability and the RAS/cyclic AMP pathway both induce expression of ribosomal protein genes in *Saccharomyces cerevisiae* but by different mechanisms. Mol Cell Biol 15:3187-3196

Nigavekar SS, Tan YS, Cannon JF (2002) Glc8 is a glucose-repressible activator of Glc7 protein phosphatase-1. Arch Biochem Biophys 404:71-79

Nikawa J, Sass P, Wigler M (1987) Cloning and characterization of the low-affinity cyclic AMP phosphodiesterase gene of *Saccharomyces cerevisiae*. Mol Cell Biol 7:3629-3636

Nishizawa M, Kawasumi M, Fujino M, Toh-e A (1998) Phosphorylation of sic1, a cyclin-dependent kinase (Cdk) inhibitor, by Cdk including Pho85 kinase is required for its prompt degradation. Mol Biol Cell 9:2393-2405

Nishizawa M, Suzuki K, Fujino M, Oguchi T, Toh-e A (1999) The Pho85 kinase, a member of the yeast cyclin-dependent kinase (Cdk) family, has a regulation mechanism different from Cdks functioning throughout the cell cycle. Genes Cells 4:627-642

Nishizawa M, Tanabe M, Yabuki N, Kitada K, Toh EA (2001) Pho85 kinase, a yeast cyclin-dependent kinase, regulates the expression of UGP1 encoding UDP-glucose pyrophosphorylase. Yeast 18:239-249

Noda T, Ohsumi Y (1998) Tor, a phosphatidylinositol kinase homologue, controls autophagy in yeast. J Biol Chem 273:3963-3966

Ogawa N, DeRisi J, Brown PO (2000) New components of a system for phosphate accumulation and polyphosphate metabolism in *Saccharomyces cerevisiae* revealed by genomic expression analysis. Mol Biol Cell 11:4309-4321

Ogawa N, Saitoh H, Miura K, Magbanua JPV, Bunya M, Harashima S, Oshima Y (1995) Structure and distribution of specific cis-elements for transcriptional regulation of PH084 in *Saccharomyces cerevisiae*. Mol Gen Genet 249:406-416

O'Neill EM, Kaffman A, Jolly ER, O'Shea EK (1996) Regulation of PHO4 nuclear localization by the PHO80-PHO85 cyclin-CDK complex. Science 271:209-212

Ostling J, Ronne H (1998) Negative control of the Miglp repressor by Snflp-dependent phosphorylation in the absence of glucose. Eur J Biochem 252:162-168

Palecek SP, Parikh AS, Kron SJ (2000) Genetic analysis reveals that FLO11 upregulation and cell polarization independently regulate invasive growth in *Saccharomyces cerevisiae*. Genetics 156:1005-1023

Palecek SP, Parikh AS, Kron SJ (2002) Sensing, signalling and integrating physical processes during *Saccharomyces cerevisiae* invasive and filamentous growth. Microbiology 148:893-907

Pan X, Harashima T, Heitman J (2000) Signal transduction cascades regulating pseudohyphal differentiation of *Saccharomyces cerevisiae*. Curr Opin Microbiol 3:567-572

Pan X, Heitman J (2000) Sok2 regulates yeast pseudohyphal differentiation via a transcription factor cascade that regulates cell-cell adhesion. Mol Cell Biol 20:8364-8372

Pan X, Heitman J (2002) Protein kinase A operates a molecular switch that governs yeast pseudohyphal differentiation. Mol Cell Biol 22:3981-3993

Pan XW, Heitman J (1999) Cyclic AMP-dependent protein kinase regulates pseudohyphal differentiation in *Saccharomyces cerevisiae*. Mol Cell Biol 19:4874-4887

Panek AC, de Araujo PS, Moura Neto V, Panek AD (1987) Regulation of the trehalose-6-phosphate synthase complex in *Saccharomyces*. I. Interconversion of forms by phosphorylation. Curr Genet 11:459-465

Pardo LA, Lazo PS, Ramos S (1993) Activation of adenylate cyclase in *cdc25* mutants of *Saccharomyces cerevisiae*. FEBS Lett 319:237-243

Parrou JL, Enjalbert B, Francois J (1999) STRE- and cAMP-independent transcriptional induction of *Saccharomyces cerevisiae* GSY2 encoding glycogen synthase during diauxic growth on glucose. Yeast 15:1471-1484

Pedruzzi I, Burckert N, Egger P, De Virgilio C (2000) *Saccharomyces cerevisiae* Ras/cAMP pathway controls post-diauxic shift element-dependent transcription through the zinc finger protein Gis1. EMBO J 19:2569-2579

Peng ZY, Trumbly RJ, Reimann EM (1990) Purification and characterization of glycogen synthase from a glycogen-deficient strain of *Saccharomyces cerevisiae*. J Biol Chem 265:13871-13877

Pernambuco MB, Winderickx J, Crauwels M, Griffioen G, Mager WH, Thevelein JM (1996) Glucose-triggered signalling in *Saccharomyces cerevisiae*: different requirements for sugar phosphorylation between cells grown on glucose and those grown on non-fermentable carbon sources. Microbiology 142:1775-1782

Popova Iu G, Padkina MV, Sambuk EV (2000) [Effect of mutations in PHO85 and PHO4 genes on utilization of proline in *Saccharomyces cerevisiae* yeasts]. Genetika 36:1622-1628

Portela P, Zaremberg V, Moreno S (2001) Evaluation of *in vivo* activation of protein kinase A under non-dissociable conditions through the overexpression of wild-type and mutant regulatory subunits in *Saccharomyces cerevisiae*. Microbiology 147:1149-1159

Pries R, Bomeke K, Irniger S, Grundmann O, Braus GH (2002) Amino acid-dependent Gcn4p stability regulation occurs exclusively in the yeast nucleus. Eukaryot Cell 1:663-672

Randez-Gil F, Herrero P, Sanz P, Prieto JA, Moreno F (1998a) Hexokinase PII has a double cytosolic-nuclear localisation in *Saccharomyces cerevisiae*. FEBS Lett 425:475-478

Randez-Gil F, Sanz P, Entian KD, Prieto JA (1998b) Carbon source-dependent phosphorylation of hexokinase PII and its role in the glucose-signaling response in yeast. Mol Cell Biol 18:2940-2948

Rayner TF, Gray JV, Thorner JW (2002) Direct and novel regulation of cAMP-dependent protein kinase by Mck1p, a yeast glycogen kinase-3. J Biol Chem 277:16814-16822

Reinders A, Burckert N, Boller T, Wiemken A, De Virgilio C (1998) *Saccharomyces cerevisiae* cAMP-dependent protein kinase controls entry into stationary phase through the Rim15p protein kinase. Genes Dev 12:2943-2955

Rittenhouse J, Moberly L, Marcus F (1987) Phosphorylation *in vivo* of yeast (*Saccharomyces cerevisiae*) fructose-1,6-bisphosphatase at the cyclic AMP-dependent site. J Biol Chem 262:10114-10119

Roberts RL, Mosch HU, Fink GR (1997) 14-3-3 proteins are essential for RAS/MAPK cascade signaling during pseudohyphal development in *S. cerevisiae*. Cell 89:1055-1065

Robertson LS, Causton HC, Young RA, Fink GR (2000) The yeast A kinases differentially regulate iron uptake and respiratory function. Proc Natl Acad Sci USA 97:5984-5988

Robertson LS, Fink GR (1998) The three yeast A kinases have specific signaling functions in pseudohyphal growth. Proc Natl Acad Sci USA 95:13783-13787

Rolland F, Wanke V, Cauwenberg L, Ma P, Boles E, Vanoni M, de Winde JH, Thevelein JM, Winderickx J (2001) The role of hexose transport and phosphorylation in cAMP signalling in the yeast *Saccharomyces cerevisiae*. FEMS Yeast Res 1403:1-13

Rolland F, Winderickx J, Thevelein JM (2002) Glucose-sensing and -signalling mechanisms in yeast. FEM Yeast Res 2:183-201

Ronne H (1995) Glucose repression in fungi. Trends Genet 11:12-17

Rose M, Albig W, Entian KD (1991) Glucose repression in *Saccharomyces cerevisiae* is directly associated with hexose phosphorylation by hexokinase-PI and hexokinase-PII. Eur J Biochem 199:511-518

Rudoni S, Colombo S, Coccetti P, Martegani E (2001) Role of guanine nucleotides in the regulation of the Ras/cAMP pathway in *Saccharomyces cerevisiae*. Biochim Biophys Acta 1538:181-189

Ruis H, Schuller C (1995) Stress signaling in yeast. Bioessays 17:959-965

Rupp S, Summers E, Lo HJ, Madhani H, Fink G (1999) MAP kinase and cAMP filamentation signaling pathways converge on the unusually large promoter of the yeast FLO11 gene. EMBO J 18:1257-1269

Santos RC, Waters NC, Creasy CL, Bergman LW (1995) Structure-function relationships of the yeast cyclin-dependent kinase pho85. Mol Cell Biol 15:5482-5491

Sanz P, Alms GR, Haystead TA, Carlson M (2000) Regulatory interactions between the Reg1-Glc7 protein phosphatase and the Snf1 protein kinase. Mol Cell Biol 20:1321-1328

Sanz P, Nieto A, Prieto JA (1996) Glucose repression may involve processes with different sugar kinase requirements. J Bacteriol 178:4721-4723

Sass P, Field J, Nikawa J, Toda T, Wigler M (1986) Cloning and characterization of the high-affinity cAMP phosphodiesterase of *S. cerevisiae*. Proc Natl Acad Sci USA 83:9303-9307

Scheid MP, Woodgett JR (2001) PKB/AKT: functional insights from genetic models. Nat Rev Mol Cell Biol 2:760-768

Schmitt AP, McEntee K (1996) Msn2p, a zinc finger DNA-binding protein, is the transcriptional activator of the multistress response in *Saccharomyces cerevisiae*. Proc Natl Acad Sci USA 93:5777-5782

Schneider KR, Smith RL, O'Shea EK (1994) Phosphate-regulated inactivation of the kinase PHO80-PHO85 by the CDK inhibitor PHO81. Science 266:122-126

Schomerus C, Kuntzel H (1992) CDC25-dependent induction of inositol 1,4,5-trisphosphate and diacylglycerol in *Saccharomyces cerevisiae* by nitrogen. FEBS Lett 307:249-252

Shemer R, Meimoun A, Holtzman T, Kornitzer D (2002) Regulation of the transcription factor Gcn4 by Pho85 cyclin PCL5. Mol Cell Biol 22:5395-5404

Shore D, Squire M, Nasmyth KA (1984) Characterization of two genes required for the position-effect control of yeast mating-type genes. EMBO J 3:2817-2823

Simon I, Barnett J, Hannett N, Harbison CT, Rinaldi NJ, Volkert TL, Wyrick JJ, Zeitlinger J, Gifford DK, Jaakkola TS, Young RA (2001) Serial regulation of transcriptional regulators in the yeast cell cycle. Cell 106:697-708

Smith A, Ward MP, Garrett S (1998) Yeast PKA represses Msn2p/Msn4p-dependent gene expression to regulate growth, stress response and glycogen accumulation. EMBO J 17:3556-3564

Stapleton D, Gao GA, Michell BJ, Widmer J, Mitchelhill K, Teh T, House CM, Witters LA, Kemp BE (1994) Mammalian 5'-AMP-activated protein kinase non-catalytic subunits are homologs of proteins that interact with yeast Snf1 protein kinase. J Biol Chem 269:29343-29346

Sugajska E, Swiatek W, Zabrocki P, Geyskens I, Thevelein JM, Zolnierowicz S, Wera S (2001) Multiple effects of protein phosphatase 2A on nutrient-induced signalling in the yeast *Saccharomyces cerevisiae*. Mol Microbiol 40:1020-1026

Sussel L, Shore D (1991) Separation of transcriptional activation and silencing functions of the RAP1-encoded repressor/activator protein 1: Isolation of viable mutants affecting both silencing and telomere length. Proc Natl Acad Sci USA 88:7749-7753

Tadi D, Hasan RN, Bussereau F, Boy-Marcotte E, Jacquet M (1999) Selection of genes repressed by cAMP that are induced by nutritional limitation in *Saccharomyces cerevisiae*. Yeast 15:1733-1745

Tan YS, Morcos PA, Cannon JF (2003) Pho85 phosphorylates the Glc7 protein phosphatase regulator Glc8 *in vivo*. J Biol Chem 278:147-153

Tanaka K, Nakafuku M, Satoh T, Marshall MS, Gibbs JB, Matsumoto K, Kaziro Y, Toh-e A (1990) *S. cerevisiae* genes *IRA1* and *IRA2* encode proteins that may be functionally equivalent to mammalian ras GTPase activating protein. Cell 60:803-807

Taylor SS, Buechler JA, Yonemoto W (1990) cAMP-dependent protein kinase: framework for a diverse family of regulatory enzymes. Annu Rev Biochem 59:971-1005

Taylor WE, Young ET (1990) cAMP-dependent phosphorylation and inactivation of yeast transcription factor Adr1 does not affect DNA binding. Proc Natl Acad Sci USA 87:4098-4102

Teem JL, Abovich N, Kaufer NF, Schwindinger WF, Warner JR, Levy A, Woolford J, Leer RJ, van Raamsdonk-Duin MM, Mager WH (1984) A comparison of yeast ribosomal protein gene DNA sequences. Nucleic Acids Res 12:8295-8312

Thevelein JM (1992) The RAS-adenylate cyclase pathway and cell cycle control in *Saccharomyces cerevisiae*. Antonie Leeuwenhoek J Microbiol 62:109-130

Thevelein JM (1994) Signal transduction in yeast. Yeast 10:1753-1790

Thevelein JM (1996) Regulation of trehalose metabolism and its relevance to cell growth and function. In: Brambl/Marzluf (ed) The Mycota III; Biochemistry and molecular biology. Springer verlag, Berlin-Heidelberg, pp 395-420

Thompson-Jaeger S, Francois J, Gaughran JP, Tatchell K (1991) Deletion of SNF1 affects the nutrient response of yeast and resembles mutations which activate the adenylate cyclase pathway. Genetics 129:697-706

Timblin BK, Bergman LW (1997) Elevated expression of stress response genes resulting from deletion of the PHO85 gene. Mol Microbiol 26:981-990

Timblin BK, Tatchell K, Bergman LW (1996) Deletion of the gene encoding the cyclin-dependent protein kinase Pho85 alters glycogen metabolism in *Saccharomyces cerevisiae*. Genetics 143:57-66

Toda T, Cameron S, Sass P, Wigler M (1988) *SCH9*, a gene of *Saccharomyces cerevisiae* that encodes a protein distinct from, but functionally and structurally related to cAMP-dependent protein kinase catalytic subunits. Gene Dev 2:517-527

Toda T, Cameron S, Sass P, Zoller M, Scott JD, McMullen B, Hurwitz M, Krebs EG, Wigler M (1987a) Cloning and characterization of BCY1, a locus encoding a regulatory subunit of the cyclic AMP-dependent protein kinase in *Saccharomyces cerevisiae.* Mol Cell Biol 7:1371-1377

Toda T, Cameron S, Sass P, Zoller M, Wigler M (1987b) Three different genes in *Saccharomyces cerevisiae* encode the catalytic subunits of the cAMP-dependent protein kinase. Cell 50:277-287

Toda T, Uno I, Ishikawa T, Powers S, Kataoka T, Broek D, Cameron S, Broach J, Matsumoto K, Wigler M (1985) In yeast, Ras proteins are controlling elements of adenylate cyclase. Cell 40:27-36

Toh-e A, Tanaka K, Uesono Y, Wickner RB (1988) PHO85, a negative regulator of the PHO system, is a homolog of the protein kinase gene, CDC28, of *Saccharomyces cerevisiae*. Mol Gen Genet 214:162-164

Treger JM, Schmitt AP, Simon JR, McEntee K (1998) Transcriptional factor mutations reveal regulatory complexities of heat shock and newly identified stress genes in *Saccharomyces cerevisiae*. J Biol Chem 273:26875-26879

Treitel MA, Carlson M (1995) Repression by SSN6-TUP1 is directed by MIG1, a repressor activator protein. Proc Natl Acad Sci USA 92:3132-3136

Uno I, Matsumoto K, Adachi K, Ishikawa T (1983) Genetic and biochemical evidence that trehalase is a substrate of cAMP-dependent protein kinase in yeast. J Biol Chem 258:10867-10872

Uno I, Mitsuzawa H, Tanaka K, Oshima T, Ishikawa T (1987) Identification of the domain of *Saccharomyces cerevisiae* adenylate cyclase associated wih the regulatory function of RAS products. Mol Gen Genet 210:187-194

Valenzuela L, Aranda C, Gonzalez A (2001) TOR modulates GCN4-dependent expression of genes turned on by nitrogen limitation. J Bacteriol 183:2331-2334

Van Aelst L, Boy-Marcotte E, Camonis JH, Thevelein JM, Jacquet M (1990) The C-terminal part of the *CDC25* gene product plays a key role in signal transduction in the glucose-induced modulation of cAMP level in *Saccharomyces cerevisiae*. Eur J Biochem 193:675-680

Van Aelst L, Jans AWH, Thevelein JM (1991) Involvement of the CDC25 gene product in the signal transmission pathway of the glucose-induced RAS-mediated cAMP signal in the yeast *Saccharomyces cerevisiae*. J Gen Microbiol 137:341-349

Van Wuytswinkel O, Reiser V, Siderius M, Kelders MC, Ammerer G, Ruis H, Mager WH (2000) Response of *Saccharomyces cerevisiae* to severe osmotic stress: evidence for a novel activation mechanism of the HOG MAP kinase pathway. Mol Microbiol 37:382-397

Vaseghi S, Macherhammer F, Zibek S, Reuss M (2001) Signal transduction dynamics of the protein kinase-A/phosphofructokinase-2 system in *Saccharomyces cerevisiae*. Metab Eng 3:163-172

Versele M, de Winde JH, Thevelein JM (1999) A novel regulator of G-protein signalling in yeast, Rgs2, downregulates glucose-activation of the cAMP pathway through direct inhibition of Gpa2. EMBO J 18:5577-5591

Vojtek AB, Fraenkel DG (1990) Phosphorylation of yeast hexokinases. Eur J Biochem 190:371-375

Vyas VK, Kuchin S, Carlson M (2001) Interaction of the repressors Nrg1 and Nrg2 with the Snf1 protein kinase in *Saccharomyces cerevisiae*. Genetics 158:563-572

Wang Z, Wilson WA, Fujino MA, Roach PJ (2001a) Antagonistic controls of autophagy and glycogen accumulation by Snf1p, the yeast homolog of AMP-activated protein kinase, and the cyclin-dependent kinase Pho85p. Mol Cell Biol 21:5742-5752

Wang Z, Wilson WA, Fujino MA, Roach PJ (2001b) The yeast cyclins Pc16p and Pc17p are involved in the control of glycogen storage by the cyclin-dependent protein kinase Pho85p. FEBS Lett 506:277-280

Ward MP, Gimeno CJ, Fink GR, Garrett S (1995) SOK2 may regulate cyclic AMP-dependent protein kinase-stimulated growth and pseudohyphal development by repressing transcription. Mol Cell Biol 15:6854-6863

Wek SA, Zhu S, Wek RC (1995) The histidyl-tRNA synthetase-related sequence in the eIF-2 alpha protein kinase GCN2 interacts with tRNA and is required for activation in response to starvation for different amino acids. Mol Cell Biol 15:4497-4506

Wera S, Bergsma JC, Thevelein JM (2001) Phosphoinositides in yeast: genetically tractable signalling. FEM Yeast Res 1:9-13

Werner-Washburne M, Brown D, Braun E (1991) Bcy1, the regulatory subunit of cAMP-dependent protein kinase in yeast, is differentially modified in response to the physiological status of the cell. J Biol Chem 266:19704-19709

Wilson RB, Tatchell K (1988) SRA5 encodes the low-Km cyclic AMP phosphodiesterase of *Saccharomyces cerevisiae*. Mol Cell Biol 8:505-510

Wilson WA, Hawley SA, Hardie DG (1996) Glucose repression/derepression in budding yeast: SNF1 protein kinase is activated by phosphorylation under derepressing conditions, and this correlates with a high AMP:ATP ratio. Curr Biol 6:1426-1434

Wilson WA, Mahrenholz AM, Roach PJ (1999) Substrate targeting of the yeast cyclin-dependent kinase Pho85p by the cyclin Pcl10p. Mol Cell Biol 19:7020-7030

Wilson WA, Wang Z, Roach PJ (2002) Analysis of respiratory mutants reveals new aspects of the control of glycogen accumulation by the cyclin-dependent protein kinase Pho85p. FEBS Lett 515:104-108

Winderickx J, de Winde JH, Crauwels M, Hino A, Hohmann S, Van Dijck P, Thevelein JM (1996) Regulation of genes encoding subunits of the trehalose synthase complex in *Saccharomyces cerevisiae*: novel variations of STRE-mediated transcription control? Mol Gen Genet 252:470-482

Woods A, Munday MR, Scott J, Yang X, Carlson M, Carling D (1994) Yeast SNF1 is functionally related to mammalian AMP-activated protein kinase and regulates acetyl-CoA carboxylase *in vivo*. J Biol Chem 269:19509-19515

Xu Z, Norris D (1998) The SFP1 gene product of *Saccharomyces cerevisiae* regulates G2/M transitions during the mitotic cell cycle and DNA-damage response. Genetics 150:1419-1428

Xue Y, Batlle M, Hirsch JP (1998) *GPR1* encodes a putative G protein-coupled receptor that associates with the Gpa2p Ga subunit and functions in a Ras-independent pathway. EMBO J 17:1996-2007

Yu L, Sabet N, Chambers A, Morse RH (2001) The N-terminal and C-terminal domains of RAP1 are dispensable for chromatin opening and GCN4-mediated HIS4 activation in budding yeast. J Biol Chem 276:33257-33264

Yun CW, Tamaki H, Nakayama R, Yamamoto K, Kumagai H (1997) G-protein coupled receptor from yeast *Saccharomyces cerevisiae*. Biochem Biophys Res Commun 240:287-292

Zahringer H, Holzer H, Nwaka S (1998) Stability of neutral trehalase during heat stress in *Saccharomyces cerevisiae* is dependent on the activity of the catalytic subunits of cAMP-dependent protein kinase, Tpk1 and Tpk2. Eur J Biochem 255:544-551

Zhang Z, Smith MM, Mymryk JS (2001) Interaction of the E1A oncoprotein with Yak1p, a novel regulator of yeast pseudohyphal differentiation, and related mammalian kinases. Mol Biol Cell 12:699-710

Zhu H, Klemic JF, Chang S, Bertone P, Casamayor A, Klemic KG, Smith D, Gerstein M, Reed MA, Snyder M (2000) Analysis of yeast protein kinases using protein chips. Nat Genet 26:283-289

Zimmermann FK, Scheel I (1977) Mutants of *Saccharomyces cerevisiae* resistant to carbon catabolite repression. Mol Gen Genet 154:75-82

Abbreviations

AMP: adenosine monophosphate
AMPK: AMP activated protein kinase
ATP: adenosine triphosphate
cAMP: cyclic 3',5'-adenosine monophosphate
CDK: cyclin dependent protein kinase
CKI: cyclin dependent protein kinase inhibitor
ESR: environmental stress response
FGM: Fermentable Growth Medium-induced pathway
FRE: filamentation and invasive response element
GDP: guanosine diphosphate
GEF: guanine nucleotide exchange factor
Glc-6-P: glucose-6-phosphate
Glk1: Glucokinase
GPCR: G-protein coupled receptor system
GSK-3: glycogen synthase kinase 3
GTP: guanosine triphosphate
Hsf: heat shock factor
Hxk1: Hexokinase P I
Hxk2: Hexokinase P II
MAPK: mitogen-activated protein kinase

NES: nuclear export signal
NLS: nuclear localisation signal
NRG: negative regulator of glucose-controlled genes
PAC: polymerase A and C box
RGS: regulator of heteromeric G-protein signalling
PDS: post-diauxic shift element
PH: pleckstrin homology domain
Pi: inorganic phosphate
PKA: protein kinase A
PKB: protein kinase B
RGS: regulator of heterotrimeric G-protein signaling
rp: ribosomal protein gene
RPPE: ribosomal RNA processing element
STRE: general stress response element
UDP: uridine diphosphate

Index

Printing: Saladruck, Berlin
Binding: Stein+Lehmann, Berlin